recent advances in phytochemistry

volume 27

Phytochemical Potential of Tropical Plants

RECENT ADVANCES IN PHYTOCHEMISTRY

Proceedings of the Phytochemical Society of North America
General Editor: Helen A. Stafford, *Reed College, Portland, Oregon*

Recent Volumes in the Series:

Volume 20 The Shikimic Acid Pathway
Proceedings of the Twenty-fifth Annual Meeting of the Phytochemical Society of North America, Pacific Grove, California, June, 1985

Volume 21 The Phytochemical Effects of Environmental Compounds
Proceedings of the Twenty-sixth Annual Meeting of the Phytochemical Society of North America, College Park, Maryland, July, 1986

Volume 22 Opportunities for Phytochemistry in Plant Biotechnology
Proceedings of the Twenty-seventh Annual Meeting of the Phytochemical Society of North America, Tampa, Florida, June, 1987

Volume 23 Plant Nitrogen Metabolism
Proceedings of the Twenty-eighth Annual Meeting of the Phytochemical Society of North America, Iowa City, Iowa, June, 1988

Volume 24 Biochemistry of the Mevalonic Acid Pathway to Terpenoids
Proceedings of the Twenty-ninth Annual Meeting of the Phytochemical Society of North America, Vancouver, British Columbia, Canada, June, 1989

Volume 25 Modern Phytochemical Methods
Proceedings of the Thirtieth Annual Meeting of the Phytochemical Society of North America, Quebec City, Quebec, Canada, August, 1990

Volume 26 Phenolic Metabolism in Plants
Proceedings of the Thirty-first Annual Meeting of the Phytochemical Society of North America, Fort Collins, Colorado, June, 1991

Volume 27 Phytochemical Potential of Tropical Plants
Proceedings of the Thirty-second Annual Meeting of the Phytochemical Society of North America, Miami Beach, Florida, August, 1992

---

recent advances in phytochemistry

volume 27

Phytochemical Potential of Tropical Plants

Edited by

Kelsey R. Downum
Florida International University
and Fairchild Tropical Garden
Miami, Florida

John T. Romeo
University of South Florida
Tampa, Florida

and

Helen A. Stafford
Reed College
Portland, Oregon

Springer Science+Business Media, LLC

Library of Congress Cataloging-in-Publication Data

Phytochemical potential of tropical plants / edited by Kelsey R. Downum, John T. Romeo, and Helen A. Stafford.
p. cm. -- (Recent advances in phytochemistry ; v. 27)
"Proceedings of an international symposium of the second joint meeting of the Phytochemical Societies of Europe and North America, as well as the Thirty-second Annual Meeting of the Phytochemical Society of North America ... held August 8-12, 1992, in Miami Beach, Florida"--T.p. verso.
Includes bibliographical references and index.
ISBN 978-1-4899-1785-0
1. Plant bioactive compounds--Tropics--Congresses. 2. Tropical plants--Composition--Congresses. 3. Botany, Economic--Tropics--Congresses. I. Downum, Kelsey R., 1952- . II. Romeo, John T. III. Stafford, Helen, 1922- . IV. Phytochemical Society of Europe. V. Phytochemical Society of North America. Meeting (32nd : 1992 : Miami Beach, Fla.) VI. Series.
QK861.R38 vol. 27
[QK898.B54]
581.19'2 s--dc20
[581.19'2'0913] 93-22417
CIP

Proceedings of an international symposium of the Second Joint Meeting of the Phytochemical Societies of Europe and North America, as well as the Thirty-Second Annual Meeting of the Phytochemical Society of North America on Phytochemical Potential of Tropical Plants, held August 8–12, 1992, in Miami Beach, Florida

ISBN 978-1-4899-1785-0 ISBN 978-1-4899-1783-6 (eBook)
DOI 10.1007/978-1-4899-1783-6

Originally published by Plenum Press, New York in 1993
Softcover reprint of the hardcover 1st edition 1993

PREFACE

Throughout the tropics, vast areas of rainforest and other biologically diverse lands are being cleared for agricultural or related uses. Rainforests, the most dramatic example of tropical habitat destruction, are estimated to be disappearing at the rate of up to 20.4 million hectares per year world-wide (based on FAO estimates; see World Resources 1990-1991, Oxford University Press)—more than 2% of the total area covered by tropical rainforests per year. Destruction of these complex habitats results in the irreversible loss of both plant and animal diversity, and dramatically illustrates the need to investigate these threatened species for potentially useful constituents—especially the identification and characterization of novel biologically-active phytochemicals with pharmacologiical and/or pesticidal properties.

This volume is based on papers presented by invited speakers at an international symposium entitled "Phytochemical Potential of Tropical Plants: held in conjunction with the second joint meeting of the Phytochemical Societies of Europe and North America, as well as the 32nd annual meeting of the latter society. The meeting was held at the Deauville Hotel, Miami Beach, Florida, USA from August 8-12, 1992. One hundred and twenty-five participants from more than 20 countries attended this meeting.

The main emphasis throughout the four-day symposium was the identification of biologically-active phytochemicals from tropical plants and discussion of their agricultural/pharmacological potential. Speakers touched on a range of topics which among others included: i)strategies for locating plants with specific biological properties; ii) use of chemical/biological techniques for the development of novel pharmaceuticals and pesticides from lead phytochemicals; iii) evaluation of plant chemicals as medicinals and insecticides; iv) studies of the phytochemical diversity of promising tropical plant families; v) use of medicinal plants by primates other than man.

The symposium was organized by Kelsey R. Downum (Florida International University) and John T. Romeo (University of South FLorida), who gratefully acknowledge advice received from Professors Peter Lea, Frank Stermitz, G.H. Neil Towers and Peter Waterman. We would like to give special thanks to Dr. William Klein Jr., Director of the Fairchild Tropical Garden, for his enthusiastic support of this meeting, and Dr. Terrence Walters who organized a day long visit to the Fairchild Tropical Garden for meeting participants. Generous Financial and/or in-kind support was provided by Florida International University (College of Arts and Sciences and Department of Biological Sciences), the U.S. Department of Agriculture (thanks to the efforts of Dr. James Saunders), the Fairchild Tropical Garden, P.C. Inc. (Potomac, MD), and Environmental Growth Chambers (Chargrin Falls, OH).

The fragile nature of tropical plant life was brought home to all meeting participants after learning of the severe devastation experienced by the Fairchild Tropical Garden during Hurricane Andrew two weeks after our visit.

February 1993

Kelsey R. Downum
John T. Romeo

CONTENTS

Chapter One

TROPICAL PLANTS AS SOURCES OF ANTIPROTOZOAL AGENTS

J. David Phillipson,[1] Colin W. Wright,[1] Geoffrey C. Kirby[2] and David C. Warhurst[2]

Department of Pharmacognosy,[1] The School of Pharmacy and
Department of Medical Parasitology[2]
London School of Hygiene and Tropical Medicine
University of London, London, U.K.

INTRODUCTION

Protozoa are the cause of a number of major diseases which spread massive misery and death, mainly throughout the tropical world. Malaria is the protozoal disease most feared by travellers to the tropics, but other protozoal diseases such as leishmaniasis, trypanosomiasis, amoebiasis and giardiasis cause havoc for many millions of people. Protozoal infections are not unique to the tropics and the AIDS epidemic has shown that other protozoa are capable of contributing to disease and death in cooler climates, as exemplified by infections of *Cryptosporidium parvum* which causes a diarrhoeal disease and *Pneumocystis carinii* which causes pneumonia. Tourists and business travellers may import tropical diseases; malaria, for example, is on the increase in non-tropical countries.[1] It is reported that there were 1987 cases of imported malaria in the UK during 1989[2] and that about 1000 cases of imported malaria are diagnosed each year in the USA.[3] It is not even necessary for a UK resident to travel abroad in order to contract malaria, because over the past decade mosquitoes have arrived on airplanes and have transmitted malaria to humans near international airports.[1,2]

WHO estimates that some 270 million people are infected with malaria and that 2.1 billion (half the world's population) are at risk because they live in areas where the disease is common.[3] There are some 12 million cases of leishmaniasis, 16-18 million cases of Chagas' disease in S. America (*Trypanosoma cruzi* infections) and 25,000 new cases of African sleeping sickness per year (caused by *Trypanosoma brucei*).[3] The problem of leishmaniasis was highlighted recently[4] when it was pointed out that an epidemic of visceral leishmaniasis (kala-azar) is adding to the death toll and suffering in war- and famine-stricken S. Sudan. There have been 40,000 deaths from kala-azar over the past five years in this area alone and 400,000 may be at risk from infection. In one part of S. Sudan 25% of the population is infected. Resistance of the sandfly vector to DDT has resulted in a serious epidemic of kala-azar in the N.E. Indian state of Bihar and in a previous epidemic, during the 1970s some 20,000 people died. Possibly, some 300,000 people in Bihar have the disease.[4] WHO estimates that 350 million are at risk of leishmaniasis, 90 millions of Chagas' disease and 50 million of African sleeping sickness.[3] These dreadful diseases cause great pain and suffering and may result in ulceration and deformation, damage to internal organs, anemia, blindness, brain damage and death. Intestinal protozoa include *Entamoeba histolytica*, the cause of amoebic dysentery and *Giardia intestinalis*, which causes giardiasis. About 42 million cases of clinical amoebiasis may occur annually; if the disease progresses it may develop into hepatic amoebiasis which is responsible for 75,000 deaths per annum.[5] *G. intestinalis* is responsible for 200 million infections annually and 10,000 deaths.[6]

The problems of treating protozoal diseases by chemotherapy are considerable since the parasites are eukaryotic like the host, and even when a successful drug is available, resistance may develop. In addition, resurgence of the disease may occur due to the development of resistance of vector insects such as mosquitoes and sandflies to insecticides. The classic example is malaria. In the 1950s it was confidently expected that the disease had been conquered, but by the late 1950s *Plasmodium falciparum* had become resistant to chloroquine in S. America and S.E. Asia. This was not a new phenomenon because the resistance of *P. falciparum* to quinine had been recognized as early as 1910. Today, not only are there strains of *P. falciparum* resistant to chloroquine but also to amodiaquine, sulphadoxine-pyrimethamine, quinine and mefloquine. It has been suggested that "if the present trend continues the affluent countries will soon arrive at the end of the drug repertoire."[7]

Only the acute stages of Chagas' disease are amenable to treatment but of the drugs available, nifurtimox and benznidazole are poorly tolerated. Suramin, melarsoprol or pentamidine may be used to treat African sleeping sickness but may be accompanied by serious adverse effects. For leishmaniasis, antimonial compounds, diamidine and amphotericin B are used. Metronidazole is used for the treatment of amoebic dysentery and giardiasis, but the drug is poorly tolerated by some patients. Hence, either because no satisfactory drug exists or because of resistance to established drugs, there is an urgent need to search for new anti-protozoal drugs, preferably those directed at different targets from existing drugs.

Higher plants provide one potential source of antiprotozoal drugs, and it should be remembered that the earliest drugs developed for the treatment of malaria and of amoebiasis were the plant products quinine and emetine, respectively. Interest in antiprotozoal drugs from higher plants has been stimulated by reports of the clinical use of the endoperoxide sesquiterpene lactone, artemisinin, isolated from the Chinese traditional antimalarial herb *Artemisia annua*. A recent update on artemisinin has noted that laboratory synthesis involves 14 separate steps and that yields are low.[8] Semisynthetic ether and ester derivatives of dihydroartemisinin, which are prepared by reduction of the lactone carbonyl, have yielded artemether, the oil soluble methyl ether, and sodium artesunate, the water soluble hemisuccinate. Both derivatives have been licensed as medicines and are used clinically in China using various dosage forms including oral, suppository and injectable; clinical experience with artemisinin and derivatives outside China is limited.[8,9]

Natural products as antiprotozoal agents and traditional medicines used for the treatment of protozoal diseases have been reviewed several times in recent years.[10-14] A range of chemical structures including terpenoids, alkaloids, quinones and phenolics have been shown to have activity against *Entamoeba*, *Plasmodium*, *Leishmania* and *Trypanosoma* species. After these reviews had been prepared, further references came to our notice, particularly for natural products with antimalarial activity. Other alkaloids with activity against *P. falciparum in vitro* include securinine from the Tanzanian antimalarial plant *Margaritaria discoidea* (Euphorbiaceae)[15] and 7-methoxy-β-carboline-l-propionic acid from *Eurycoma longifolia* (Simaroubaceae).[16] A series of steroidal taccalonolides from the Chinese plant *Tacca plantagina*[17] and japonicine A (a hexadienone) from *Hypericum japonicum*[18] have been reported to be active against *P. berghei* in mice. Of particular interest is the report of *in vitro* antimalarial activity from compounds isolated from the red alga *Portieria hornemannii*..[19] Aplystatin and 5,R-aplystatin were active *in vitro* against *P. falciparum* but were cytotoxic. One compound isolated, however, had medium antimalarial activity and lacked toxicity. The potential of marine organisms as antiprotozoal agents is untapped.

The purpose of this chapter is to focus on some of the findings which have been made by a collaborative research program between the London School of Hygiene and Tropical Medicine and The School of Pharmacy, two independent institutes within the University of London.

QUASSINOID TERPENOIDS

The most extensive screening of higher plants for antimalarial activity was published in 1947 and some 600 plants representing 126 families were extracted and tested against three avian malarias *in vivo*.[20] Thirty genera yielded active extracts and two families in particular, Simaroubaceae and Amaryllidaceae, were the most active in terms of the number of extracts with activity. Many of the plants screened were selected on the basis of their traditional use for the treatment of malaria. Species of Simaroubaceae have reputations for the treatment of malaria and amoebic dysentery in areas as far apart as Asia, India, Africa, S. and C. America. Antiprotozoal activities have been reported for quassinoids and for indole alkaloids isolated from simaroubaceous species, but it is the former group of compounds which is the more potent. Quassinoids are terpenoids biosynthetically produced from triterpenoids. Their name is derived from quassin, the first of its class to be characterized. Individual quassinoids have been shown to have *in vitro* activity against *Plasmodium falciparum*,[11-14,21,22] *Entamoeba histolytlica* [23-25] and *Leishmania donovani*.[26]

Following a lead provided in the 1947 antimalarial screening programme,[20] five plants that have traditional reputations for antimalarial or other antiprotozoal activities were selected for further investigations; namely, *Brucea javanica* (S.E. Asia, China, Thailand), *Eurycoma longifolia* (Malaysia), *Ailanthus altissima* (India), *Simarouba amara* (C. America) and *Picramnia antidesma* (C. America). Prepared extracts were tested for their *in vitro* activity against *P. falciparum* (multi-drug resistant strain, Kl) using a modification of the *in vitro* microplate assay developed by Desjardins *et al.* (1979).[27] Bioassay guided fractionation led to the isolation of a series of antimalarial quassinoids from *Brucea javanica*,[28,29] *Eurycoma longifolia*,[30] *Ailanthus altissima*[31] and *Simarouba amara*.[32] *In vitro* testing of 40 individual quassinoids against *P. falciparum* (Kl) revealed that ten of them had IC_{50} values of less than 0.02 µg ml^{-1} as compared with an IC_{50} value for chloroquine diphosphate of 0.21 µg ml^{-1}, under the same test conditions.[33]

The antimalarial quassinoids possess a tetracyclic skeleton with a lactone ring; more potently active quassinoids possess a further ring in which the C-8 methylene is linked through an oxygen atom to either C-ll or C-13 (Fig. 1). The *in vitro* IC_{50} values against *P. falciparum* (Kl) for some selected quassinoids are given in Figure 1. The state of oxidation and substitution of ring A is important for activity and the most potent compounds have a hydroxyl substituent at either C-l or C-3 and an adjacent carbonyl conjugated to a double bond.[13,33] Many of the naturally-occurring quassinoids possess ester functions, usually at C-15 or occasionally at C-6. Loss of the esterifying acid leads to significant loss of antimalarial activity. However, an ester function is not necessarily a prerequisite for antimalarial activity and several quassinoids with 14,15-diol functions and no ester moiety are active against *P. falciparum in vitro* (Fig. 1). Only a very small number of quassinoid glycosides has been tested for antimalarial activity and these are very inactive in comparison to their corresponding aglycones.[29]

During the course of these investigations into the identification and structure determination of quassinoids and their *in vitro* antimalarial activities, a number of questions arose naturally from the work:

1. Do the compounds active *in vitro* possess *in vivo* activity?
2. Are the quassinoids active *in vitro* responsible for the antimalarial activity of species of Simaroubaceae used in traditional medicine?

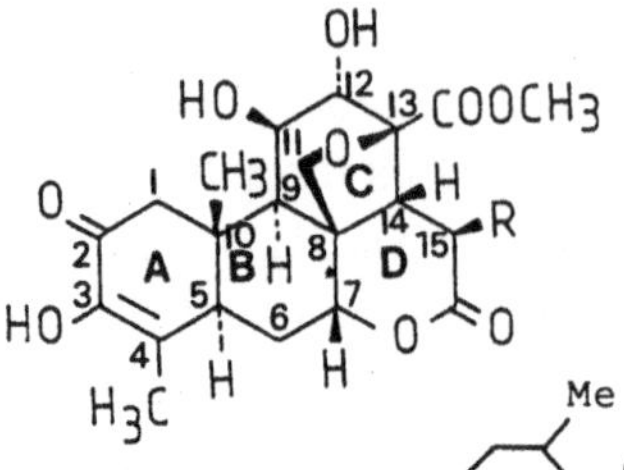

bruceine A, R = OCO Me Me (0.01)
bruceolide, R = OH(0.50)

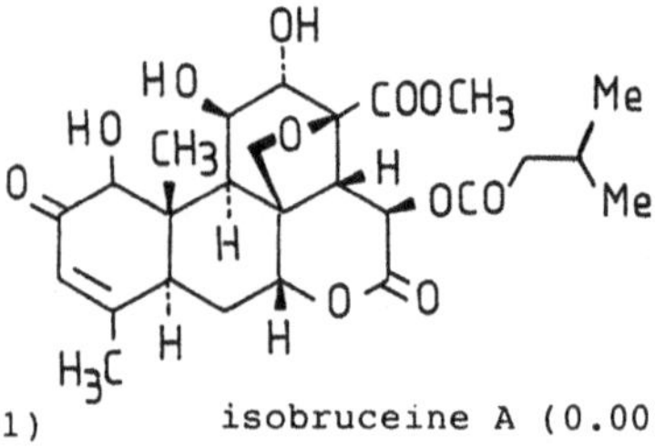

isobruceine A (0.002)

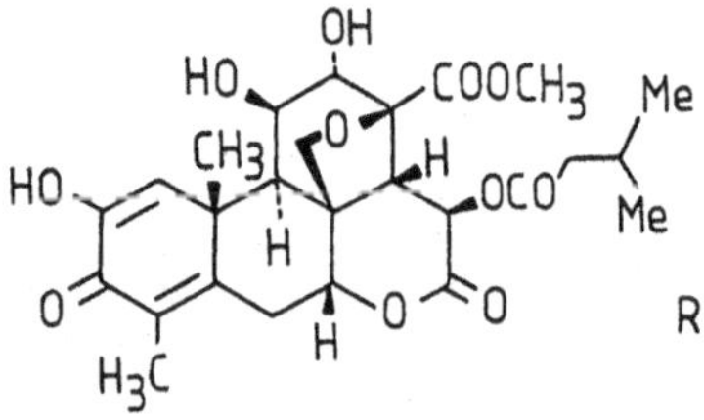

dehydrobruceine A (0.046)

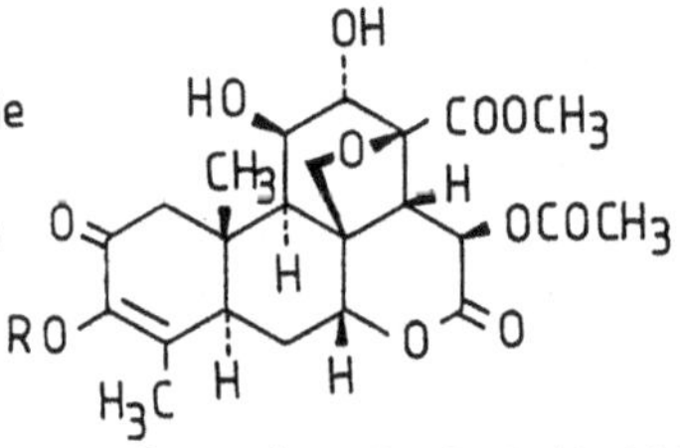

bruceine B, R=H (0.01)
yadanzioside I, R=glucose (22.04)

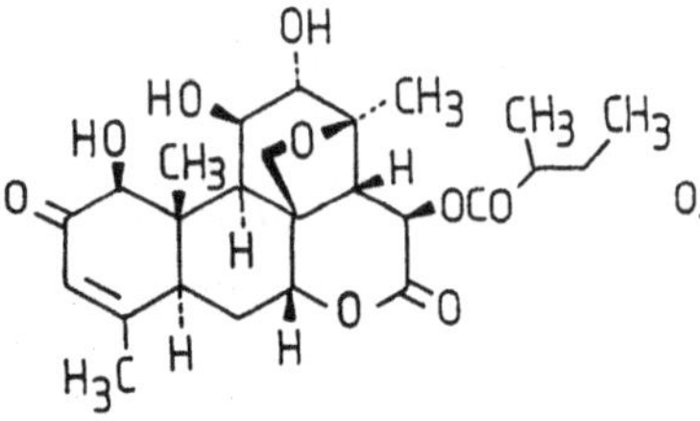

simalikalactone D (0.0009)

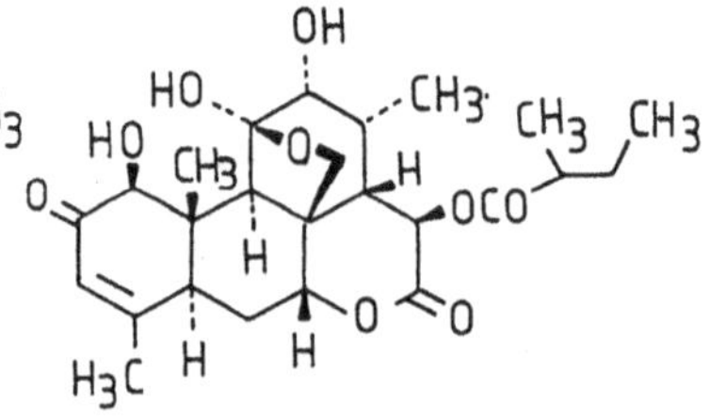

ailanthinone (0.009)

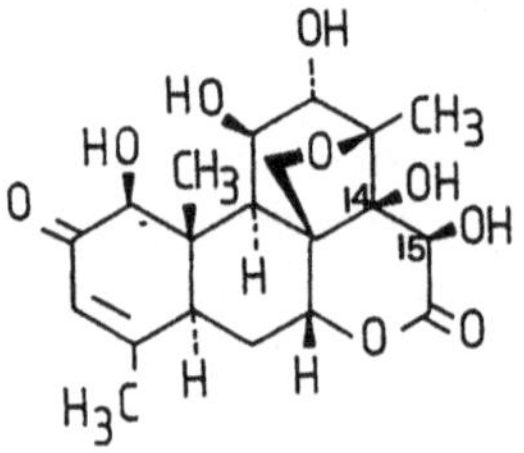

bruceine D (0.015)

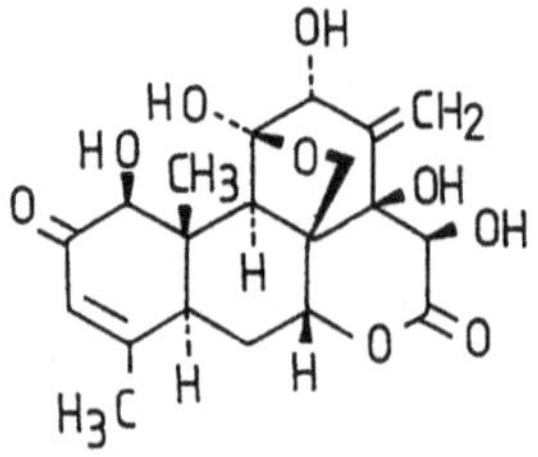

eurycomanone (0.11)

Fig. 1. Antimalarial quassinoids, with *in vitro* antiplasmodial activity (IC values in $\mu g\ ml^{-1}$ against *Plasmodium falciparum* K-1 strain; chloroquine diphosphate IC_{50} value 0.2 $\mu g\ ml^{-1}$). (IC = 50% inhibitory concentration).

3. Does the C-15 ester group simply contribute to activity by providing a lipid soluble 'pro-drug'?
4. Do quassinoids act as general cytotoxic agents or do they possess some selectivity to *P. falciparum*?
5. Is it possible to modify the chemical structure of quassinoids in order to improve activity and selectivity?
6. How do quassinoids exert their effects against *P. falciparum*?
7. Do quassinoids possess activity against other protozoa?

Table 1. *In vivo* antimalarial activity of quassinoids against *P. berghei* (N strain) in mice

Species	Compound	Oral activity against *P. berghei* (mg $kg^{-1}day^{-1}$) ED_{50}	ED_{90}	No. of toxic deaths (from 5 mice)	Dose tested
Brucea javanica	Bruceine A	3.36	26.7	1	9
	Bruceine B	0.90	2.80	5	9
	Brusatol	1.27	3.03	4	3
	Bruceine D	2.79	8.19	0	3
Ailanthus altissima	Ailanthone	0.7	3.05	5	9
				0	3
	Ailanthinone	1.25	4.13	5	9
				1	3
Simarouba amara	Glaucarubinone	0.86	3.43	5	9
				1	3
	2'Acetoxy-glaucarubinone	2.19	11.02	0	9
	Holacanthone	1.70	4.50	1	18
------	Chloroquine diphosphate	2.27	6.02	0	30

The following paragraphs address some of the aspects of these questions and indicate the type of experimental work undertaken to increase our knowledge of the antiprotozoal effects of quassinoids.

In Vivo Antimalarial Activity of Quassinoids

In the Peters' 4-day suppressive test using *Plasmodium berghei* in mice[34] orally administered chloroform and butanol extracts of Thai *Brucea javanica* fruits had ED_{50} values of 17.9 and 73.8 mg $kg^{-1}day^{-1}$ (E.D. = effective dose). [29]*In vivo* activities of crude extracts of five other simaroubaceous species have been reported.[35] The ED_{50} and ED_{90} values for some nine individual quassinoids isolated from three simaroubaceous species, *Brucea javanica, Ailanthus altissima* and *Simarouba amara* are given in Table 1.[29,31,32,35,36] The ED_{50} values ranged from 0.76 to 3.36 mg $kg^{-1}day^{-1}$ (Table 1) whilst the ED_{90} value for chloroquine diphosphate was 2.27 mg $kg^{-1}day^{-1}$, under the same test conditions. The ED_{90} values of the nine quassinoids ranged from 2.80 to 26.7 mg $kg^{-1}day^{-1}$ (Table 1) in comparison to the ED_{90} value for chloroquine diphosphate of 6.02 mg $kg^{-1}day^{-1}$. Seven of the nine quassinoids caused toxic deaths in mice within dose ranges of 3 to 18 mg $kg^{-1}day^{-1}$, whereas chloroquine diphosphate given at a level of 30 mg $kg^{-1}day^{-1}$ caused no toxic deaths (Table 1). Bruceine D, 2' acetoxyglaucarubinone and holacanthone were found to be the least toxic of the quassinoids tested.

In vivo antimalarial activities of some selected quassinoids have been compared to quinine in mice infected with *P. berghei*.[36] The quinine index used for comparison purposes is calculated by dividing the ED_{90} value in mg $kg^{-1}day^{-1}$ for quinine by the ED_{90} of the individual quassinoid. The results are given in Table 2 and indicate that the nature of the esterifying acid at C-15 markedly affects ED_{90} values (contrast quinine index of bruceantin, 0.88 with that of brusatol, 30.55).[36] Comparison of the quinine indices of glaucarubinone and 6 α-senecioyloxychaparrinone suggests that esterification at C-15 results in higher activity than esterification at C-6, but presence of two ester groups at C-6 and C-15 (undulatone) results in loss of activity when compared to the corresponding C-15 ester (holacanthone).

Quassinoids and the Antimalarial Activity of Traditional Medicines

The highly potent antimalarial quassinoids active *in vitro* are mainly chloroform soluble and the few glycosides tested to date against *P. berghei* in

Table 2. Comparison of *in vivo* activity of selected quassinoids with quinine in *P. berghei* infected mice

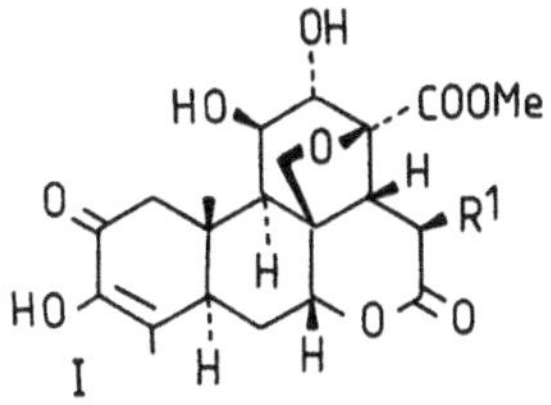

I

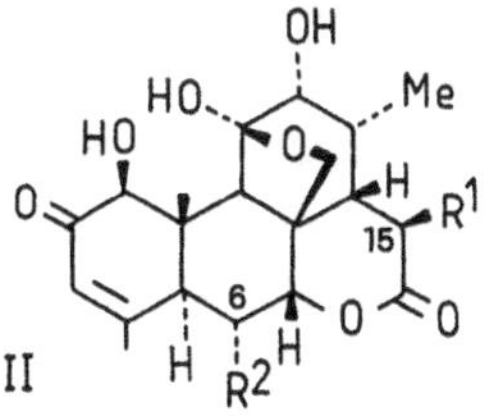

II

	Quassinoid	R^1	R^2	Quinine Index[a]
I	brusatol	$OCOCH{=}CMe_2$	-	30.35
	bruceine A	$OCOCH_2CH(Me)_2$	-	3.44
	bruceantin	$OCOCH{=}C(Me)CHMe_2$	-	0.88
II	glaucarubinone	OCOC(Me)(OH)Et	H	26.81
	ailanthinone	OCOCH(Me)Et	H	22.26
	2'-acetoxy-glaucarubinone	OCOC(Me)(OAc)Et	H	7.91
	holacanthone	OAc	H	20.43
	undulatone	OAc	OCOC(Me)=CHMe	5.78
	6α-senecioylox-chaparrinone	H	$OCOCH{=}CMe_2$	2.08
	quinine dihydrochloride			1.00

[a]Quinine Index = $\dfrac{ED_{90}\ (mg\ kg^{-1}day^{-1})(quinine)}{ED_{90}\ (mg\ kg^{-1}day^{-1})(quassinoid)}$

mice have been inactive. Since the plants used in traditional medicines are frequently given as aqueous extracts (i.e. used in the form of a 'tea'), the question arises as to whether or not quassinoids are the active ingredients. Fruits of *Brucea javanica* were extracted in boiling water and the filtered aqueous extract was freeze dried.[37,38] Chromatographic investigation indicated that only small quantities of potent lipophilic quassinoids were present in the freeze dried preparation but that much larger quantities of polar quassinoids had been extracted. Acid hydrolysis of the freeze dried aqueous extract followed by partition with chloroform yielded a chloroform extract which was rich in lipophilic quassinoids. The *in vitro* IC_{50} values of the dried aqueous extract and the chloroform extract from the hydrolysed portion were 0.43 and 0.011 μg ml^{-1}, respectively, against *P. falciparum* (K1). In contrast, the *in vivo* activities of the dried aqueous tea extract and its chloroform extacted hydrolysate were 500 and 112 mg $kg^{-1}day^{-1}$ respectively against *P. berghei* in mice. Hence the chloroform extracted hydrolysate of *B. javanica* fruits was 39 times more potent *in vitro* than the dried aqueous tea extract, whereas the *in vivo* activity of the hydrolysate was 4.5 times more potent than the aqueous tea extract. These results indicated that acid hydrolysis increases both *in vitro* and *in vivo* antiplasmodial activity. The decreased ratio for *in vivo* activity may be due to more potency *in vivo* or to partial hydrolysis of glycosidic quassinoids to potent aglycones in mice.[37,38]

The species of Simaroubaceae which are used to treat malaria also contain β-carboline alkaloids which have *in vitro* activity against *P. falciparum in vitro.* [16,39] The IC_{50} values against *P. falciparum* are in the order of 2 to 3 μg ml^{-1}. It is not known whether the β-carbolines exert any synergistic activity on quassinoids and whether or not this would be of significance in the antimalarial activity of traditional medicines. Such medicines may also act by more than one mechanism and it is possible that their activities may be due to stimulation of the immune system or by the lowering of body temperature in fevers. Experimental investigations for such effects need to be carried out.

The Stability of the C-15 Ester of Bruceine A

Bruceine A, one of the major quassinoids of *Brucea javanica* fruits has an *in vitro* IC_{50} value of 0.011 μg ml^{-1} (21 nM) against *P. falciparum* (K1) making it more potent than chloroquine diphosphate (IC_{50} value of 0.2 μg ml^{-1}; 407 nM) (Fig. 1). *In vivo* activities of bruceine A and chloroquine diphosphate given orally against *P. berghei* in mice were 3.36 and 2.27 mg $kg^{-1}day^{-1}$ respectively (Table 1). In order to explain the correlations between the *in vitro* and the *in vivo* potencies of bruceine A and chloroquine diphosphate, it has been postulated that *in vivo* hydrolysis of the C-15 ester of bruceine A may occur, resulting in the formation of the less active C-15 alcohol, bruceolide (IC_{50} 0.502 μg ml^{-1}; 1146 nM) (Fig. 1).[40] Although the C-15 esterifying acid is readily removed by alkali under laboratory conditions, it has been shown that it is not lost from bruceine A after 90 minutes incubation with either human plasma esterases or rabbit-liver carboxylic acid esterase. The close structural similarity of bruceine A and other C-15 esterified bruceolides is an indication that ester hydrolysis does not take place *in vivo*. Whatever the reason for lack of *in vivo* potency of bruceine A against *P. berghei* in comparison to its *in vitro* activity against *P. falciparum*, it is not likely to be due to hydrolysis of the C15 ester function.[40]

Cytotoxic and Antimalarial Activities of Quassinoids

A microdilution assay with KB cells (human epidermoid cancer of the mouth) has been developed in order to compare *in vitro* cytotoxicity of quassinoids and other antimalarial natural products with *in vitro* activity against *P. falciparum*.[41,42] The IC_{50} values for seven *B. javanica* quassinoids against KB cells and against *P. falciparum* are given in Table 3. The parent alcohol bruceolide has an IC_{50} value in the KB cell test of 7.57 μg ml^{-1} whilst five of its naturally occurring esters had IC_{50} values ranging from 0.008 (bruceantin) to 0.102 (brucitol) μg ml^{-1} (Table 3). The nature of the C-15 esterifying acid of the bruceolides markedly affects the toxicity to KB cells. Comparison of cytotoxicity to KB cells and activity against *P. falciparum* (Ratio A/B, Table 3) shows that this ratio varies from 4.2 to 34. The 14,15 diol, bruceine D, with comparable activity to bruceine B against *P. falciparum* (0.015, 0.011 μg ml^{-1}, respectively) is markedly less toxic to KB cells and the A/B ratios of cytotoxic to antimalarial activity of bruceine D was 77 in comparison to 5 for bruceine B. It has been reported that the 14,15-diol, eurycomanone, isolated from *Eurycoma longifolia* has an IC_{50} value of 0.048 μg ml^{-1} against *P. falciparum* (W-2, chloroquine-resistant strain)[16] and that its IC_{50} value against KB cells is 1.9 μg ml^{-1}. The cytotoxic to antiplasmodial ratio of eurycomanone is calculated as 39.6. The cytotoxic to antiplasmodial ratios of some 30 quassinoids and their derivatives have been compared and the results show that cytotoxicity does not necessarily parallel antiplasmodial activity.[41,42]

Under the same conditions, chloroquine diphosphate, quinine dihydrochloride and artemisinin have cytotoxic to antimalarial ratios of 347, 600 and 13,900. When the chloroquine-sensitive strain of *P. falciparum*, T9-96, is used the cytotoxic to antiplasmodial ratio of chloroquine diphosphate is calculated as 30,000. Nine of the quassinoids tested had cytotoxic to antiplasmodial ratios of between 20 and 284 whilst the remainder had ratios ranging from 0.1 to 18.[42] The use of cytotoxic to antimalarial ratios as an *in vitro* guide to therapeutic ratios has been proposed as a method for selecting natural products for *in vivo* evaluation.[41]

Structural Modifications of Quassinoids

Lipidic amino acid conjugation has been used to modify drug action, e.g., conjugates have increased blood-brain penetration in comparison to the parent drug, and may deliver a product which produces the active drug on hydrolysis. α-Tertiarybutoxycarbonylaminodecanoic acid has been coupled to

Table 3. *In vitro* activity against KB cells and *in vitro* activity against *Plasmodium falciparum* of *Brucea javanica* quassinoids

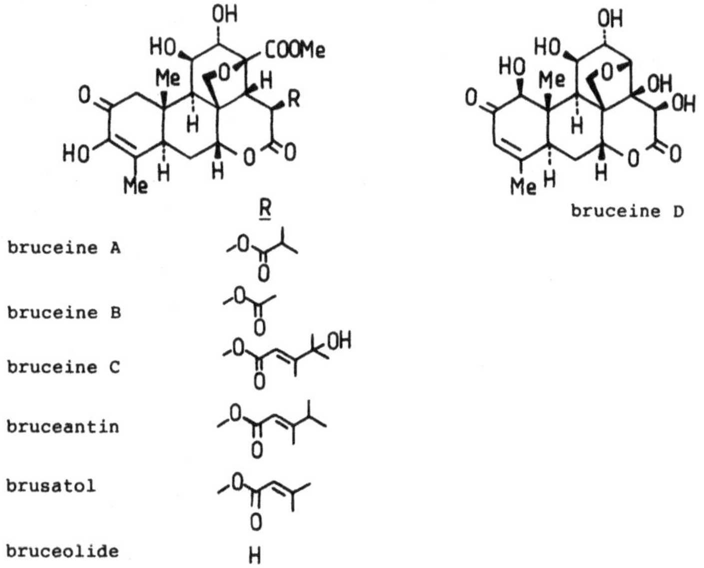

Compound	A KB cells (IC_{50} $\mu g ml^{-1}$)	B P. falciparum(K1) (IC_{50} $\mu g ml^{-1}$)	A/B
bruceine A	0.098	0.011	8.9
bruceine B	0.055	0.011	5.0
bruceine C	0.021	0.005	4.2
bruceantin	0.008	0.0008	10
brusatol	0.102	0.003	34
bruceolide	7.57	0.451	17
bruceine D	1.16	0.015	77
chloroquine diphosphate	72.8	0.210	347

brusatol by standard methods.[43] The lipidic amino acid moiety is conjugated by an ester linkage to C-3 of brusatol as indicated by ^{1}H-NMR spectroscopy and by a negative ferric chloride reaction. The C-3 acetate was prepared for comparison purposes. Mild alcoholic ammonolysis of brusatol resulted in the C-13 amide analogue of brusatol and not the C-15 alcohol, bruceolide, as anticipated. The C-3 t-butoxycarbonylamino-decanoate of the C13 amide derivative of brusatol was also prepared. The *in vitro* activities of these semisynthetic derivatives of brusatol (**1**-**4**) were compared against KB cells and *P. falciparum* (K1 strain). The results are given in Table 4.

Selective etherification at C-3 of brusatol has been reported to enhance *in vitro* antimalarial activity[44] and this approach has been extended into C-3 ester derivatives. The C-3 *t*-butoxycarbonyldecanoate ester of brusatol (**1**) and the C-3 acetyl (**2**) possess similar IC_{50} values to brusatol against *P. falciparum* whereas both esters are markedly less active against KB cells (Table 4). These results suggest that esterification at C-3 reduces toxicity to mammalian cells and not to *P. falciparum*. The cytotoxic to antimalarial ratios of brusatol, its C-3 lipidic amino acid conjugate (**1**) and its C-3 acetate (**2**) were 6.0, 11.7 and 25.0, respectively (Table 4). These findings may be due to differences at the binding site (possibly ribosomal), between the malaria parasite and the human KB cells. This is of interest because it is generally assumed that parasite and host ribosomes are sufficiently similar to preclude selective toxicity.

The C-13 amide of brusatol (**3**) was 250 times less active than brusatol against *P. falciparum in vitro* and further loss of activity was observed for the corresponding C-3 *t*-butoxycarbonyldecanoate ester (**4**) (Table 4). Amidation at C-13 also resulted in less activity against KB cells. It is concluded that the C-13 antimalarial activity and that this part of the quassinoid may be involved in binding at the site of action. Other C-3 and C-13 derivatives of quassinoids should be prepared in order to ascertain whether antimalarial activity is retained and cytotoxic action to mammalian cells is reduced.[43]

Mode of Action of Quassinoids Against *P. falciparum*

A number of quassinoids has been shown to disrupt normal ribosome function and cause irreversible inhibition of protein synthesis in eukaryotic cells.[44-50] Ailanthinone, bruceantin, bruceine B, glaucarubinone and holacanthone (Fig. 1, Table 2) were compared for their ability to affect incorporation of ^{3}H-isoleucine and ^{3}H-hypoxanthine into *P. falciparum in vitro* as measures of protein- and nucleic acid-synthetic ability, respectively.[51]

Rate of incorporation of radiolabels into parasite protein or nucleic acids was observed to be linear for up to 6 hours. These five quassinoids at doses ten times their respective 48 hr *in vitro* IC_{50} values (Table 5), all caused marked inhibition of protein synthesis within 20-40 min. of application. Effects upon

Table 4. *In vitro* cytotoxicity against KB cells and against *P. falciparum* (K1) of some semi-synthetic quassinoids

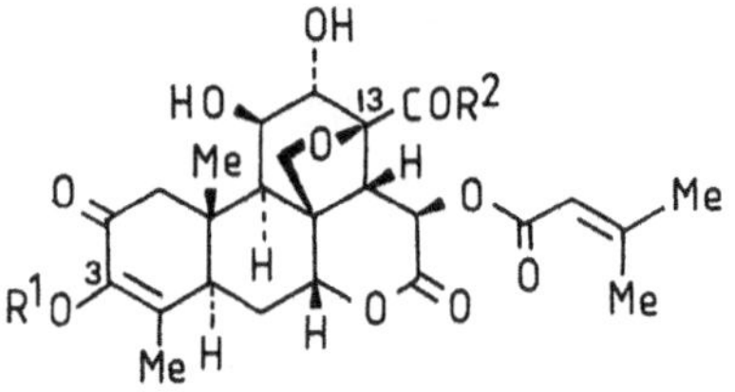

Compound	R^1	R^2
brusatol	H	OCH_3
1	$(CH_3)_3COOCNHCH\text{-}CO$ with $(CH_2)_7CH_3$ on CH	OCH_3
2	CH_3CO	OCH_3
3	H	NH_2
4	$(CH_3)_3COOCNHCH\text{-}CO$ with $(CH_2)_7CH_3$ on CH	NH_2

Compound	A KB cells (IC_{50} nM)	B *P. falciparum* (IC_{50} nM)	A/B
brusatol	28.8	4.81	6.0
1	68.4	5.83	11.7
2	102.3	4.09	25.0
3	18,000	1208	14.9
4	34,121	5203	6.6

Table 5. Effects of selected quassinoids at ten times their IC_{50} values on protein- and nucleic acid-synthesis in *Plasmodium falciparum*

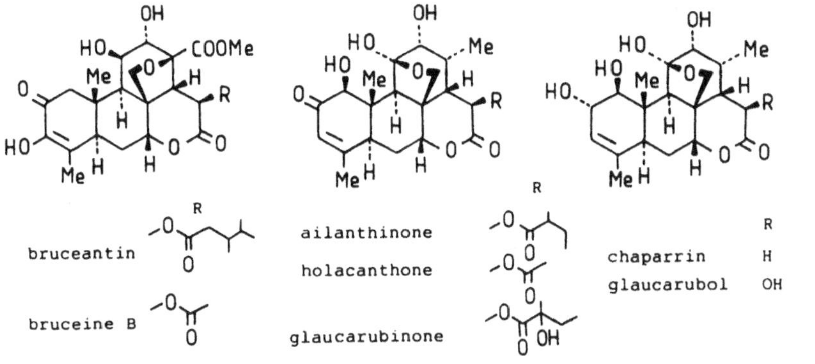

Compound	t_{50} (min) Protein synthesis (isoleucine)	Nucleic acid synthesis (hypoxanthine)	t_{50} ratio (isoleucine/ hypoxanthine)
bruceantin	122	233	0.52
bruceine B	73	299	0.24
ailanthinone	71	184	0.38
holacanthone	66	195	0.33
glaucarubinone	108	343	0.31
chaparrin	inactive	-	-
chaparrin[a]	93	257	0.36
glaucarubol	inactive	-	-
glaucarubol[b]	239	833	0.29
chloroquine diphosphate	inactive	-	-
cycloheximide[c]	70	184	0.38

concentration tested [a] 209 x IC_{50} value
[b] 114 x IC_{50} value
[c] 423 x IC_{50} value

nucleic acid synthesis were delayed; the time to which nucleic acid synthesis was reduced to 50% of control (t_{50}) was directly proportional to the t_{50} for inhibition of protein synthesis suggesting that failure of nucleic acid synthesis was a consequence of inhibition of protein synthesis.[51] Two other quassinoids, chaparrin and glaucarubol (Table 5), with relatively low *in vitro* antiplasmodial activities, were also tested but at concentrations ten times their 48 hr *in vitro*

IC_{50} values, they failed to inhibit significantly, within the time course of the experiment, synthesis of either protein or nucleic acids. However, when used at concentrations 209 and 114 times their respective IC_{50} values their observed effects were identical to those of the other quassinoids studied. Increasing the concentrations of ailanthinone or holacanthone above 10 times their *in vitro* IC_{50} values had no further effect upon the inhibitory effects of either quassinoid upon either of the parameters measured. It was concluded that in *P. falciparum*, as in eukaryotic cells, quassinoids are potent inhibitors of protein synthesis and that this is most likely due to direct effects upon the ribosome, rather than secondarily through action on nucleic acid synthesis.

The three quassinoids ailanthinone, bruceantin and chaparrin do not differ significantly in their IC_{50} values against a chloroquine-sensitive strain of *P. falciparum* (T9-96) and a chloroquine-resistant strain (Kl).[52] Similarly there were no differences between these two strains of *P. falciparum* in their sensitivities to the protein synthesis inhibitors anisomycin, desacetyl-anisomycin, cephalotaxine, homoharringtonine, cycloheximide, puromycin and puromycin aminonucleoside. Anisomycin and the quassinoid bruceantin compete for binding at a site within the eukaryotic 60 S ribosomal subunit[45] and cause inhibition of ribosomal peptidyl transferase activity.

Ailanthinone, bruceine B and glaucarubinone at concentrations of ten times their IC_{50} values were investigated for effects on lactic acid production by *P. falciparum*. The results indicated that these quassinoids had little significant effect on glycolysis.[53]

The antimalarial effects *in vitro* of some protein synthesis inhibitors in combination have been investigated.[54] IC_{50} values were obtained for a series of individual compounds alone and in the presence of four fixed concentrations of another drug. Isobolograms were drawn in order to assess the nature of any interactions. Bruceantin and glaucarubinone, glaucarubinone and anisomycin, and bruceantin and anisomycin were shown to be potentiating combinations. Such findings warrant further investigations particularly since traditional medicines containing quassinoids may well be taken in combination with allopathic medicine. It is not yet known whether these findings have any clinical significance.

Activity of Quassinoids Against Other Protozoa

A number of species of Simaroubaceae have been reported to be used in traditional medicine for the treatment of amoebic dysentery.[10,23,55] The active principles have been identified as quassinoids.[55,56] A micro-dilution assay was used to assess plant extracts and isolated natural products for activity against *Entamoeba histolytica in vitro*.[24,25] Bruceantin is about ten times more potent than metronidazole against *E. histolytica in vitro*. Comparison of the *in vitro* IC_{50} values of 25 quassinoids against *P. falciparum* and *E. histolytica* has shown that, in general, the activities against both protozoa parallel each other.[24]

Several quassinoids have been demonstrated to have activity against *Leishmania donovani* promastigotes but the most active compounds, simalikalactone D and 15 β-heptylchaparrinone were also toxic to macrophages.[26] A microplate assay for determination of *in vitro* activity against *Giardia intestinalis* has also been developed.[57] Bruceantin and bruceine A have IC_{50} values of 1.20 and 8.84 μM, respectively, and under the same test conditions metronidazole, anisomycin and furazolidine have IC_{50} values of 2.98, 0.24 and 4.14 μM, respectively. The test provides a convenient method of searching for new antigiardial agents from natural products.

ALKALOIDS

Bisbenzylisoquinoline Alkaloids

Crude extracts from eighteen species of plants used in Sierra Leone for the treatment of fevers and malaria were extracted and the crude extracts tested for *in vitro* activity against *P. falciparum* (Kl). The most active extract was that of *Triclisia patens* (Menispermaceae) wood which had an IC_{50} value of 5 μg ml^{-1}; further purification yielded an alkaloid-rich extract with an IC_{50} value of 0.5 μg ml^{-1}.[58] Five bisbenzylisoquinoline alkaloids previously isolated from *T. patens*[59-61] were tested for activity against *P. falciparum* (Kl) *in vitro*. Phaeanthine, pycnamine and aromoline (Fig. 2) had IC_{50} values ranging from 0.15-1.43 μg ml^{-1} comparing with an IC_{50} value for chloroquine diphosphate of 0.14 μg ml^{-1}, under the same test conditions. The dibenzo-1,4-dioxin alkaloids cocsuline and trigilletimine (Fig. 2) were considerably less active against *P. falciparum*. *T. dictyophylla* is used as an antimalarial remedy in Ghana and *in vitro* activity of the alkaloidal constituents against *P. falciparum* has been investigated.[62] Independent investigation of *Tiliacora triandra* (Menispermaceae), which is used as an antimalarial in Thailand, resulted in the isolation of the three bisbenzylisoquinoline alkaloids tiliacorine, tiliacorinine and nortiliacorinine A, with IC_{50} values against *P. falciparum* of between 0.56 and 3.5 μg ml^{-1}.[63] These findings lend support to the folkloric reputations of these species in Sierra Leone, Ghana and Thailand as remedies for the treatment of malaria.

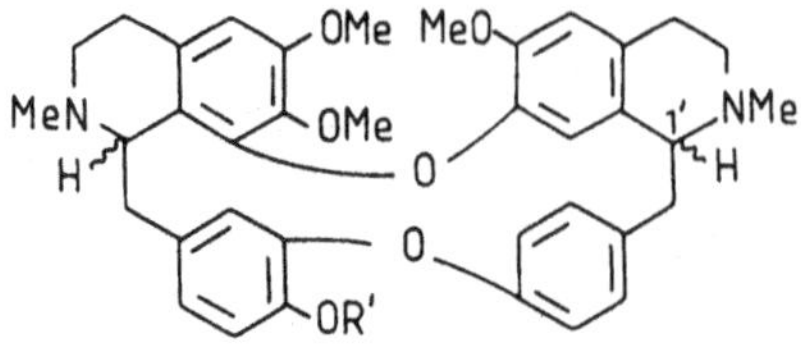

phaeanthine (R,R) R'=Me
tetrandrine (S,S) R'=Me
pycnamine (R,R) R'=H
berbamine (R,S) R'=H

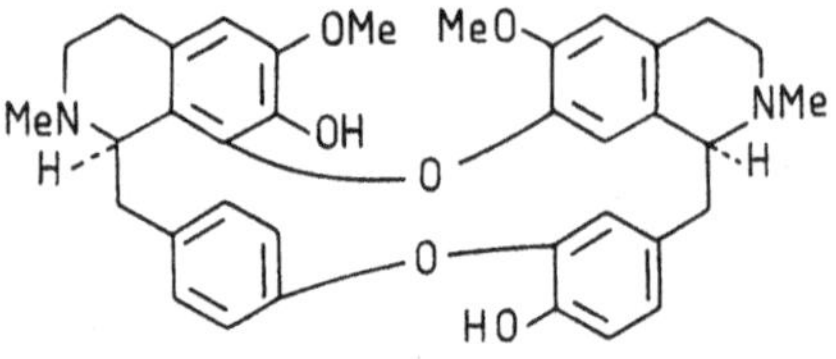

aromoline (R,S)

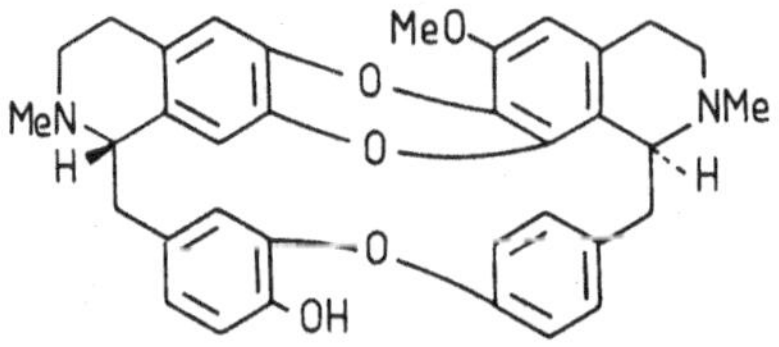

cocsuline (S,S)

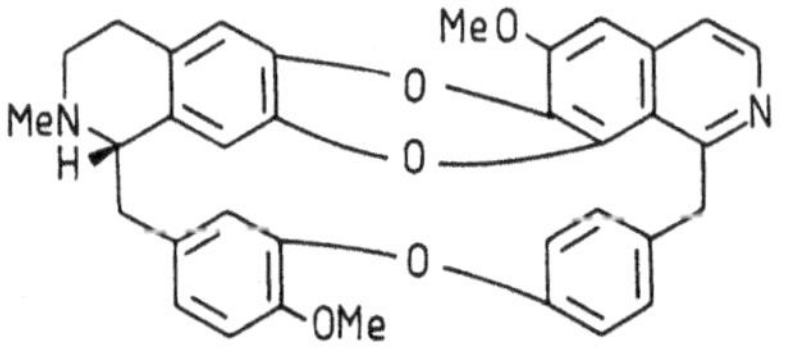

trigilletimine

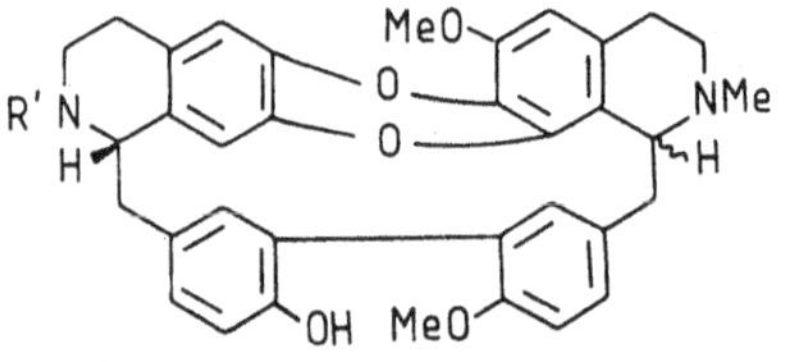

tiliacorine (C-1'R) R'=Me
tiliacorinine (C-1'S) R'=ME
nortiliacorinine A (C-1'S) R'=H

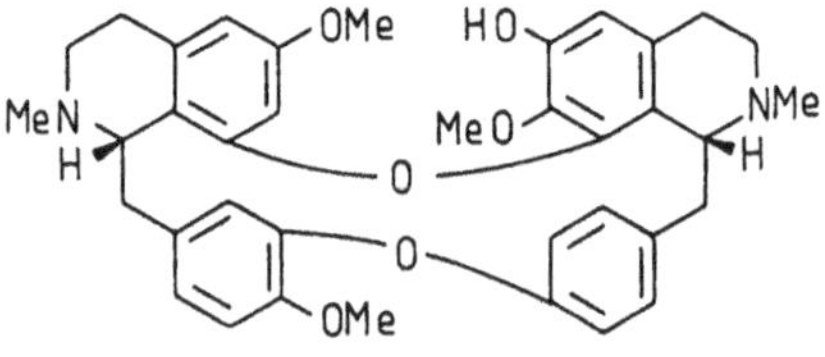

gyrocarpine

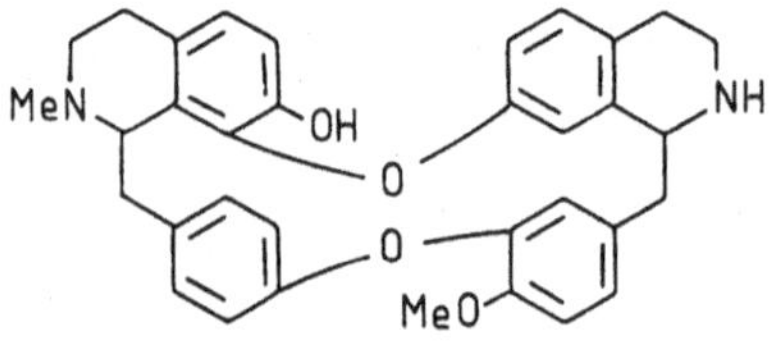

daphandrine

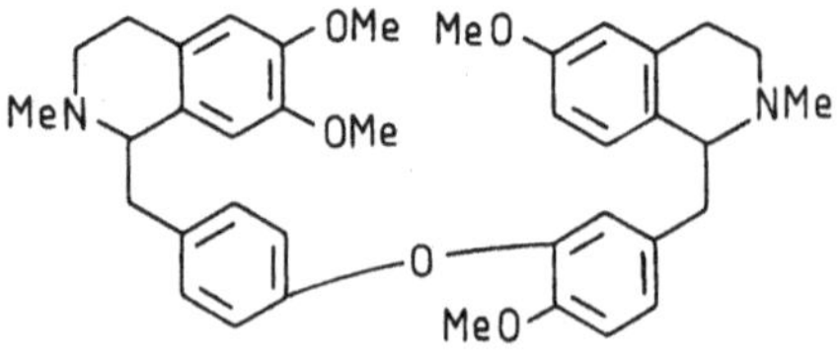

obaberine

Figure 2. Examples of bisbenzylisoquinoline alkaloids which have been tested for antiprotozoal activities.

Some 24 bisbenzylisoquinoline alkaloids have been tested for *in vitro* activity against *P. falciparum*, *E. histolytica* and KB cells.[64] We are grateful to Professor P.L. Schiff of Pittsburgh for the generous supply of these alkaloids.

Eight of the alkaloids had IC_{50} values of less than 1 μM against multi-drug resistant *P. falciparum* (K1). Activity against *E. histolytica* was less pronounced with only one alkaloid, aromoline, showing significant activity with an IC_{50} value of 5 μM compared with an IC_{50} value of 2.2 μM for emetrine. None of the 24 alkaloids tested possessed significant activity against KB cells in a 48 hr microplate assay.[42,64] The most cytotoxic alkaloid was berbamine (Fig. 2) with an IC_{50} value against KB cells of 17.8 μM. The IC_{50} value of berbamine against *P. falciparum* (Kl) was 0.45 μM and hence the *in vitro* cytotoxic to antimalarial ratio of the most cytotoxic bisbenzylisoquinoline alkaloid tested was 40. Phaeanthine was found to be twice as potent against a chloroquine-resistant *P. falciparum* (Kl), as against a chloroquine-sensitive clone (T9-96), with IC_{50} values of 365.85 and 704.87 nM, respectively (Table 6).[65] Tetrandrine, the enantiomer of phaeanthine shows a similar behaviour, being more active against chloroquine-resistant than chloroquine-sensitive strains of *P. falciparum*.[66] T9-96 is more sensitive to the action of chloroquine than is Kl, by a factor of 16.6 and chloroquine diphosphate exhibits 52.9 times greater activity than phaeanthine against T9-96 whereas for Kl the difference is only 1.65 fold (Table 6).[65] In addition, a combination of chloroquine and phaeanthine against the chloroquine-sensitive strain T9-96 of *P. falciparum* was found to show antagonism whilst an additive effect was observed against the chloroquine-resistant strain Kl. Phaeanthine exhibits selectivity of action against *P. falciparum* as compared to KB cells with cytotoxic to antimalarial ratios of 96.4 and 185 for sensitive clone T9-96 and resistant strain K1, respectively (Table 6).[65]

Tetrandrine and structurally related alkaloids are effective in reversing resistance in chloroquine-resistant *P. falciparum*; when used in combination with artemisinin (qinghaosu), tetrandrine is said to provide a long-acting and synergistic activity.[67] Gyrocarpine, daphnandrine and obaberine (Fig. 2) have been shown to be highly active against promastigotes of *Leishmania amazonensis*, *L. braziliensis* and *L. donovani*.[68] Clearly, this group of alkaloids needs further investigation, in particular *in vivo* assessment, as potential antiprotozoal agents.

Berberine and Related Alkaloids

Berberine (Table 7) is widely distributed in nine plant families including Annonaceae and Menispermaceae, and is used widely in the treatment of malaria, amoebiasis and leishmaniasis.[69] Three closely related alkaloids, berberine, palmatine and jatrorrhizine isolated from *Enantia chlorantha* have *in vitro* activity against two chloroquine-resistant strains (D-6 and W-2) of *P. falciparum*

Table 6. IC_{50} values (nM) of phaeanthine and chloroquine diphosphate against sensitive (T9-96) and resistant (K1) strains of *P. falciparum* and against KB cells

	T9-96	K1	Ratio KI/T9-96	KB	Ratio KB/T9-96	Ratio KB/KI
Phaeanthine	704.87	365.85	0.52	67,949	96.40	185.72
Chloroquine diphosphate	13.325	221.5	16.62	117,000	8,783.8	528.2
Ratio Phaeanthine/ Chloroquine	52.9	1.65				

(Table 7).[64,70] Berberine, jatrorrhizine and columbamine have been shown to have *in vitro* activity against *P. falciparum* (K1).[69] The results taken from the two different laboratories show that the IC_{50} values obtained against *P. falciparum* are between 0.14 and 0.36 $\mu g\ ml^{-1}$ (3 strains) for berberine, 0.16-0.28 $\mu g\ ml^{-1}$ (2 strains) for palmatine, 0.42-1.6 $\mu g\ ml^{-1}$ (3 strains) for jatrorrhizine and 0.72 $\mu g\ ml^{-1}$ (1 strain) for columbamine.

The *in vitro* IC_{50} values for four protoberberine alkaloids have been compared for their effects against *P. falciparum*, *E. histolytica* and KB cells (Table 8). Berberine was the only one of the four alkaloids tested which showed appreciable toxicity to KB cells (Table 8) and it proved to be the most active alkaloid against *P. falciparum* (IC_{50} 0.97 μM). Thalifendine (a des-methyl berberine) was approximately ten times less active against *P. falciparum* in comparison with berberine. Although thalifendine, jatrorrhizine and columbamine are less active than berberine against *P. falciparum*, they possess more favourable cytotoxic to antimalarial ratios (Table 8). None of the four alkaloids showed any appreciable activity against *E. histolytica*.[70]

These results lend some support to the use of plants containing protoberberine alkaloids for the treatment of malaria but not for the treatment of amoebiasis. Intriguingly, *in vivo* activity could not be demonstrated for berberine against *P. berghei* in mice[69] whereas, in contrast, berberine has *in vivo* activity against *E. histolytica* in mice.[71]

Table 7. *In vitro* activities of protoberberine alkaloids against three different strains of *Plasmodium falciparum*

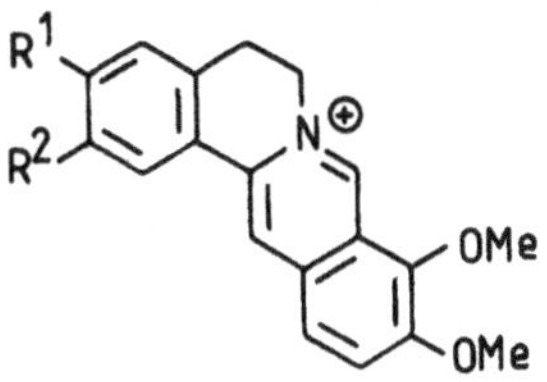

Alkaloid[a]			IC_{50} values (μg ml^{-1}) against P. falciparum		
	R^1	R^2	D-6[b]	W-2[b]	K1[a]
berberine	$-OCH_2O-$		0.14	0.15	0.36
palmatine	OMe	OMe	0.28	0.16	-
jatrorrhizine	OH	OMe	0.42	1.61	1.18
columbamine	OMe	OH	-	-	0.72

[a]as chlorides
[b]Reference 69
[c]Reference 70

Alstonia Alkaloids

A number of species of *Alstonia* (Apocynaceae) are used in traditional medicine throughout S.E. Asia for the treatment of malaria and dysentery.[72,73] *Alstonia* was the subject of a monograph in the 1934 issue of the British Pharmaceutical Codex and was defined as the bark of either *Alstonia scholaris* R.Br. from trees growing in India, Burma and the Philippine Islands or of *A. constricta* F.Muell, which grows in Australia. The action and use of *Alstonia* was regarded as "a tonic in malarial conditions and as a remedy for chronic diarrhoea."[74] It was pointed out in the monograph that its value in malaria treatment could not be compared with that of cinchona bark although it was noted that it produced no bad effects such as cinchonism. Echitamine is the most common alkaloid of *Alstonia* species and it has been asumed that either echitamine or another alkaloid must be the antimalarial principle.

Alstonia angustifolia Wall. is a medium sized tree of the Malaysian peninsula and some nine alkaloids have been isolated from the leaves and 27 alkaloids from the bark.[75] Nine alkaloids isolated from roots of *A. angustifolia* have been assessed for their *in vitro* activities against *P. falciparum* and *E. histolytica*.[76] The results for five of these alkaloids are given in Table 9 and their chemical structures are illustrated in Figure 3. The monomeric alkaloids alstonerine and pleiocarpamine have only weak activities against *P. falciparum* (Kl) and *E. histolytica* whereas the dimeric alkaloids villastonine and macrocarpamine (both are dimers of alstonerine and pleocarpamine) and macralstonine acetate (a dimer of alstonerine) have IC_{50} values of 2.9, 9.3 and 3.4 μM against *P. falciparum* (Table 9). Macrocarpamine was the most active alkaloid against *E. histolytica* with an IC_{50} value of 8.12 μM (Table 9). Two of the dimeric alkaloids were also assessed for their activities against *Giardia intestinalis* and for cytotoxicity against KB cells. The results are given in Table 9 and show some evidence for selection of activity. Villastonine and macralstonine acetate are the most active against *P. falciparum* whilst macrocarpamine was the most active against *E. histolytica* and villastonine the most active against *G. intestinalis*. Villastonine and macralstonine acetate were 80 and 130 times less toxic to KB cells than was emetine hydrochloride (Table 9). It has been reported that at least one *Alstonia* species, *A. coriacea* from New Caledonia, contains a quinoline alkaloid, corialstonine.[77] The structure of another quinoline alkaloid, corialstonidine, has recently been established (G. Massiot, personal communication) and since both alkaloids are structurally related to quinine (Fig. 3), it was of interest to test them for antimalarial activity. We are grateful to Professor G. Massiot for supplying these alkaloids and a sample of echitamine. The *in vitro* activities of the two monomeric quinoline alkaloids against *P. falciparum* are given in Table 10, and comparison is made with echitamine which is one of the most common alkaloids of *Alstonia* species and which has been reported to be the major alkaloid of *A. angustifolia* stem bark from Indonesia.[78] The results show that echitamine has only weak activity which is comparable with the other monomeric alkaloids tested but that corialstonine and corialstonidine have similar IC_{50} values to the dimeric alkaloids tested (Table 9) for antimalarial activity.

There are many other alkaloids from *Alstonia* species which have not been tested for antiprotozoal activities and although the values obtained *in vitro* are low compared to some of the quassinoids, they do tend to lend some support to the traditional use of *Alstonia* species. It is not known whether these alkaloids potentiate the action of one another or of other antiprotozoal drugs and nothing appears to be known of their *in vivo* activities or mode of action.

Table 8. *In vitro* antimalarial, antiamoebic and cytotoxic activities of four protoberberine alkaloids

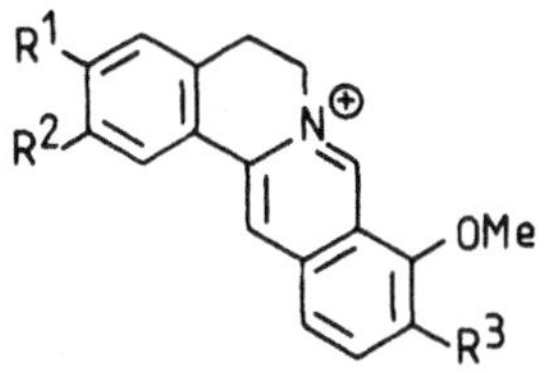

				IC_{50} values (μM)			
Alkaloid[a]	R^1	R^2	R^3	A. *P. falciparum*	*E. histolytica*	B. KB cells	B/A ratio
berberine	$-OCH_2O-$		OMe	0.97	111	7.3	8
thalifendine	$-OCH_2O-$		OH	7.9	115	>698	>88
jatrorrhizine	OH	OMe	OMe	3.1	83	>334	>106
columbamine	OMe	OH	OMe	1.9	156	78	41

[a]as chloride salts

Table 9. *In vitro* antiprotozoal and cytotoxic activities of *Alstonia angustifolia* alkaloids

Compound	Type of Alkaloid	*Plasmodium falciparum* (Kl)	*Entamoeba histolytica*	*Giardia*[a] *intestinalis*	KB[a] cells
Alstonerine (A)	monomer	46.3	75.3	—[b]	—
Pleiocarpamine (P)	monomer	20.5	47.4	—	—
Villastonine	dimer (A+P)	2.9	11.8	8.32	16
Macralstonine acetate	dimer (A+A)	3.4	15.5	31.7	26
Macrocarpamine	dimer (A+P)	9.3	8.12	—	—
Chloroquine diphosphate		0.1	—	—	—
Emetinehydrochloride		—	2.04	5.8	0.2
Metronidazole		—	1.8	2.9	—

[a]unpublished results.

[b]indicates not tested.

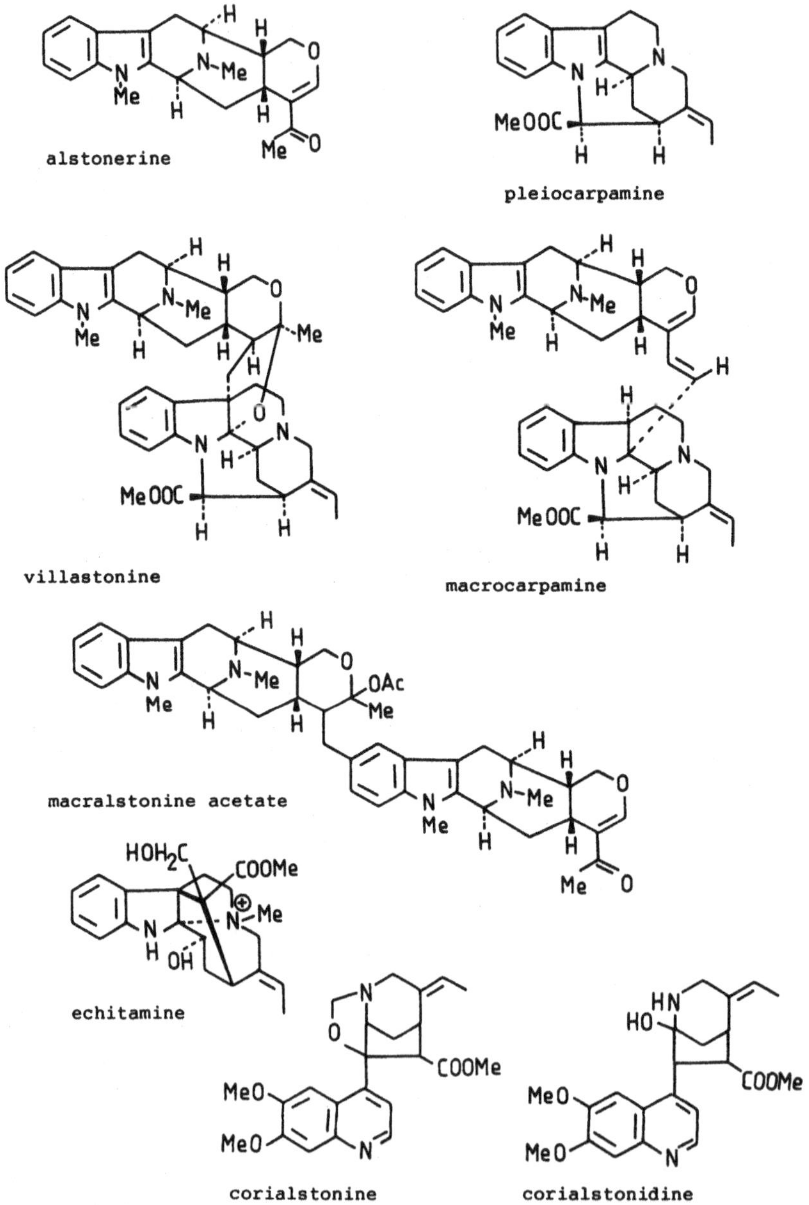

Fig. 3. Some *Alstonia* alkaloids which have been assessed for *in vitro* antiprotozoal activities.

Table 10. *In vitro* activity of echitamine, corialstonine and corialstonidine against *Plasmodium falciparum* (K1)

	In vitro IC_{50} value (μM)[a]
Echitamine	42.6
Corialstonine	5.71
Corialstonidine	5.45

[a]unpublished results.

Cinchona, Cephaelis and Strychnos Alkaloids

Quinine and emetine were the first natural products to be utilized as antiprotozoal drugs. Quinine, the quinoline alkaloid from the barks of some species of *Cinchona* is effective against malaria whilst emetine, the isoquinoline alkaloid of the roots and rhizomes of some *Cephaelis* species, has proved to be effective in the treatment of amoebic dysentery. Quinine and emetine are produced in plants by related biosynthetic pathways. Each molecule of quinine is derived from one molecule of the iridoid glycoside secologanin and one molecule of tryptamine whilst each molecule of emetine is derived from one molecule of secologanin and two molecules of dopamine (Fig. 4). Cinchophylline-type alkaloids are found in the leaves of some *Cinchona* species, each molecule being derived from one molecule of secologanin and two of tryptamine (Fig. 4). Hence, cinchophylline-type alkaloids may be considered as indole analogues of emetine.

In vitro amoebicidal and cytotoxic activities have been reported for some 14 cinchophylline-type alkaloids which were available in our laboratories. Some of these alkaloids showed pronounced *in vitro* amoebicidal activity and 3α,17β-cinchophylline, which possesses the same overall configuration as emetine, was found to be potently amoebicidal but also cytotoxic.[79] Alkaloids which are structurally similar to the cinchophyllines are found in some species of *Strychnos*. One such species, *S. usambarensis*, is used traditionally as an arrow poison by people of the Banyambo tribe who live on the Rwanda-Tanzania border. One of its alkaloids, 3',4'-dihydrousambarensine, has been reported to be active against *E. histolytica* and *Trichomonas vaginalis in vitro*.[80] Seven alkaloids have been isolated from *S. usambarensis* and four of them have been

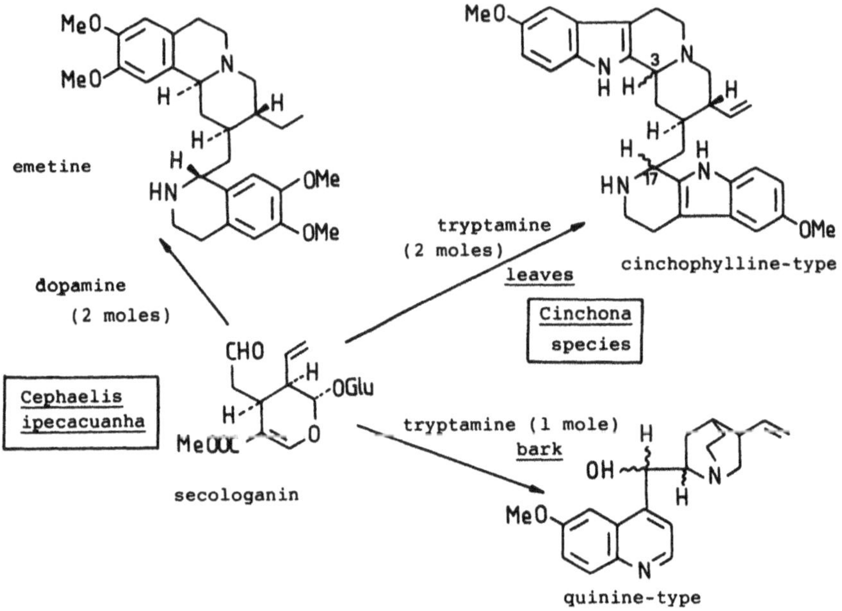

Fig. 4. Biosynthetic relationships between quinine, emetine and cinchophylline-type alkaloids.

tested for *in vitro* activity against *P. falciparum*, *E. histolytica*, *G. intestinalis* and KB cells.[81-83] The chemical structures of these alkaloids are given in Figure 5 and the IC_{50} values are recorded in Table 11. Cytotoxic to antiprotozoal ratios are calculated as the ratio of IC_{50} against KB cells divided by IC_{50} for a specific protozoan; hence, the higher the value, the greater the selectivity for antiprotozoal activity.

The IC_{50} values of usambarensine against the three protozoa are close, within the range of 0.88-1.4 μM and the cytotoxicity to antiprotozoal ratios are between 15 and 24 (Table 11). In contrast, 3',4'-dihydrousambarensine is highly potent against *P. falciparum* with an IC_{50} value of 0.023 μM and with a high cytotoxicity to antiplasmodial ratio of 1474. 3',4'Dihydrousambarensine is equipotent against *E. histolytica* and *G. intestinalis* (IC_{50} values of 4.84 and 3.85 μM, respectively) and with cytotoxicity to antiprotozoal ratios of 7.0 and 8.8, respectively (Table 11). Usambarine has a similar activity against *E. histolytica* (IC_{50} value of 1.02 μM) to usambarensine and its cytotoxicity to

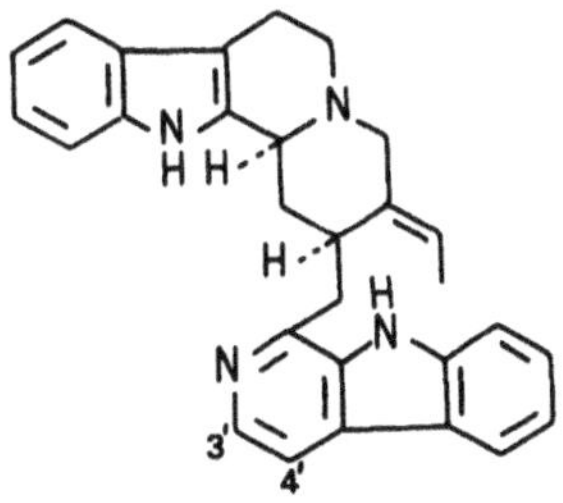

usambarensine

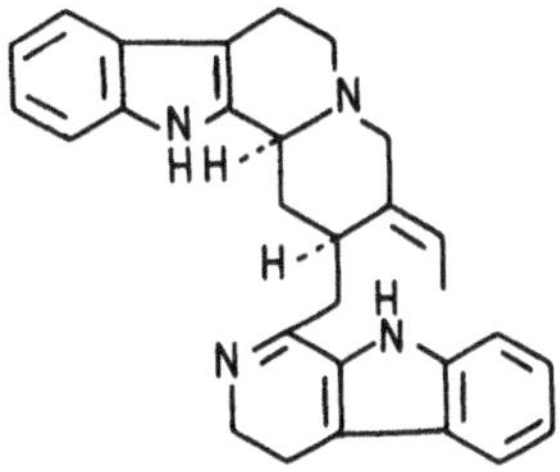

3',4'-dihydrousambarensine

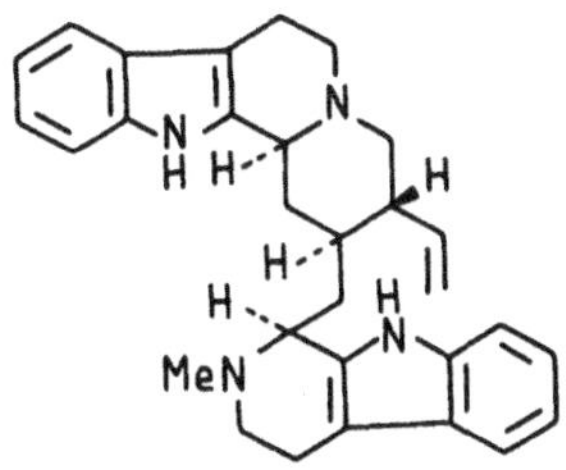

usambarine

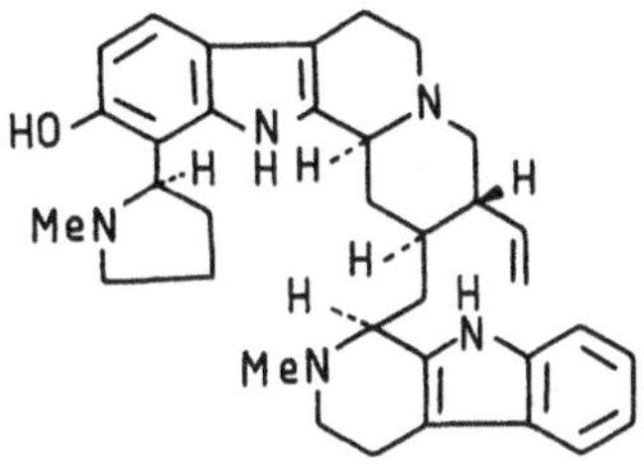

strychnopentamine

Fig. 5. Some alkaloids of *Strychnos usambarensis* which have been assessed for *in vitro* antiplasmodial and cytotoxic activities.

Table 11. *In vitro* antiprotozoal and cytotoxic activities of four alkaloids isolated from *Strychnos usambarensis* and of emetine hydrochloride

Alkaloid	IC50 values (μM)			
	A. *Plasmodium falciparum* (ratio D/A)	B. *Entamoeba histolytica* (ratio D/B)	C. *Giardia*[a] *intestinaliss* (ratio D/C)	D. KB cells[a]
Usambarensine	0.88 (24)	1.13 (18.8)	1.40 (15.1)	21.2
3',4'-Dihydrousambarensine	0.023 (1474)	4.84 (7.0)	3.85 (8.8)	33.9
Usambarine	4.1 (5)	1.02 (20.4)	12.6 (1.6)	20.8
Strychnopentamine	0.09 (213)	7.70 (4.5)	3.01 (11.7)	35.0
Emetine hydrochloride	0.106 (2.57)	2.04 (0.13)	5.81 (0.05)	0.272

[a]unpublished results.

Table 12. *In vitro* antimalarial activity of cryptolepine

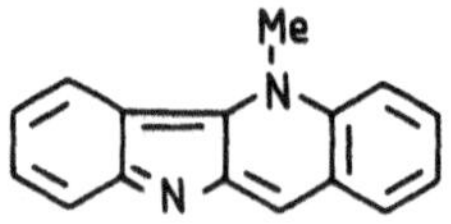

	IC_{50} value against P. falciparum (K1)	
	μgml^{-1}	μM
cryptolepine	0.03	0.134

amoebicidal ratio (20.4) is also similar. The IC_{50} values of usambarine against *P. falciparum* and *G. intestinalis* are 4.1 and 12.6 μM, respectively, with cytotoxicity to antiprotozoal ratios of 5 and 1.6, respectively. Strychnopentamine proved to be most active against *P. falciparum* (IC_{50} 0.09 μM,cytotoxicity to antiplasmodial ratio of 213), whilst its activities against *E. histolytica* and *G. intestinalis* were similar (IC_{50} values 7.70 and 3.01 μM, respectively: cytotoxicity to antiprotozoal ratios of 4.5 and 11.7, respectively).

The activities of emetine hydrochloride against the three protozoa and against KB cells are given in Table 11 for comparison. Emetine hydrochloride is active against the three protozoa in the range of 0.1 to 5.8 μM but its cytotoxicity to antiprotozoal ratios are poor. The four alkaloids from *S. usambarensis*, which differ only slightly in their chemical structures, show interesting selectivity against the three protozoa. 3',4'-Dihydrousambarensine and strychnopentamine are more selective for *P. falciparum* and have favourable cytotoxicity to antiplasmodial ratios. Usambarensine and usambarine were the most active against *E. histolytica* whereas usambarensine was the most active against *G. intestinalis*. All four alkaloids were much less toxic to KB cells (IC_{50} values 20-35 μM) than was emetine hydrochloride (IC_{50} value 0.272 μM).

Dihydrousambarensine and strychnopentamine were tested orally and by subcutaneous injection against *P. berghei* in mice using the four day suppressive test[34] and although no deaths were observed at the highest dose tested (30 mg kg^{-1}) there was no antiplasmodial activity.[81] The *in vivo* effects of these four alkaloids need to be investigated in more detail. The mode of action of the *S. usambarensis* alkaloids may well differ from that of emetine and this also requires further investigation.[83] Molecular modelling of usambarensine, tubulosine and emetine has shown that it is not possible for these alkaloids to adopt the planar conformation which has been one proposal used to explain the activity of emetine on protein synthesis.[84] A common conformation can be adopted by the three alkaloids; this may point to a common receptor which could be slightly different in amoebae which are more sensitive to usambarensine.

Cryptolepine

The root of *Cryptolepis sanguinolenta* (Lindl.) Schlecter is used in Ghana for the treatment of malaria. In one reported clinical trial 4 kg of root was boiled with 32 l of water and the final volume of aqueous extract reduced to 20 l. Doses of two dessertspoons of aqueous extract were given three times daily to 100 patients diagnosed as having *P. falciparum* malaria. Seventy patients completed the trial and it was claimed that complete elimination of parasitaemia resulted after 5 days and no serious side effects were observed.[85] Root decoctions of *Cryptolepis sanguinolenta* continue to be used for the treatment of malaria and in one large clinic at Mampong Akwapim malaria patients are treated regularly with such extracts. It is highly likely that the majority of these patients have acquired some degree of immunity to malaria.

Cryptolepine, the major alkaloid of *C. sanguinolenta* roots has been tested for antimalarial activity against *P. falciparum* (Kl) and found to be highly active (Table 12). The *in vitro* IC_{50} value was 0.03 $\mu g\ ml^{-1}$ (0.134 μM) comparing with chloroquine base under the same test conditions as 0.073 $\mu g\ ml^{-1}$ (0.23 μM).[86] It is speculated that intercalation with DNA is the target for cryptolepine because of its structural similarity to 9-aminoacridine and the bathochromic shift in the UV spectrum of cryptolepine on addition of DNA.

When cryptolepine was administered subcutaneously to mice infected with *P. berghei* no reduction in parasitaemia was observed.[87] In view of the continued clinical use of *Cryptolepis sanguinolenta* as an antimalarial and the potent *in vitro* action of cryptolepine on *P. falciparum*, further investigations of *in vivo* effects are required. Cryptolepine has anti-inflammatory activity as evidenced by its ability to reduce carageenan-induced oedema in rat paw[88] and this effect may also contribute to the antimalarial activity of the root extracts.

CONCLUSION

Not all of our research was included; reference should be made to our work on other antiprotozoal natural products including limonoids,[89] which are closely related biosynthetically to the quassinoids. Other groups include*Cephalotaxus* alkaloids[90,91] and sesquiterpene lactones such as parthenolide and its semi-synthetic analogues.[92] Another area of interest has been the flavonoids, particularly highly methoxylated ones such as artemetin and casticin, which act synergistically with artemisinin against *P. falciparum in vitro* at concentrations which have no intrinsic antimalarial activity.[93,94] As *P. falciparum* matures within erythro-cytes, selective changes in permeability of the erythrocyte membrane occur. Casticin and artemetin are able to inhibit the influx of L-glutamine and myo-inositol across the erythrocyte membrane and this may contribute towards their synergistic activity on artemisinin. It is not known whether flavonoids present in aqueous teas used in traditional medicine exert any synergistic effects on artemisinin in humans.

Our aim has been to illustrate how a combined interdisciplinary approach can be used to further knowledge of natural products with antiprotozoal activities. Most of the compounds discussed come from plants used in traditional medicine and it is salutary to realise that much of the World's population relies on systems of indigenous medicine for the treatment of such severe diseases. Interest in other compounds has resulted from an appreciation of biosynthetic pathways as exemplified by the section on *Cinchona*, *Cephaelis* and *Strychnos* alkaloids. A number of natural products with antiprotozoal activity also exhibit anticancer properties and we propose that all compounds which have been shown to have anticancer activity should be screened for antiprotozoal activity. This may reveal new natural product molecules with activity against protozoa. The main approach discussed here has been to try to identify candidates for *in vivo* studies which have favourable cytotoxic to antiplasmodial ratios.

The research results presented here open up a series of questions, some of which have been addressed in the section on quassinoids. None of these questions can yet be answered fully. One in particular concerns the nature and mechanism of action of the active ingredients within a traditional medicine. Activity may be due to direct antiprotozoal action and this has been demonstrated for a number of isolated natural products. However, plant extracts contain complex mixtures of chemicals and other actions such as anti-inflammatory, antipyretic or immunostimulatory actions may contribute to the overall curative effect. Water soluble extracts used in traditional medicine do not necessarily contain the active principles identified as a result of *in vitro* bioassay-guided screening of organic solvent extracts. There may be much to learn from thorough investigation of aqueous extracts of plants used in traditional medicine.

In vitro screening is not necessarily ideal for this work, but it is one approach to searching for biological activity in plant extracts and is used in laboratories of the developed countries because of economic and ethical considerations. In contrast, *in vivo* assays are often preferred by laboratories in developing countries because of the practical difficulties which may be encountered with *in vitro* test methods. Our findings show some disturbing differences between *in vitro* and *in vivo* test results and with laboratory correlations of alleged efficacy of plant extracts in humans. This can be illustrated by reference to the section on cryptolepine in which the extract of *Cryptolepis sanguinolenta* is used clinically with evidence of effect against *P. falciparum* infections in humans. The major alkaloid cryptolepine is highly potent against *P. falciparum in vitro* whereas it is inactive against *P. berghei* in mice. Berberine poses a similar dilemma because it is active against *P. falciparum in vitro* but is inactive in mice infected with *P. berghei*. In contrast, berberine is inactive *in vitro* against *E. histolytica* but is reportedly active *in vivo*. These differences may reflect the test procedures used and it is possible that the *P. berghei* mouse model has limitations as a reliable indicator of activity against *P. falciparum*. Furthermore, compounds may act as 'pro-drugs' being metabolised *in vivo* to active compounds; conversely, activity may be lost due to lack of absorption or subsequent rapid metabolism and deactivation.

These findings suggest that, wherever possible, both *in vitro* and *in vivo* tests for antiprotozoal activity should be carried out for each compound. This may be accomplished by establishing closer collaborative links between laboratories of developed and developing countries in joint programmes of research.

To date, the main input into antiprotozoal natural products has been concerned with malaria, specifically with *P. falciparum*, and also with *E. histolytica*; comparatively little has been done in searching for natural products with activity against *Leishmania*, *Trypanosoma* or *Giardia*.[11-13] Such extended research programmes also require collaboration and, if it could be properly co-ordinated, it would seem sensible that each plant extract under investigation should be tested for activity against a range of protozoa.

Our results on natural products with antiprotozoal activity show considerable promise, and indicate that there is a vast amount of research which should be undertaken to hopefully lead to the selection and standardization of plants used in traditional medicine, and the development of new drugs.

ACKNOWLEDGMENTS

C.W. Wright thanks the C.W. Maplethorpe Trust; G.C. Kirby the Wellcome Trust; and D.C. Warhurst the Public Health Laboratory Service for financial support. The research described in this chapter was supported by the Medical Research Council, Science and Engineering Research Council, The

Wellcome Trust, the European Economic Community, Royal Pharmaceutical Society, Sir Halley Stewart Trust, the British Council and the Association of Commonwealth Universities.

REFERENCES

1. CONLON, C. 1991. Imported Malaria. The Practitioner 234:841-843.
2. BRADLEY, D.J., WARHURST, D.C., BLAZE, M., SMITH, V. 1991. Malaria imported into the United Kingdom. Communicable Disease Report 1:R45-R48.
3. LAITMAN, C. 1990. Tropical diseases in media spotlight. UNDP/World Bank/WHO Special programme for research and training in tropical diseases. TDR News No 31:3.
4. MAURICE, J. 1991. Kala-azar surges on two fronts. UNDP/World Bank/WHO Special programme for research and training in tropical diseases. TDR News No 37:1-2.
5. GUERRANT R.L. 1986. The global problem of amoebiasis; current status, research needs and opportunity for progress. Rev. Inf. Dis. 8:218-227.
6. BOREHAM, P.F.L. 1991 Giardiasis and its control. Pharm J. August 31, 271-274.
7. WERNSDORFER, W.H. 1991. The development and spread of drug resistant malaria. Parasitol. Today 7:297-303.
8. DAVIDSON, D.E. 1991. Qinghaosu drugs: a new generation of antimalarials. UNDP/World Bank/WHO Special programme for research and training in tropical diseases. TDR News No. 35:7.
9. ANON. 1992. Rediscovering wormwood: quinghaosu for malaria. The Lancet 339:649-651.
10. PHILLIPSON, J.D., O'NEILL, M.J. 1987. Antimalarial and amoebicidal natural products. In: Biologically Active Products. (K. Hostettmann, P.J. Lea, eds.). Clarendon Press, Oxford, pp 50-64.
11. WRIGHT, C.W., PHILLIPSON, J.D. 1990. Natural products and the development of selective antiprotozoal drugs. Phytotherapy Res. 4:127-139.
12. PHILLIPSON, J.D., WRIGHT, C.W. 1991. Medicinal plants against protozoal diseases. Trans. Roy. Soc. Trop. Med. Hyg. 85:18-21.
13. PHILLIPSON, J.D., WRIGHT, C.W. 1991. Antiprotozoal agents from plant sources. Planta Med. 57:S53-S59.

14. PHILLIPSON, J.D., WRIGHT, C.W. 1991. Can ethnopharmacology contribute to the development of antimalarial agents? J. Ethnopharmacol. 32:155-165.

15. WEENEN, H., NKUNYA, M.H.H., BRAY, D.H., MISASUMBI, L.B., KINABOS, L.S., KILIMALI, V.A.E.B., VIJNBERG, J.B.P.A. 1990. Antimalarial compounds containing an α,β-unsaturated carbonyl moiety from Tanzanian medicinal plants. Planta Med. 56:371-373.

16. KARDONO, L.B.S., ANGERHOFER, C.K., TSAURI, S., PADMAWINATA, K., PEZZUTO, J.M., KINGHORN, A.D. 1991. Cytotoxic and antimalarial constituents of the roots of *Eurycoma longifolia*. J. Nat. Prod. 54:1360-1367.

17. CHEN, Z., WANG, B., CHEN, M. 1987. Steroidal bitter principles from *Tacca plantaginea*: structures of taccalonolide A and B. Tet. Lett. 1673-1676.

18. GU, G., FENG, S., WANG, X., ZHOU, Y., LI, G. 1984. Studies on the active principles of D-Er CaO: isolation and structure of japonicine A. Kexue Tong-bao 29:548-549.

19. KONIG, G.M., WRIGHT, A.D., STICHER, O., JURCIC, K., OFFERMANN, F., REDL, K., WAGNER, H., ANGERHOFER, C.K., PEZZUTO, J.M. 1991. Abstract, International Research Congress on Natural Products. 32nd Annual Meeting American Society of Pharmacognosy. Chicago, Illinois, July 21-26.

20. SPENCER, C.F., KONIUSZY, F.R., ROGERS, E.F., SHAVEL, J., Jr., EASTON, N.R., KACZKA, E.A., KUEHL, F.A., Jr., PHILLIPS, R.F., WALTI, J., FOLKERS, K., MALANGA, C., SEELER, A.O. 1947. Survey of plants for antimalarial activity. Lloydia 10:145-174.

21. TRAGER, W., POLONSKY, J. 1981. Antimalarial activity of quassinoids against chloroquine-resistant *Plasmodium falciparum in vitro*. Am. J. Trop. Med. Hyg. 30:531-537.

22. GURU, P.Y., WARHURST, D.C., HARRIS, A., PHILLIPSON, J.D. 1983. Antimalarial activity of bruceantin *in vitro*. Ann. Trop. Med. Parasitol. 77:433-435.

23. GILLIN, F.D., REINER, D.S., SUFFNESS, M. 1982. Bruceantin, a potent amoebicide from a plant, *Brucea antidysenterica*. Antimicrob. Agents Chemother. 22:342-345.

24. WRIGHT, C.W. 1989. Antimoebic and antimalarial natural products. Ph.D. thesis, University of London.

25. WRIGHT, C.W., O'NEILL, M.J., PHILLIPSON, J.D., WARHURST, D.C. 1988. Use of microdilution to assess *in vitro* antiamoebic activities of *Brucea javanica* fruits, *Simarouba amara* stem and a number of quassinoids. Antimicrob. Agents Chemother. 32:1725-1729.

26. ROBERT-GERO, M., BACHRACH, U., BHATNAGAR, S., POLONSKY, J. 1985. Inhibition *in vitro* de la croissance des promastigotes de *Leishmania donovani* par des quassinoides. C.R. Acad. Sc. Paris 300. II:803-806.

27. DESJARDINS, R.E., CANFIELD, C.J., HAYNES, J.D., CHULAY, J.D. 1979. Quantitative assessment of antimalarial activity *in vitro* by a semi-automated microdilution technique. Antimicrob. Agents Chemother. 16:710-718.

28. O'NEILL, M.J., BRAY, D.H., BOARDMAN, P., PHILLIPSON, J.D., WARHURST, D.C. 1985. Plants as sources of antimalarial drugs. 1. *In vitro* test method for the evaluation of crude extracts from plants. Planta Med. 394-397.

29. O'NEILL, M.J., BRAY, D.H., BOARDMAN, P., CHAN, K.L., PHILLIPSON, J.D., WARHURST, D.C., PETERS, W. 1987. Plants as sources of antimarial drugs. 4. Activity of *Brucea javanica* fruits against chloroquine-resistant *Plasmodium falciparum in vitro* and *Plasmodium berghei* in vivo. J. Nat. Prod. 50:41-48

30. CHAN, K.L., O'NEILL, M.J., PHILLIPSON, J.D., WARHURST, D.C. 1986. Plants as sources of antimalarial drugs. 3. *Eurycoma longifolia.* Planta Med. 52:105-107.

31. BRAY, D.H., BOARDMAN, P., O'NEILL, M.J., CHAN, K.L., PHILLIPSON, J.D., WARHURST, D.C., SUFFNESS, M. 1987. Plants as sources of antimalarial drugs. 5. Activities of *Ailanthus altissima* stem constituents and some related quassinoids. Phytotherapy Res. 1:22-24.

32. O'NEILL, M.J., BRAY, D.H., BOARDMAN, P., WRIGHT, C.W., PHILLIPSON, J.D., WARHURST, D.C., GUPTA, M.P., CORREYA, M., SOLIS, P. 1988. Plants as sources of antimalarial drugs. 6. Activities of *Simarouba amara* fruits. J. Ethnopharmacol. 22:183-190.

33. O'NEILL, M.J., BRAY, D.H., BOARDMAN, P., PHILLIPSON, J.D., WARHURST, D.C., PETERS, W., SUFFNESS, M. 1986. Plants as sources of antimalarial drugs. 2. *In vitro* antimalarial activities of some quassinoids. Antimicrob. Agents Chemother. 30:101-104.

34. PETERS, W., PORTUS, J.H., ROBINSON, B.L. 1975. The chemotherapy of rodent malaria XXII. The value of drug resistant strains of *Plasmodium berghei* in screening for blood schizonticidal activity. Ann. Trop. Med. Hyg. Parasitol. 60:155-171.

35. PHILLIPSON, J.D. 1991. Assays for antimalarial and amoebicidal activities. In: Methods in Plant Biochemistry. 6. Assays for Bio-activity. (P.M. Dey, J.B. Harborne, eds.). Academic Press. pp. 135-152.

36. BRAY, D.H., O'NEILL, M.J., PHILLIPSON, J.D., WARHURST, D.C. 1987. *In vivo* antimalarial activity of quassinoids. J. Pharm. Pharmac. 39:85P.

37. ANDERSON, M.M., O'NEILL, M.J., PHILLIPSON, J.D., BRAY, D.H., KIRBY, G.C., WARHURST, D.C. 1988. Antimicrobial activity of an aqueous tea prepared from *Brucea javanica* fruits. J. Pharm. Pharmac.40:52P.

38. ANDERSON, M.M., O'NEILL, M.J., PHILLIPSON, J.D., BRAY, D.H., KIRBY, G.C., WARHURST, D.C. 1989. Antiplasmodial activity of an aqueous infusion prepared from *Brucea javanica* fruits against *Plasmodium falciparum* and *P. berghei*. Trans. Roy. Soc. Trop. Med. Hyg. 83:421.

39. PAVANAND, K., YONGVANITCHIT, K., WEBSTER, H.K., DECHATIWONGSE, T., NUTAKUL, W., JEWVACHDAM-RONGKUL, Y., BANSIDDHI, J. 1988. *In vitro* antimalarial activity of a Thai medicinal plant *Picrasma javanica*. Phytotherapy Res. 2:33-36.

40. ALLEN, D., PHILLIPSON, J.D., WRIGHT, C.W., KIRBY, G.C., WARHURST, D.C. 1991. Bruceine A is not hydrolysed *in vitro* by human plasma esterases or rabbit liver carboxylic acid esterase. J. Pharm. Pharmac. 43: 50P.

41. ANDERSON, M.M., O'NEILL, M.J., PHILLIPSON, J.D., WAR-HURST, D.C. 1991. *In vitro* cytotoxicity of a series of quassinoids from *Brucea javanica* fruits against KB cells. Planta Med. 57:62-64.

42. ANDERSON, M.M. 1992. Cytotoxic and antimalarial natural products. Ph.D. thesis, University of London, pp 1-165.

43. ALLEN, D., TOTH, I., WRIGHT, C.W., KIRBY, G.C., WARHURST, D.C., PHILLIPSON, J.D. 1992. *In vitro* antimalarial and cytotoxic activities of semisynthetic derivatives of brusatol. European J. Med. Chem., in press.

44. PATEL, G., O'NEILL, M.J., BOARDMAN, P., PHILLIPSON, J.D.,

KIRBY, G.C., WARHURST, D.C. 1989. Selective etherification of quassinoids and their *in vitro* antimalarial activity. J. Pharm. Pharmac. 41:90P.

45. GALE, E.F., CUNDLIFFE, E., REYNOLDS, P.E., RICHMOND, M.H., WARING, M.J. 1981. Antibiotic inhibitors of ribosome function. In: The Molecular Basis of Antibiotic Action. (E.F. Gale, E. Cundliffe, P.E. Reynolds, M.H. Richmond, M.J. Waring, eds.). John Wiley and Sons Ltd, London. pp. 402-457
46. LIAO, L.L., KUPCHAN, S.M., HOROWITZ, S.B. 1976. Mode of action of the antitumour compound bruceantin, an inhibitor of protein synthesis. Mol. Pharmacol. 12:167-176.
47. FRESNO, M., GONZALES, A., VASQUEZ, D., JIMENEZ, A. 1978. Bruceantin, a novel inhibitor of peptide bond formation. Biochem Biophys. Acta. 518:104-112.
48. BERAN, Y.M., BENZIE, C.R., KAY, J.E. 1980. Bruceantin, an inhibitor of the initiation of protein synthesis in eukaryotes. Biochem. Soc. Trans. 8:357-359.
49. WILLINGHAM, W., STAFFORD, E.A., REYNOLDS, S.H., CHANEY, S.G., CHANG, S.G., LEE, K-H., OKANO, M., HALL, I.H. 1981. Mechanisms of the eukaryotic protein synthesis inhibition by brusatol. Biochem. Biophys. Acta. 654:169-174.
50. HALL, I.H., LIOU, Y.F., LEE, K-H., CHANEY, S.G., WILLINGHAM, W. 1983. Antitumour agents LIX: effects of quassinoids on protein synthesis of a number of murine tumors and normal cells. J. Pharm. Sci. 72:626-630.
51. KIRBY, G.C., O'NEILL, M.J., PHILLIPSON, J.D., WARHURST, D.C. 1989. *In vitro* studies on the mode of action of quassinoids with activity against chloroquine-resistant *Plasmodium falciparum*. Biochem. Pharmacol. 38:4367-4374.
52. EKONG, R.M., KIRBY, G.C., PATEL, G., PHILLIPSON, J.D., WARHURST, D.C. 1990. Comparison of the *in vitro* activities of quassinoids with activity against *Plasmodium falciparum*, anisomycin and some other inhibitors of eukaryotic protein synthesis. Biochem. Pharmacol. 40:297-301.
53. KIRBY, G.C., WARHURST, D.C., PHILLIPSON, J.D. 1992. Mode of action of antimalarial quassinoids: limited effects *in vitro* upon glycolysis in *Plasmodium falciparum*. Trans. Roy. Soc. Trop. Med. Hyg. 86:343.
54. AL-KHAYAT, A.H.A., KIRBY, G.C., WARHURST, D.C.,

PHILLIPSON, J.D. 1990. *In vitro* antimalarial effects of combinations of protein synthesis inhibitors. Trans. Roy. Soc. Trop. Med. Hyg. 85:310.

55. KEENE, A.T., HARRIS, A., PHILLIPSON, J.D., WARHURST, D.C. 1986. *In vitro* amoebicidal testing of natural products. Part 1. Methodology. Planta. Med. 48:278-284.
56. CASINOVI, C.G., FARDELLA, G., GRANDOLINI, G., BURINATO, C. 1981. Propreta antiamebiche di alcuni derivati dell'ailanthone a della quassina. Il Farmaco 36:116-122.
57. WRIGHT, C.W., MELWANI, S.I., PHILLIPSON, J.D., WARHURST, D.C. 1992. Determination of *in vitro* antigiardial activity by means of soluble formazon production. Trans. Roy. Soc. Trop. Med. Hyg. 86: 517-519.
58. PARTRIDGE, S.J., RUSSELL, P.F., KIRBY, G.C., BRAY, D.H., WARHURST D.C., PHILLIPSON, J.D., O'NEILL, M.J., SCHIFF, P.L., Jr. 1988. *In vitro* antimalarial activity of *Triclisia patens* and of some of its constituent alkaloids. J. Pharm.Pharmac. 40:53P.
59. TACKIE, A.N., DWUMA-BADU, D., OKARTER, T., KNAPP, J.E., SLATKIN, D.J., SCHIFF, P.J., Jr. 1974. Constituents of some West African medicinal plants. II. The isolation of alkaloids from selected *Triclisia* species. Lloydia 37:1-5.
60. DWUMA-BADU, D., AYIM., J.S.K., TACKIE., A.N. 1975. Additional alkaloids of *Triclisia patens* and *Triclisia subcordata*. Phytochemistry 14:2524-2525.
61. DWUMA-BADU, D., AYIM, J.S.K., TACKIE, A.N., SOHLY, M.A., KNAPP, J.E., SLATKIN, D.J., SCHIFF, P.L., Jr. 1975. Trigilletimine: A new bisbenzylisoquinoline alkaloid from *Triclisia* species. Experientia 31:1251-1252.
62. PARTRIDGE, S.J., RUSSELL, P.F., KIRBY, G.C., WARHURST, D.C., PHILLIPSON, J.D., O'NEILL, M.J., SCHIFF, P.L. Jr. 1989. *In vitro* antimalarial activity of *Triclisia dictophylla* and some of its constituents. J. Pharm. Pharmac. 41:92P.
63. PAVANAND, K., WEBSTER, H.K., YONGVANICHIT, K., DECHATIWONGSE, T. 1989. Antimalarial activity of *Triclisia triandra* Diels against *Plasmodium falciparum in vitro*. Phytotherapy Res. 3:215-217.
64. MARSHALL, S.J. 1991. Antimalarial activities of some West African plants. Ph.D. thesis, Council for National Academic Awards Department of Pharmacy, Brighton Polytechnic and The School of

Pharmacy, University of London.

65. EKONG, R., PARTRIDGE, S.J., ANDERSON, M.M., KIRBY, G.C.,WARHURST, D.C., RUSSELL, P.F., PHILLIPSON, J.D. 1991. *Plasmodium falciparum* effects of phaeanthine, a naturally occurring bisbenzylisoquinoline alkaloid, on chloroquine-resistant and -sensitive parasites *in vitro*, and its influence on chloroquine activity. Ann. Trop. Med. Parasit. 85:205-213.
66. YE, Z.-G., VAN DYKE, K. 1989. Selective antimalarial activity of tetrandrine against chloroquine-resistant *Plasmodium falciparum* Biochem. Biophys. Res. Commun. 159:242-248.
67. VAN DYKE, K., 1991. Use of tetrandrine and its derivatives to treat malaria. U.S. Patent 5,025,020.
68. FOURNET, A., MUNOZ, V., MANJON, A.M., ANGELO, A., HOCQUEMILLER, R., CORTES, D., CAVÉ, A., BRUNETON, J. 1988. Activité antiparasitaire d'alcaloides bisbenzylisoquinoleiques. 1. Activité *in vitro* sur des promastigotes de trois souches de Leishmania. J. Ethnopharmacol. 24:327-335.
69. VENNERSTROM, J.L., KLAYMAN, D.L.1988. Protoberberine alkaloids as antimalarials. J. Med. Chem. 31:1084-1087.
70. PARTRIDGE, S.J., RUSSELL, P.F., ANDERSON, M.M., WRIGHT, C.W., PHILLIPSON, J.D., KIRBY, G.C., WARHURST, D.C., SCHIFF, P.L. Jr. 1990. *In vitro* cytoxic antimalarial and antimoebic activities of protoberberine alkaloids. J. Pharm. Pharmac. 42:97P.
71. SUBBAIAH, T.V., AMIN, A.H. 1967. Effect of berberine sulphate on *Entamoeba histolytica*. Nature 215:527-528.
72. BURKHILL, I.M. 1935. A Dictionary of Economic Products of the Malay Peninsular, 2 vols. p.113. Crown Agents for the Colonies, London (Reprint 1966, Ministry of Agriculture and Co-operatives, Kuala Lumpur).
73. PERRY, L.M., METZGER, J. 1980. Medicinal Plants of East and Southeast Asia. MIT Press, Cambridge. pp. 22.
74. ANONYMOUS. 1934. *Alstonia*. British Pharmaceutical Codex, The Pharmaceutical Press, London. pp. 94-95.
75. GHEDIRA, K., ZECHES-HANROT, M., RICHARDS, B., MASSIOT, G., LE MEN-OLIVIER, L., SEVENET, T., GOH, S.H. 1988. Alkaloids of *Alstonia angustifolia*. Phytochemistry 27:3955-3962.
76. WRIGHT, C.W., ALLEN, D., CAI, Y., PHILLIPSON, J.D., SAID, I.M., KIRBY, G.C., WARHURST, D.C. 1992. *In vitro* antiamoebic and antiplasmodial activities of alkaloids isolated from *Alstonia*

angustifolia roots. Phytotherapy Res. 6:121-124.

77. CHERIF, A., MASSIOT, G., LE-MEN-OLIVIER, L., PUSSET, J., LABARRE, S. 1989. Alkaloids of *Alstonia coriacea*. Phytochemistry 28:667-670.

78. HU, W., ZHU, J., HESSE, M. 1989. Indole alkaloids from *Alstonia angustifolia*. Planta Med. 55:463-466.

79. KEENE, A.T., PHILLIPSON, J.D., WARHURST, D.C., KOCH, M., SEGUIN, E. 1987. *In vitro* amoebicidal testing of natural products. Part 2. Alkaloids related to emetine. Planta Med. 49:201-206.

80. OHIRI, F.C., VERPOORTE, R., BAARHEIM-SVENSEN, A. 1983. The African *Strychnos* species and their alkaloids: a review. J. Ethnopharmacol. 9:167-223.

81. WRIGHT, C.W., BRAY, D.M., O'NEILL, M.J., WARHURST, D.C., PHILLIPSON, J.D., QUETIN-LE CLERQ, J., ANGENOT, L. 1991. Antiamoebic and antiplasmodial activities of alkaloids isolated from *Strychnos usambarensis*. Planta Med. 57:337-340.

82. WRIGHT, C.W., O'NEILL, M.J., PHILLIPSON, J.D., WARHURST, D.C., ANGENOT, L., QUETIN-LE CLERQ, J. 1990. Antiamoebic activity of indole analogues of emetine with *in vitro* potency greater than that of emetine. J. Pharm. Pharmacol. 40:94P.

83. WRIGHT, C.W., ALLEN, D., CAI, Y., CHEN, Z.P., PHILLIPSON, J.D., KIRBY, G.C., WARHURST, D.C., SAID, I.M., QUETIN-LE CLERQ, J., ANGENOT, L. 1991. Comparison of the antiplasmodial and cytoxic activities of emetine with those of some related alkaloids isolated from *Alstonia* and *Strychnos* species. Abstract Booklet, British Society of Parasitology, 4th Malaria Meeting, 13-14 February pp 2-3.

84. QUETIN-LE CLERQ, J., DUPONT, J., WRIGHT, C.W., PHILLIPSON, J.D., WARHURST, D.C., ANGENOT, L. 1991. Modelisation moleculaire de usamarensine, de la tubulosin et de l'emetine, alcaloides cytotoxiques et antiambiens. J. Pharm. Belg. 46:85-92.

85. BOYE, G.L., OKU-AMOFO 1983. Proceedings of the first international seminar on cryptolepine. (K. Boakye-Yiadom, S.O.A. Bamgbose, eds.) Kumasi University of Science and Technology, Ghana. pp 1-82.

86. NOAMESI, B.K., PAINE, A., KIRBY, G.C., WARHURST, D.C., PHILLIPSON, J.D. 1991. *In vitro* antimalarial activity of cryptolepine, an indoloquinoline. Trans. Roy. Soc. Trop. Med. Hyg; 85:315.

87. KIRBY, G.C., NOAMESI, B.K., PAINE, A., WARHURST, D.C.,

PHILLIPSON, J.D. 1992. Antimalarial activity *in vivo* of cryptolepine. Trans. Roy. Soc. Trop. Med. Hyg. 86:343.

88. BAMGBOSE, S.O.A., NOAMESI, B.K. 1981. Studies on cryptolepine II. Inhibition of carageenan induced oedema by cryptolepine. Planta Med. 41:392-396.

89. BRAY, D.H., WARHURST, D.C., CONNOLLY, J.D., O'NEILL, M.J., PHILLIPSON, J.D. 1990. Plants as sources of antimalarial drugs. Part 7. Activity of some species of Meliaceae plants and their constituent limonoids. Phytotherapy Res. 4:29-35.

90. EKONG, R.M., KIRBY, G.C., PATEL, G., PHILLIPSON, J.D., WARHURST, D.C. 1990. Comparison of the *in vitro* activities of quassinoids with activity against *P. falciparum*, anisomycin, and some other inhibitors of eukaryotic protein synthesis. Biochem. Pharmacol. 40:297-301.

91. WRIGHT, C.W., TOTH, I., PHILLIPSON, J.D., WARHURST, D.C. 1991. Antiamoebic testing of some natural and synthetic compounds structurally related to berberine. J. Pharm. Pharmac. 43:18P.

92. HOOPER, M., KIRBY, G.C., KULKARNI, M.M., KULKARNI, S.N., NAGASAMPAGI, B.A., O'NEILL, M.J., PHILLIPSON, J.D., ROJATKAR, S.R., WARHURST, D.C. 1990. Antimalarial activity of parthenin and its derivatives. Eur. J. Med. Chem. 25:717-723.

93. ELFORD, B.C., ROBERTS, M.F., PHILLIPSON, J.D., WILSON, R.J.M. 1987. Potentiation of the antimalarial activity of qinghaosu by methoxylated flavones. Trans. Roy. Soc. Trop. Med. Hyg. 81:434-436.

94. YANG, S.L. 1992. Phytochemical studies of *Artemisia annua* L. Ph.D. thesis. University of London. pp. 1-265.

Chapter Two

CHEMICAL STUDIES AND BIOLOGICAL ASPECTS OF SOME MEXICAN PLANTS USED IN TRADITIONAL MEDICINE*

Rachel Mata

Facultad de Quimica, Universidad Nacional Autonoma de Mexico,
Coyoacan, 04510, Mexico D.F., Mexico

INTRODUCTION

Mexican folk medicine has its roots in the traditions of the ancient Mesoamerican civilizations, which according to early post-conquest accounts possessed a rich medicinal herbolaria. The use of herbal remedies still remains in modern Mexico, where an important segment of the population relies on traditional medicine for primary health care. Perhaps the best testimony of their efficacy and cultural value lies in the persistence of medicinal plants in Mexican markets.

***Dedicated to Professor Jerry McLaughlin, Purdue University, Indiana, who introduced me to the interesting and challenging world of medicinal plants.**

To show the richness and variety of the Mexican green medicine it would be necessary to cite the works written from the sixteenth century to modern times. However, the interested reader is referred elsewhere for this information.[1] Regardless of its richness and variety, only a small percentage of the Mexican medicinal flora has been investigated from the phytochemical point of view, and the fraction subjected to biological or pharmacological screening is even smaller. Thus the potential of Mexican plants used in folk medicine as a source for new drugs remains largely unexplored. Therefore, in order to promote further knowledge about these plants, we have instituted a program to investigate their chemical composition and some biological properties. In this paper I review some of the results of our work along this lines.

CHEMICAL AND BIOLOGICAL STUDIES ON *TELOXYS GRAVEOLENS*

The leaves and above-ground parts of *Teloxys graveolens* (Willd.) Weber (Chenopodiaceae), commonly known as "epazote de zorrillo," are highly valued as medicinal by Mexicans. A tea is ingested to alleviate gastrointestinal ailments, to act as a vermifugue, and to relieve headaches and fevers.[2-4]

Chromatographic fractionation of a chloroform extract of the aerial parts allowed the isolation of two new sesquiterpenoids: (+)-8-α-hydroxyelemol (**1**) and (+)-8-α-acetoxycryptomeridiol (**2**).[5] Cryptomeridiol (**3**) and the flavonoids pinocembrin (**4**), pinostrobin, (**5**) and chrysin (**6**) were also obtained.

More recently, bioactivity- directed fractionation of an acetone extract based on the brine shrimp lethality test (BS LT),[6,7] led to the isolation of (**4**) as the only active compound.[8] The toxic acetone extract (BS LC_{50}=157.25 μg/ml) was fractionated by column chromatography on silica gel to yield two toxic major fractions: F-1 (BS LC_{50}=8.06 μg/ml) and F-2 (BS LC_{50}=81.02 μg/ml). Further chromatography of both active fractions afforded pinocembrin as the only toxic compound (BS LC_{50}=4.25 μg/ml).

Pinocembrin also exhibited fasciolicide, ovicide and larvicide activities on newly excysted *Fasciola hepatica* L., on infective eggs of *Ascaridia galli* Dujardin, and on stage III larvae of *Stomoxys calcitrans* Geoffrey, respectively.[8] In all the cases, the effect of graded concentrations of the flavanone was determined. The *in vitro* activity of (**4**) against newly excysted *F. hepatica* was assessed at 10, 5, 1 and 0.1 μg/ml, and the percentage of mortality at these concentrations were 100, 100, 100 and 41 %, respectively. At 10 μg/ml the activity was much better than that of the standard (diamphenetide free amine)

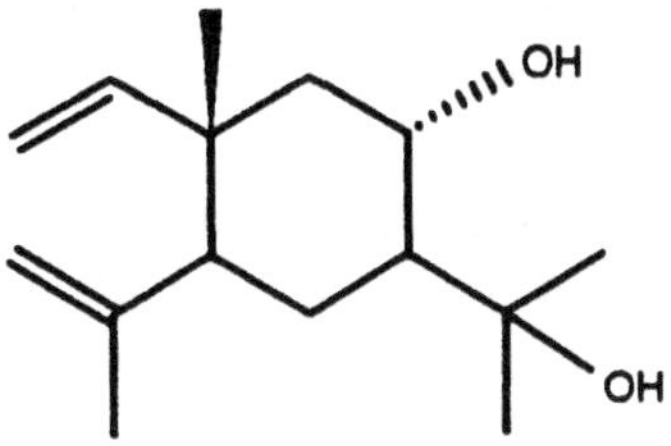

1

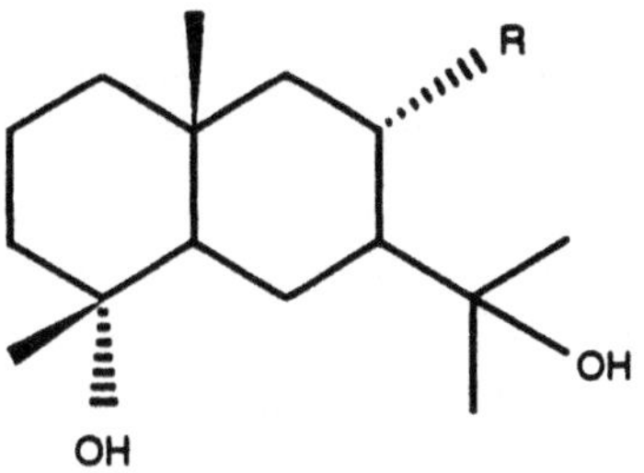

R = - OAc 2

R = - H 3

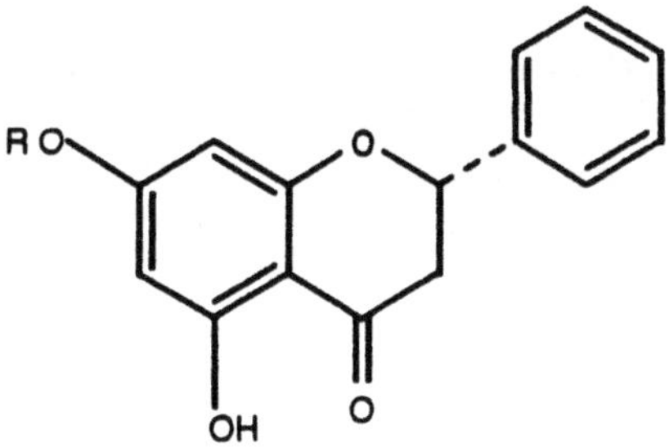

R = - H 4

R = - Me 5

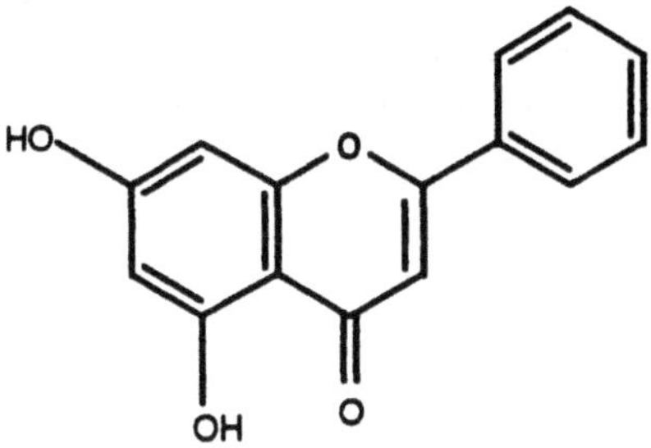

6

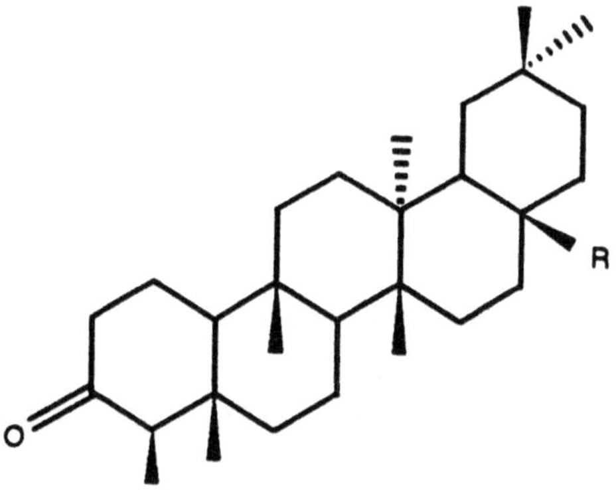

R = - Me 7

R = - CH_2OH 8

R = - CHO 9

R = - COOH 10

because the small flukes were killed and disintegrated.[8] The same effect has been observed for diamphenetide free amine but at higher doses. The ovicide effect on *A. galli* (LC_{50}=623.49 µg/ml), which is a nematode similar to *Ascaris lumbricoides*, and the effect on *S. calcitrans* stage III larvae (LC_{50}=418.69 µg/ml) were rather moderate.

The related flavonoid compounds (**5**) and (**6**) were devoid of any of the anthelmintic or larvicide properties of pinocembrin, indicating that the hydroxyl group at C-7 and the absence of the double bond between C-2 and C-3 are essential for the activities. The anthelmintic properties demonstrated by pinocembrin could be related to the use of *T. graveolens* in folk medicine for the treatments of worms.

Finally, the potential antimicrobial properties of the crude methanol extract of the plant against *Candida albicans* and selected Gram-positive and Gram-negative bacteria were determined. The extract exhibited marginal activity against *Bacillus subtilis*, *Pseudomonas aeruginosa* and *C. albicans*.[9] Of the isolated compounds from *T. graveolens*, (**6**) showed a moderate activity against *C. albicans* [MIC (minimum inhibitory concentration)=25 µg/ml] and *P. aeruginosa* (MIC=31.25 µg/ml), whereas (**4**) was active against *C. albicans* (MIC=25 µg/ml).[9]

ALKALOIDS AND TERPENOIDS OF *HIPPOCRATEA EXCELSA*

Throughout Mexico the reddish-brown bark of the roots and stem of *Hippocratea excelsa* HBK (Hippocrateaceae), known under the name of "cancerina," is used for treating skin cancer, gastric ulcers, kidney illnesses and menstruation disorders. In rural Mexico the plant is known as "mata piojo" (louse killer); as this name suggests, the pesticidal properties are valued by farmers and ranchers.[10] Previous chemical work on the genus dealt mainly with the isolation and structural elucidation of flavonoids[11] and friedelane type of triterpenoids.[12]

A hexane extract prepared from the roots and stem bark of *H. excelsa* showed moderate activity in the brine shrimp lethality test.[13] From this extract a non-toxic mixture of friedelanes precipitated, and the mixture was resolved chromatographically to yield friedelin (**7**), canophyllol (**8**), canophyllal (**9**) and canophyllic acid (**10**). The remaining portion of the extract (BS LC_{50}=650 µg/ml) was further fractionated by column chromatography on silica gel to render a mixture of triterpenoid quinone methides as the only active fraction (BS

LC_{50}=298 μg/ml); preparative TLC allowed the isolation of tingenone (**11**) [BS LC_{50}=235 μg/ml], celastrol (**12**), pristimerin (**13**) and excelsine (**14**), which is the first triterpenoid quinone methide possessing a hydroxymethylene group at C-29.

The presence of tingenone (**11**) could be related to some of the uses of the stem bark of *H. excelsa* in traditional medicine, since it has been demonstrated that topical application of this compound is effective for the treatment of epidermal carcinoma.[14] Tingenone is also active against Yoshida sarcoma, sarcoma 180,[14] and KB [human nasopharyngeal carcinoma] (ED_{50}=2.7 x 10^{-1} μg/ml).[15] Compound (**9**) possesses specific but moderate activity on *P. aeruginosa* (MIC=75 μg/ml).[9]

From the methanol extract obtained from the roots and stem barks of another collection of the plant, three sesquiterpene evoninoate alkaloids, (**15-17**), were isolated.[16] Emarginatine A (**15**) was previously found in *Maytenus emarginata* [17] and displayed significant cytotoxicity (ED_{50}=4 μg/ml) against KB cells. Hippocrateines I (**16**) and II (**17**) are new naturally occurring substances. The structure of (**16**) was unequivocally established by X-ray crystallographic analysis, and it showed marginal activities in the brine shrimp assay (LC_{50}=212 μg) and in the 9 PS (*in vitro* murine leukaemic cells) cytotoxicity test (ED_{50}=1.85 x 10^{-1} μg/ml), but it was inactive in the A-549, HT-29 and MCF-7 human solid tumor cell culture systems (ED_{50} > 50 μg/ml).[16]

It is interesting to point out that the stem and root barks of this plant may be regarded as a practical source of gutta.[10] The overall yield of gutta based on the plant material dry weight is 11 %.

LONG CHAIN PHENOLS FROM *AMPHIPTERYGIUM ADSTRINGENS*

Amphipterygium adstringens Shiede ex Schlecht (Julianiaceae), local name "cuachalalate," is a small tree indigenous to Mexico. The stem bark of this plant is considered effective for the treatment of cholelithiasis, cancer of the gastrointestinal tract, fevers, hypercholesterolemia, gastritis, ulcers, skin burns and fresh wounds.[18] As "epazote" and "cancerina," the stem bark of "cuachalalate" is commonly sold in the Mexican markets, alone or mixed with other medicinal plants.

Some years ago the antitumor activity of a methanolic extract prepared from the stem bark of *A. adstringens* was described by Gonzalez *et al.*[19] The chemistry of the stem bark of this tree has been investigated by several authors.

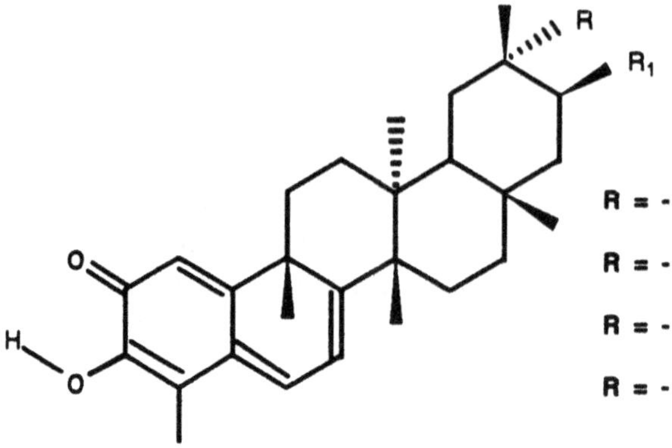

R = - H, R_1 = =O <u>11</u>

R = - COOH, R_1 = - H <u>12</u>

R = - COOMe, R_1 = - H <u>13</u>

R = - CH_2 OH, R_1 = - OH <u>14</u>

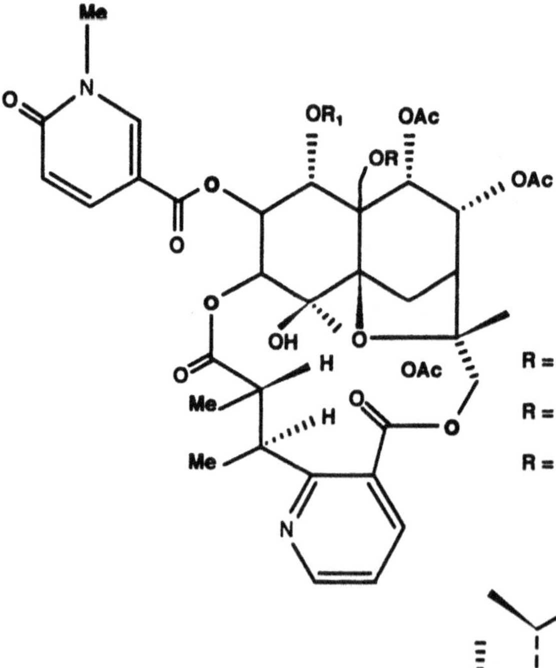

R = R_1 = - Ac <u>15</u>

R = - Ac, R_1 = - OBz <u>16</u>

R = - CO-CH (Me)-CH_2 -Me <u>17</u>

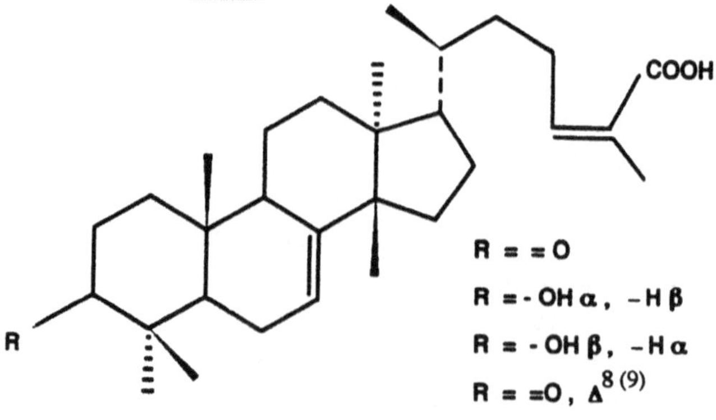

R = =O <u>18</u>

R = - OH α, – H β <u>19</u>

R = - OH β, – H α <u>20</u>

R = =O, $\Delta^{8(9)}$ <u>21</u>

First Gonzalez *et al.*, who conducted a general phytochemical screening on the basis of its promising anticancer activity, demonstrated the presence of sarsapogenin by means of IR spectroscopy and thin layer chromatography.[20] Later, Dominguez *et al.* reported the isolation of instipolinacic[21] and cuachalic[22] acids. Finally, Soriano-Garcia *et al.*[23] described the presence of masticadienonic acid (**18**), whose structure was determined by single crystal X-ray analysis.

In our preliminary study[18] of this plant, column chromatography on silica gel of an hexane extract of the stem bark yielded four known tyrucalane derivatives, (**18-21**), and a mixture of anacardic acids. In subsequent work,[24] we demonstrated that the hexane extract of the stem bark of "cuachalalate," at a dose of 100 mg/Kg subcutaneously, exhibited significant hypocholesterolemic effects on 24 h fasted rats, lowering the cholesterol levels by 31%, an effect similar to that seen for estrone (positive control) at 15 mg/kg given by the same route.

Table 1. Long chain phenolic acids of *A. adstringens* [24]

Compound	Concentration (%)
6-Pentadecyl-salicylic acid (**22**)	50.00
6-Hexadecyl-salicylic acid (**23**)	6.41
6-Heptadecyl-salicylic acid (**24**)	28.84
6-Nonadecyl-salicylic acid (**25**)	8.50
6-[15' (Z)-Nonadecenyl]-salicylic acid (**26**)	6.48

Table 2. Long chain phenolic aldehydes of *A. adstringens* [24]

Compound	Concentration (%)
6-Octadecyl-salicylaldehyde (**27**)	50.00
6-Eicosyl-salicylaldehyde (**28**)	35.00
6-Docosyl-salicylaldehyde (**29**)	15.00

Chromatographic fractionation of the active extract allowed the isolation of compounds (**18**) and (**19**), and two mixtures of long chain phenols, one of them phenolic acids (Table 1) and the second phenolic aldehydes (Table 2). The anacardic aldehydes were not previously described as natural products, although they have been regarded as possible precursors of cardanols.[25] The primary technique for the identification of the individual compounds present in both mixtures was GC/MS as well as ^{1}H and ^{13}C NMR. In the case of the phenolic acid mixture, the components were identified as their methyl ester methyl ether derivatives. The location of the double bond in compound (**26**) was accomplished by epoxidation of the olefin present in the methylated mixture, followed by GC/MS analysis of the crude reaction products. Cleavage between carbons bearing the epoxide function yielded fragments at *m/z* 403, 71 and 43, indicating that the double bond of the parent compound was between C-15' and C-16' of the side chain. The Z configuration of the double bond was assumed from the J values (J= 7 Hz) exhibited by the olefinic protons at d 5.30 in the ^{1}H NMR.[24]

The presence of aldehyde groups for constituents of the second mixture of long chain phenols was indicated initially by a positive reaction with Tollen's reagent. In addition, the ^{1}H NMR spectra exhibited the characteristic aldehyde signal at d 10.92, which did not exchange with D_2O. Neither of the long chain phenolic mixtures or the triterpenoids significantly affected serum cholesterol levels in rats at doses of 15 mg/kg. The acidic mixture slightly lowered the levels (6 %) whereas the aldehydes slightly increased them. The pharmacologic action shown by the hexane extract of *A. adstringens* tends to support the use of the plant as a hypocholesterolemic agent in folk medicine. The presence of phenolic acids as well as other substances from the plant, yet to be isolated, could account for the hypocholesterolemic effect of the hexane extract. On the other hand, the presence of anacardic acids with demonstrated antimicrobial activity,[25] tends to substantiate the ethnomedical use of "cuachalalate' as an antiseptic agent in the treatment of fresh wounds.

CHEMICAL AND BIOLOGICAL STUDIES ON SOME MEXICAN RUBIACEAE

In Mexico, as well as in other parts of the world, traditional healers treat the recurrent fevers typical of malaria with herbal preparations. The stem barks of several Rubiaceae such as *Exostema caribaeum* Jacq., *Exostema mexicanum* Gray, *Hintonia latiflora* (Sesse et Mociño ex D.C.) Bullock (syn.

Coutarea latiflora Sesse et Mociño ex D.C.) and *Simira mexicana* (Bullock) Steyermark, are among the most widely prescribed. All of these species are popularly known by the name of "copalchi," and like *Cinchona* species they possess extremely bitter stem barks.

The genus *Exostema* Richards includes twenty-six species found mainly in Mexico, The West Indies, Costa Rica and northern South America. In

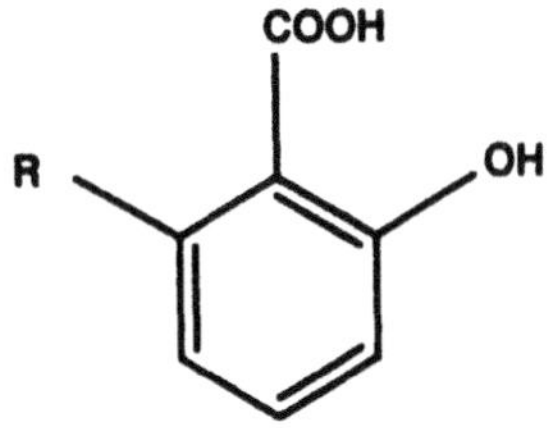

R = - $(CH_2)_{14}$ - Me **22**

R = - $(CH_2)_{15}$ - Me **23**

R = - $(CH_2)_{16}$ - Me **24**

R = - $(CH_2)_{18}$ - Me **25**

R = - $(CH_2)_{14}$ - CH = CH - $(CH_2)_2$ - Me **26**

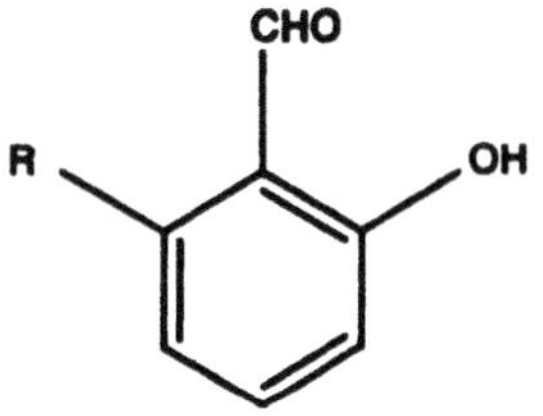

R = - $(CH_2)_{17}$ - Me **27**

R = - $(CH_2)_{19}$ - Me **28**

R = - $(CH_2)_{21}$ - Me **29**

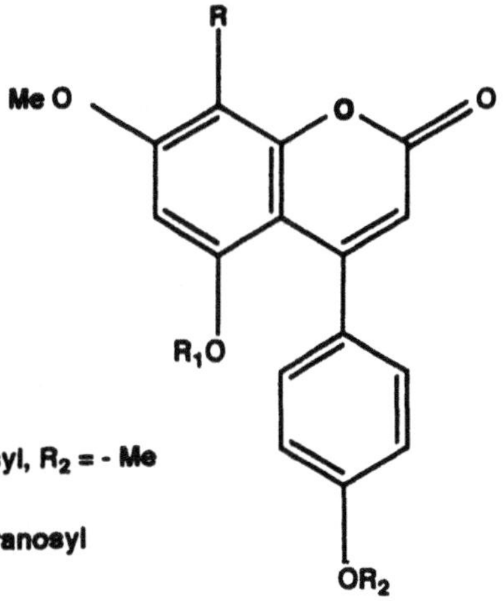

30 R = - OH, R_1 = R_2 = - Me

32 R = - H, R_1 = R_2 = - Me

39 R = - H, R_1 = - β – D - galactopyranosyl, R_2 = - Me

46 R = R_2 = - H, R_1 = - β – D - galactopyranosyl

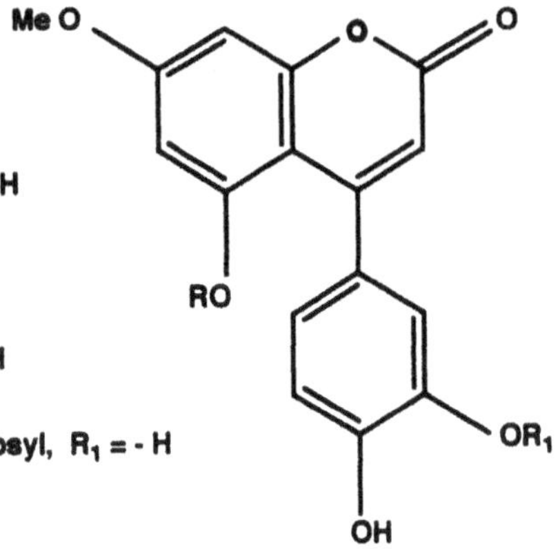

33 R = - β – D - galactopyranosyl, R_1 = - H

34 R = - H, R_1 = - Me

37 R = - β – D - glucopyranosyl, R_1 = - H

38 R = 6'' - acetyl - β – D -galactopyranosyl, R_1 = - H

40 R = R_1 = - H

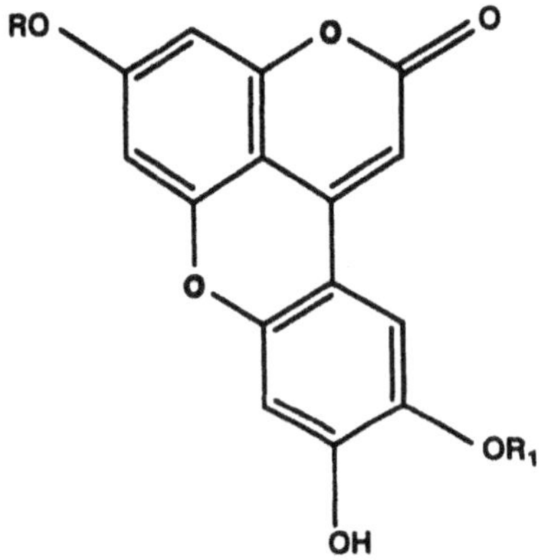

R = - Me , R_1 = - H **31**

R = R_1 = - Me **35**

R = R_1 = - H **36**

Mexico, four species have been described, namely *Exostema coulteri* Hook, *Exostema indutum* Standley, *E. caribaeum* and *E. mexicanum*,[26] but only the last two species are widely used in Mexican folk medicine as substitutes for *Cinchona*.[4,27] *E. caribaeum* is also known locally by the names of "copalchi de Jojutla," "falsa quina" and "quina de Michoacan," among others.[4,28] Previous chemical work on the species resulted in the isolation of exostemin[29] (**30**) and mannitol.[30] From the stem bark of this plant we isolated nine phenylcoumarins, (**31-39**).[31,32] The ground plant material was extracted with methanol. After removal of the solvent, the residue was partitioned between water and ethyl acetate. Chromatographic separation of the organic fraction led to the isolation of the coumarins. Compounds (**31**) and (**32**) were previously isolated from the related species *C. latiflora* [33] and *C. hexandra* Jacq.,[34] respectively, but the other coumarins were new natural products. In each case, the presence of a 4-phenylcoumarin skeleton was deduced by the UV and IR spectra, along with the highly diagnostic resonance of H-3, which usually appeared as a sharp singlet at δ 5.76-6.11; ^{1}H and ^{13}C NMR analysis as well as chemical or enzymatic transformations allowed us to fully elucidate the structures.[31,32] Recently, the structure of 5-*O*-β-D-galactopyranosyl-3',4'-dihydroxy-7-methoxy-4-phenyl-coumarin (**33**) was confirmed by single crystal X ray analysis.[35]

During the course of our investigation on *E. caribaeum*, we demonstrated that 4-phenylcoumarins undergo oxidative cyclization under aerobic, alkaline conditions to give 4-phenyl-5,2'-oxido-coumarins.[31,32] 5,3',4'-Trihydroxy-7-methoxy-4-phenylcoumarin (**40**), obtained by acid hydrolysis of (**33**), was converted into 4',5'-dihydroxy-7-methoxy-4-phenyl-5,2'-oxido-coumarin (**31**) by treatment with KOH in methanol. Also the natural product, 5,3'-dihydroxy-7,4-dimethoxy-4-phenylcoumarin (**34**) yielded 5'-hydroxy-7,4-dimethoxy-4-phenyl-5,2'-oxidocoumarin (**35**) upon reaction with alkali (KOH/MeOH) in the presence of $K_3[Fe(CN)_6]$. Since the reaction took place only in basic conditions and in the presence of air, it might proceed *via* an oxidative phenol coupling process. Previously Bhanu *et al.*[36] achieved this kind of transformation by treatment of 5,2' dihydroxyl 4-phenylcoumarins with boiling HI. In this case the oxide ring formation involved the loss of water between the hydroxyl groups located at positions 2' and 5'. By this method they claimed to obtain several oxido-coumarins, all of them without any substituent in ring B, which were used as intermediates for the synthesis of xanthones. Unfortunately, they did not provide any chemical or spectral data for the synthetic products regarded to be oxido-coumarins.

As previously speculated by Bhanu *et al.*,[36] the coexistence of both 4-arylcoumarins and oxido-coumarins provides evidence of their biogenetic inter-relationship. The biosynthesis of compound (**31**) from (**33**) or (**34**) might involve an oxidative phenol coupling after hydrolysis with a suitable enzyme.

None of the compounds isolated from *E. caribaeum* displayed significant activity against *Plasmodium falciparum in vitro*, nor were they toxic to *A. salina*.[6,7] However, the galactoside (**33**) and 7,4',5'-trihydroxy-4-phenyl-5,2'-oxido-coumarin (**36**) exhibited a selective activity against *C. albicans* with MIC values of 12.5 μg/ml and 15 μg/ml, respectively.[9]

E. mexicanum, in addition to its reputed antimalarial properties, is considered effective for the cure of dengue,[37] and it is known with the popular names of "quina," "melena de leon," "quina blanca" and "sabacche".[28] Bioactivity-guided fractionation of a methanol extract of the stem bark of *E. mexicanum*, based on the brine shrimp lethality test, led to the isolation of two cytotoxic constituents, namely cucurbitacin F (**41**) [BS LC_{50}=16 μg/ml] and cucurbitacin F-25 acetate (**42**) [BS LC_{50}=17.3 μg/ml].[37] The non-toxic glucosides (**43**) and (**37**) were also isolated from inactive fractions of the extract.

Cucurbitacins (**41**) and (**42**) displayed considerable cytotoxicity in several *in vitro* tumor cell culture test systems (Table 3), but in the preliminary test for determining the anti-*Plasmodium falciparum in vitro* (Table 4) they did not exhibit significant activity.

The stem bark of *H. latiflora*, popularly known as "copalchile," "falsa quina," "palo amargo" and "copalquin,"[28] is highly valued in Mexican traditional medicine, not only for the cure of malaria, but also for treating wounds[3] and diabetis mellitus.[38,39] In 1960 Paris and Bastien showed that the extract of *H. latiflora* had no influence on the blood sugar of normal rabbits. However, if the extract was given orally and followed by an oral dose of glucose, the peak of hyperglycemia produced was lower in the test animals than in controls. Coutareoside, a non identified product isolated from this plant, exhibited a similar effect on glycemia. [38] In 1984 Reher and Kraus reported the isolation and structure elucidation of the oxido-coumarin (**31**).[33]

In 1987 we described the isolation of 23,24-dihydrocucurbitacin F (**44**) and 23,24-dihydrocucurbitacin F-25-acetate (**45**) in high yields, from a chloroform extract of the stem bark of this species, collected in the state of Chihuahua, northern Mexico.[40] This was the first time this type of compound had been found in the Rubiaceae family. This finding was of chemotaxonomic significance when we demonstrated later that many related species of the genera *Coutaportla, Portlandia, Exostema, Coutarea* and *Cigarrilla*, with bitter stem barks, contained cucurbitacins as common phytochemical components.[37,41,42]

Table 3. Cytotoxicity of cucurbitacin F (**41**) and cucurbitacin F-25-acetate (**42**) [37]

Compound	Tumor System	ED_{50} (µg/ml)
41	A-549	3.44×10^{-1}
	MCF-75	2.3×10^{-1}
	HT-29	3.27×10^{-1}
42	A-549	$< 10^{-2}$
	MCF-73	1.4×10^{-2}
	HT-29	$< 10^{-2}$

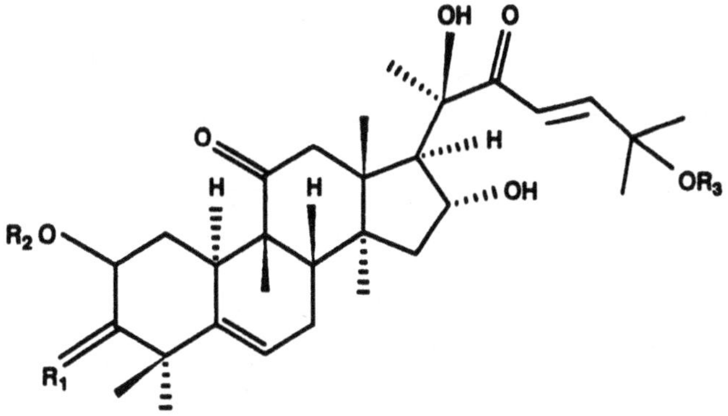

R_1 = - OH α , - H β, R_2 = R_3 = - H **41**

R_1 = - OH α , - H β, R_2 = - H, R_3 = - Ac **42**

R_1 = - OH α , – H β, R_2 = β – D - glucopyranosyl, R_3 = - Ac **43**

R_1 = - OH α , - H β, R_2 = R_3 = - H, 23, 24 - dihydro **44**

R_1 = - OH α , – H β, R_2 = - H, R_3 = - Ac, 23, 24 - dihydro **45**

R_1 = - O - β – D - glucopyranosyl α , - H β, R_2 = R_3 = - H, 23, 24 - dihydro **48**

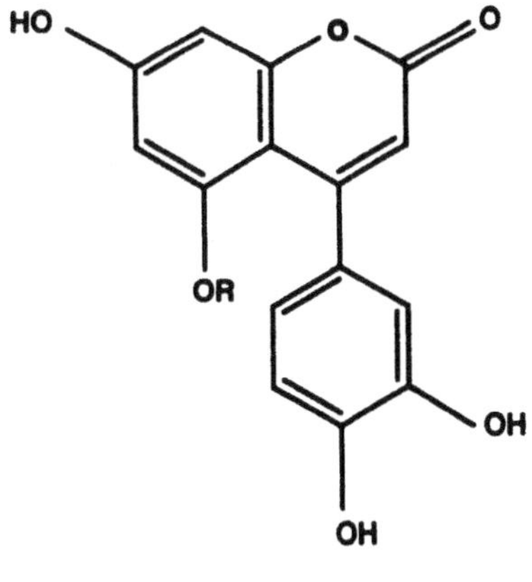

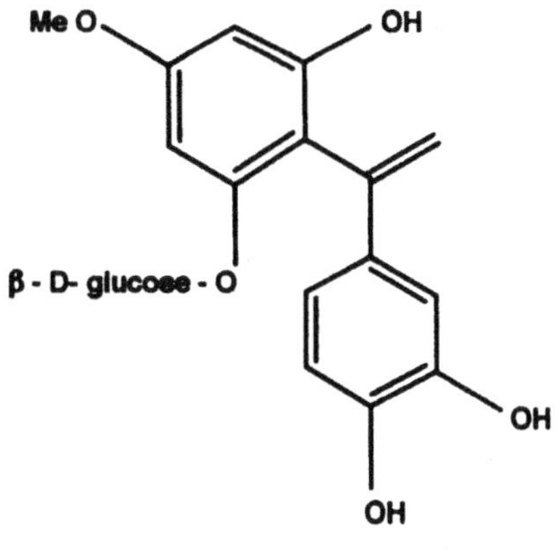

R = 6''- acetyl - β - D - glucopyranosyl **47**

R = - β – D - glucopyranosyl **50**

49

More recently, we found that methanolic extracts of the defatted stem bark of *H. latiflora* contained four new glycosides which include the phenylcoumarins (**46**) and (**47**), the cucurbitacin (**48**)[43] and the novel phenylstyrene, 6-*O*-β-D-glucopyranosyl-2,3',4'-trihydroxy-4-methoxy-α-phenylstyrene (**49**). The latter represents the first example of this type of compound found in nature.[44] Additionally, the known phenyl coumarins (**33**), (**37**) and (**50**) were obtained.[43] The proposed structure of compound (**49**) was based on spectroscopy and chemistry.

The coexistence of (**49**) with phenylcoumarins suggest a common biogenesis which may proceed via an intermediate such as (**51**). According to Seshadri's hypothesis,[45] intermediate (**51**) could generate the lactone ring to give 4-phenylcoumarins. Alternatively it might be decarboxylated to yield phenylstyrenes.

Although compound (**33**) possesses moderate activity against *C. albicans*, the crude methanol extract of *H. latiflora* was inactive against the yeast, gram-positive and gram-negative bacteria. None of the new glycosides (**46-49**) exhibited antiseptic properties against the several organisms tested.[9] The methanol extract and compounds isolated from *H. latiflora* were devoid of *in vitro* anti-*Plasmodium falciparum* activity.

The last "copalchi" we have investigated is *Simira mexicana*, also known as "quina roja." The infusion of the red stem bark, alone or with other "quinas," is not only employed as a remedy for malaria but also for dengue.[46] From this plant the known alkaloids harman and strictosamide were isolated. Harman was toxic to brine shrimp (LC_{50}=25 μg/ml) and exhibited moderate *in vitro* anti-*P. falciparum* activity (IC_{50}=0.97 μg/ml).[46]

In summary, it is probable that none of "copalchis" we have studied so far contains compounds with significant antimalarial properties. However, the

Table 4. Anti-*Plasmodium falciparum in vitro* activity of cucurbitacin F (**41**) and cucurbitacin F-25 (**42**)

Compound	IC_{50} (μg/ml)
41	7.8
42	3.1

traditional preparations made with these plants may alleviate the symptoms of the disease and, in some cases, the cure might be achieved immunologically. It is interesting to point out that the only species containing alkaloids, but no quinine, was *S. mexicana* . It has been claimed that *H. latiflora* possesses quinine,[3] but, in all the collections we investigated alkaloids were not detected.

CONSTITUENTS OF *RATIBIDA LATIPALEARIS*

Ratibida latipalearis Richards (Asteraceae) is a yellow-flowered perennial herb restricted to the pine-oak forest of the Sierra Madre Occidental of western Chihuahua, Mexico. The roots and leaves are heated and applied topically to skin wounds and inflammations. An infusion of the roots is imbibed to alleviate headaches.[47] Chemical investigation of the related species, *R. columnifera* and *R. peduncularis*, led to the isolation of several sesquiterpene lactones[48,49] acetylenes[48] and aromatic compounds including flavonoids and simple phenylpropanoids.[49]

An extract of the whole plant of *R. latipalearis*, prepared with a mixture of methanol and chloroform 1:1, exhibited moderate activity in the brine shrimp lethality test [LC_{50}=366.61 µg/ml]. Successive column chromatography on silica gel of the active extract, monitoring the fractions with the brine shrimp bioassay, led to the isolation of ratibinolide (**52**)[47] [LC_{50}=62.75 µg/ml) and ratibinolide II (**53**)[50] [LC_{50}=69.53 µg/ml]. The structures of both compounds were unequivocally established by spectral and X-ray crystallographic analyses.

Table 5. Cytotoxicity of a methanol extract of *Ratibida latipalearis*

Human Tumor system	ED_{50} (µg/ml)
Sarcoma	4.6
Breast	2.4
Lung	7.7
Colon	4.0
KB (epidermoid carcinoma)	6.2

It was subsequently demonstrated that the methanol extract of *R. latipalearis* exhibited marked activities against *B. subtilis* and *C. albicans*.[9] These results support the use of the plant as an anti-infective agent. On the other hand, the extract showed significant cytotoxicity in several *in vitro* cell systems (Table 5).

CONSTITUENTS OF *DODONAEA VISCOSA* AND *ANREDERA SCANDENS*

Dodonaea viscosa (L.) Jacq. (Sapindaceae) is a shrub widely distributed in tropical and subtropical areas of the world. The plant is widespread through Mexico and reputed to possess several medicinal properties. The leaves are used against several kinds of aches, rheumatism, skin infections, fevers and swellings, whereas the roots are employed as an astringent. Some popular names of this species are "chapuliztle," "granadina," "cuerno de cabra," "jirumi," "hierba de la cucaracha" and "ocotillo".[4]

From the aerial parts of this species several flavonoids,[51-57] the diterpenoid- derived hautriwaic (**54**)[58] and dodonoic[59] acids, condensed tannins,[51] some triterpenoids[60] and the essential oil[61] were isolated and characterized. Additionally, two biologically active saponins, dodonosides A and B, were obtained from the seeds.[62] The saponin mixture exhibited antiexudative, phagocytosis enhancement and molluscicidal activities.[62]

The significant antimicrobial activity of a methanol extract[9] against *C. albicans* and Gram-positive and Gram-negative bacteria prompted us to investigate this plant. Column chromatography on silica gel of the active extract led to the isolation of a new *ent*-labdane diterpene [*ent*-15,16-epoxy-9αH-labda-13(16),14-diene-3β,8a-diol (**55**)] and a novel *p*-coumaric acid ester of L-*myo*-inositol [1-L-1-*O*-methyl-2-acetyl-3-*p*-coumaryl-*myo*-inositol (**56**)].[63] Additionally, the known flavonoids, sakuranetin, 6-hydroxykaempferyl-3-7-dimethyl ether, not previously described in the plant, and hautriwaic acid were obtained. The structures were determined by spectroscopic analysis and chemical reactions. The NMR assignments of compound (**55**) were based on 2D heteronuclear chemical shift correlation , APT and DEPT spectra of the natural product and some derivatives. The *ent*-type absolute configuration was indicated by the negative Cotton effect of the CD curve of the ketone derivative. Alkaline hydrolysis of (**56**) with NH_4OH afforded *p*-coumaric acid and (+)bornesitol. The stereochemistry and relative disposition of the substituents in the cyclitol moiety were deduced from the natural substance's NMR data and its acetyl derivative.

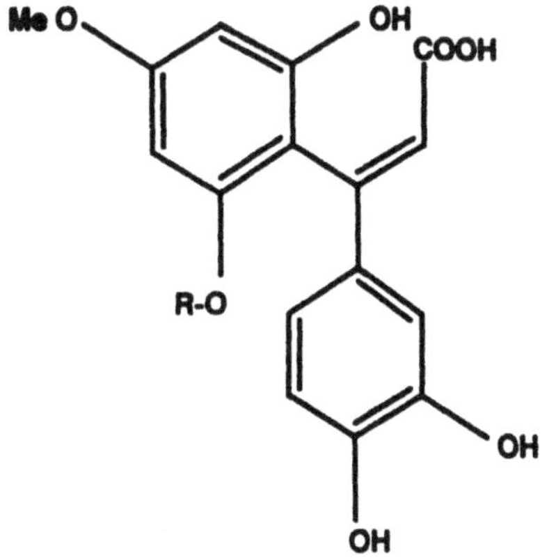

51

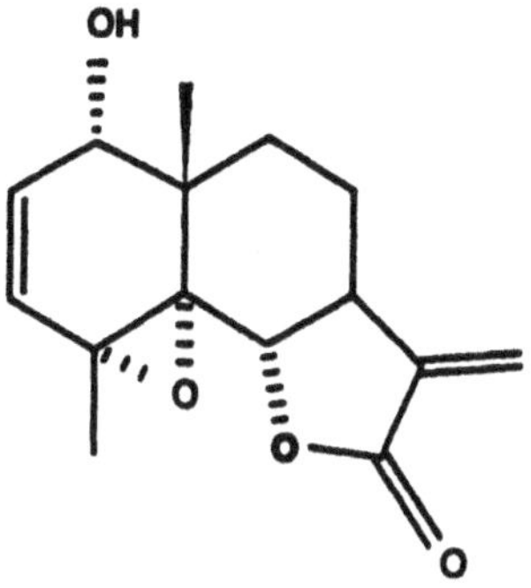

52

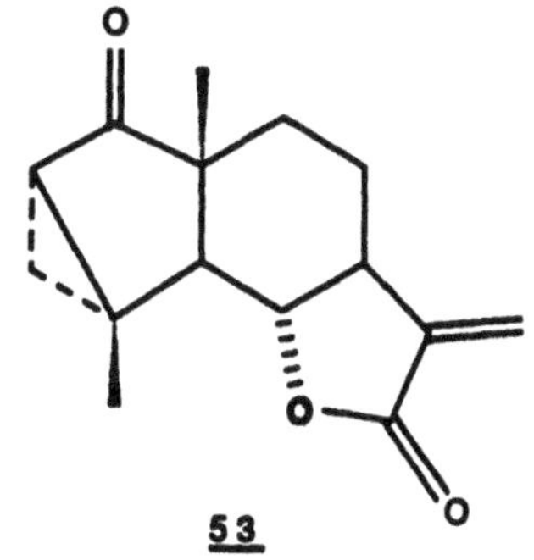

53

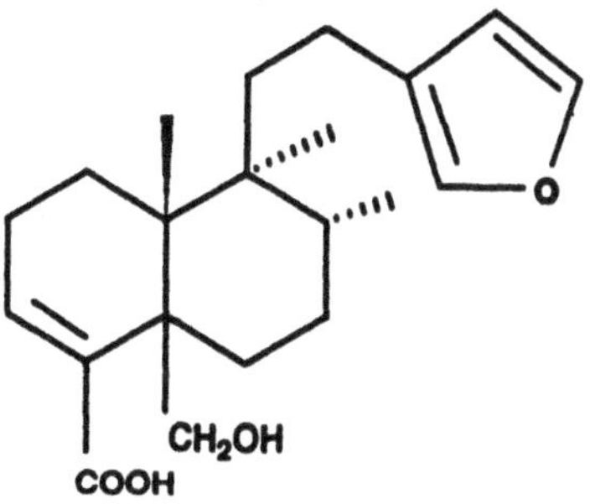

54

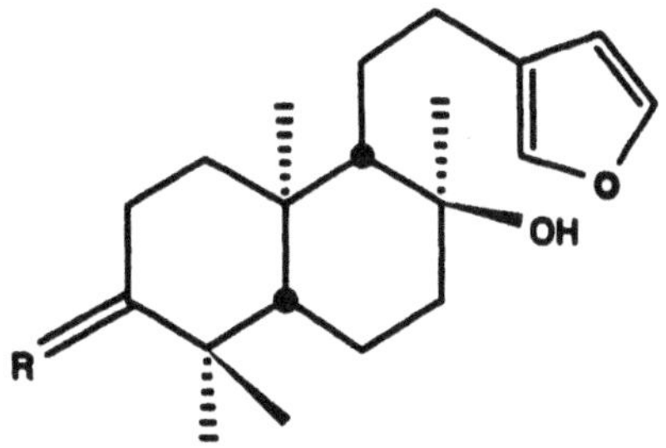

55

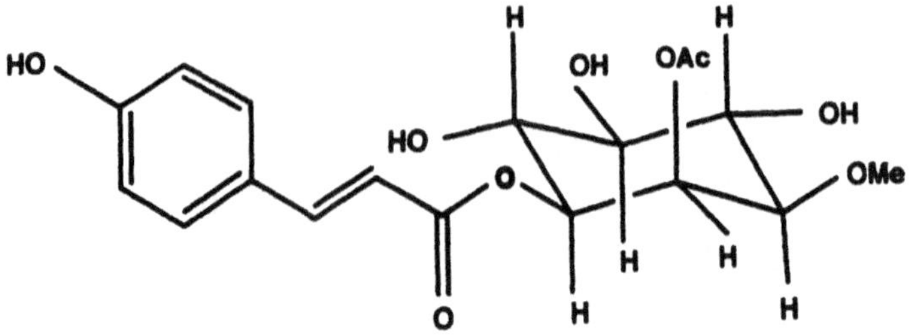

56

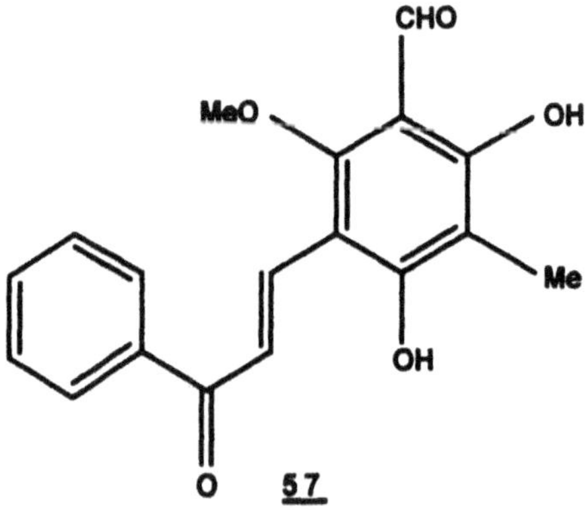

57

Although the preliminary screening of the crude extract supported the use of *D. viscosa* as anti-infective agent in folk medicine, none of the isolated compounds exhibited significant antimicrobial activity; only the two diterpenoids showed marginal activity against *Staphylococcus aureus*.[9]

The rhizomes of *Anredera scandens* (L.) Moq. (Basellaceae), commonly known as "sacasil," are used traditionally for the treatment of broken bones and flesh wounds.[64] The new naturally occurring retrochalcone, 2,4-dihydroxy-6-methoxy-5-formyl-3-methylchalcone (**57**), was isolated from a methanol extract of dried rhizomes of "sacasil" by column chromatography, and was characterized by spectral means.[64] This product exhibited moderate activity against *S. aureus* (MIC=13.5 μg/ml) and *P. aeruginosa* (MIC=15 μg/ml).

CONCLUSION

The results of our chemical and biological investigations on some Mexican plants used in traditional medicine have been reviewed. Although the wide range of properties attributed to each plant remains to be investigated, it has been possible to provide scientific evidence supporting some of the ethnomedical uses of *Teloxys graveolens*, *Amphipterygium adstringens*, *Hippocratea excelsa*, *Ratibida latipalearis* and *Dodonaea viscosa*. Bioassay-guided fractionation of four species, based on the brine shrimp lethality test, led to the isolation of several bioactive compounds. Upon further biological exploration, two cytotoxic triterpenoids from *Exostema mexicanum* and one anthelmintic flavanone from *T. graveolens* were discovered. Also, a few new antimicrobial substances were obtained from *Hintonia latiflora*, *Exostema caribaeum* and *Anredera scandens*.

During the course of the investigation of several medicinal species of the Rubiaceae, we found that 4-phenyl coumarins undergo oxidative cyclization, under alkaline conditions to give 4-phenyl-5,2'-oxidocoumarins. Additionally, the first naturally occurring phenylstyrene was obtained from *Hintonia latiflora*.

This short survey clearly indicates that the Mexican medicinal flora shows a considerable potential for the discovery of new or known substances with valuable biological activities.

ACKNOWLEDGEMENTS

This review was carried out during a sabbatical leave sponsored by DGAPA, UNAM; their support is gratefully acknowledged. The author is also grateful to all her students; without them these investigations could not have been initiated.

REFERENCES

1. LOZOYA, X. 1984. Bibliografia Basica Sobre Herbolaria Medicinal de Mexico. Secretaria de Desarrollo Urbano y Ecologia, Mexico.
2. BYE, R. 1985. Medicinal plants of the Tarahumara Indians of Chihuahua, Mexico. In: Two Mummies from Chihuahua: A Multidisciplinary Study. (R.A. Tyson, D.V. Elerick, eds.), San Diego Museum Papers, No 19, San Diego, California, pp.77-104.
3. BYE, R. 1986. Medicinal plants of the Sierra Madre: Comparative study of the Tarahumara and Mexican market plants. Economic Botany, 40:103-124.
4. MARTINEZ, M. 1989. Las Plantas Medicinales de Mexico. Ediciones Bota, Mexico D.F., pp 657.

5. MATA, R., NAVARRETE, A., ALVAREZ, L., PEREDA-MIRANDA, R., DELGADO, G., ROMO DE VIVAR, A. 1987. Flavonoids and terpenoids of *Chenopodium graveolens*. Phytochem. 26:191-193.
6. MEYER, B.N., FERRIGNI, N.R., PUTNAM, J.E., JACOBSEN, L.B., McLAUGHLIN, J.L. 1982. Brine shrimp: A convenient general bioassay for active plant constituents. Planta Medica. 45:31-34.
7. ANDERSON, J.E., GOETZ, C.M., McLAUGHLIN, J.L., SUFFNESS, M. 1991. A blind comparison of simple bench-top bioassays and human tumor cell cytotoxicities as antitumor prescreens. Phytochem. Anal. 2:107-111.
8. CAMACHO, M. del R., SANCHEZ, B., QUIROZ, H., CONTRERAS, J.L., MATA, R. 1991. Pinocembrine, a bioactive flavanone from *Teloxys graveolens*. J. of Ethnopharmacology. 31:383-389.
9. ROJAS, A., HERNANDEZ, L., PEREDA-MIRANDA, R., MATA, R. 1992. Screening for antimicrobial activity of crude drug extracts and pure natural products from Mexican medicinal plants. Journal of Ethnopharmacology. 32:275-283.
10. PALACIOS, J., MATA, R., LOPEZ, R., LINARES, E., BYE, R. 1989. *Hippocratea excelsa* (Hippocrateaceae), a new source of *trans* polyisoprene. Econ. Bot. 43:508-509.
11. ILYAS, M., ILYAS, N. 1986. Flavonoids from the flowers of *Hippocratea mucronata* . Nigerian Journal of Pharmacy. 2:99-101.
12. BRUNING, R., WAGNER, H. 1978. Ubersicht die Celastraceen-inhalsstoffe: Chemie, chemotaxonomie, biossynthese, pharmakologie. Phytochem. 17:1821-1858.
13. CALZADA, F., MATA, R., LOPEZ, R., LINARES, E., BYE, R., BARRETO, V.M., DEL RIO, F. 1991. Friedelanes and triterpenoid quinone methides from *Hippocratea excelsa*. Planta Medica. 57:194-195.
14. MARINI-BETOLO, G.B. 1979. Un grupo particular de sustancias naturales. Las fenoldienonas triterpenicas. Revista Latinoamericana de Quimica. 10:97-104.
15. KUTNEY, J.P., BEALE, M.H., SALISBURY, P.J., STUART, K.L., WORTH, B.R., TOWNSLEY, P.M., CHALMERS, W.T., NILSSON, K., JACOLI, G.G. 1981. Isolation and characterization of natural products from plant tissue cultures of *Maytenus buchananii*. Phytochem. 20:653-657.
16. MATA, R., CALZADA, F., DIAZ, E., TOSCANO, R.A. 1990. Chemical studies on Mexican plants used in traditional medicine XV.

Sesquiterpene evoninoate alkaloids from *Hippocratea excelsa.* J. of Natural Products. 53:1212-1219.

17. KUO, Y.-H., CHEN, C.-H., YANGKUO, L.-M., KING, M.-L., WU, T.-S., LU, S.-T., CHEN, I.-S., McPHAIL, D.R., McPHAIL, A.T., LEE, K.-H. 1989. Structure and stereochemistry of emarginatine-A, a novel cytotooxic sesquiterpene pyridine alkaloid from *Maytenus emarginata* : X-ray crystal structure of emarginatine-A monohydrate. Heterocycles. 29:1465-1468.
18. NAVARRETE, A., MATA, R., DELGADO, G. 1989. Alkylanacardic acids from *Amphipterygium adstringens.* Planta Medica. 55:579.
19. GONZALEZ, E.E., McKENNA, G.P., DELGADO, J.N. 1962. Anti-cancer activity of *Amphipterygium adstringens.* J. of Pharmaceut. Sci. 51:901-905.
20. GONZALEZ, E.E., DELGADO, J.N. 1962. Phytochemical investigation of *Amphipterygium adstringens.* Journal of Pharmaceut. Sci. 51:786-790.
21. DOMINGUEZ, X., FRANCO, R., GARCIA, S., PORRAS, M.E., VAZQUEZ, G., AMEZCUA, B. 1983. Plantas medicinales Mexicanas XLVIII: Estructura del acido instipolinacico separado de la corteza del cuachalalate *(Amphipterygium adstringens).* Revista Latinoamericana de Quimica. 14:99-100.
22. WATSON, W.H., DOMINGUEZ, X.A., VAZQUEZ, G., GARCIA, S. 1987. Cuachalic acid, a new triterpene from *Amphipterygium adstringens.* Revista Latinoamericana de Quimica. 18:89-90.
23. SORIANO-GARCIA, M., TOSCANO, R.A., ORTIZ, B., NAVARRETE, A., SANCHEZ-OBREGON, G.R., BARRIOS, H., YUSTE, F. 1987. Structure and stereochemistry of the methyl ester of (5a, 13a, 14b, 17a, 20 S,24Z) 3-oxalanosta-7,24-dien-26-oic acid (masticadienonic acid). Acta Crystallographyca C. 43:990-992.
24. MATA, R., CALZADA, F., NAVARRETE, A., DEL RIO F., DELGADO, G. 1991. Long chain phenols from the bark of *Amphipterygium adstringens.* J. of Ethnopharmacology. 31:147-154.
25. TYMAN, J.H.P. 1979. Non-isoprenoid long chain phenols. Chem. Soc. Reviews. 8:499-537.
26. STANDLEY, W. 1975. Flora de Guatemala. Field Museum of Natural History, New York.
27. AIELLO, A. 1979. A reexamination of *Portlandia* (Rubiaceae) and associated taxa. J. of the Arnold Arrboretum. 60:38-126.
28. MARTINEZ, M. 1987. Catalogo de Nombres Vulgares y Cientificos de

Plantas Mexicanas. Fondo de Cultura Economica, Mexico D.F. 1116 pp.

29. SANCHEZ-VIESCA, F. 1969. The structure of exostemin, a new 4-phenylcoumarin isolated fron *Exostema caribaeum*. Phytochem. 8:1821-1823.
30. KREBS, K.G., GRIESINGER, E. 1960. Inhaltsstoffe der rinde von *Exostema caribaeum*. Arzneimittel-Forschung. 10:32-33.
31. MATA, R., CALZADA, F., GARCIA, M. del R., REGUERO, M.T. 1987. Chemical studies on Mexican plants used in traditional medicine, III: New 4-phenylcoumarins from *Exostema caribaeum*. Journal of Natural Products. 50:866-871.
32. MATA, R., CALZADA, F., GARCIA, M. del R. 1988. Chemical studies on Mexican plants used in traditional medicine, VI. Additional new 4-phenylcoumarins from *Exostema caribaeum*. J. of Natural Products. 51:851-856.
33. REHER, G., KRAUS, J. 1984. New neoflavonoids from *Coutarea latiflora*. J. of Natural Products. 42:172-174.
34. DELLE MONACHE, G., BOTTA, B., SERAFIM, A., ALVES de LIMA, R. 1983. 4-Arylcoumarins from *Coutarea hexandra*. Phytochem. 22:1657-1658.
35. SORIANO-GARCIA, M., VILLENA-URIBE, R., MENDOZA-DIAZ, S., CAMACHO, M. del R., MATA, R. 1992. Structure of 5-*O*-β-D-galactopyranosyl-7-methoxy-3',4'-dihydroxy-4-phenylcoumarin trihydrate. Acta Crystallographyca C. 48. In press.
36. BHANU, S., SAROJA, T., SESHADRI, T.R., MUKERJEE, S.K. 1972. Conversion of 4-phenylcoumarins into xanthones. Biogenetic implications. Indian J. of Chem. 10:577-580.
37. MATA, R., ALBOR, C., PEREDA-MIRANDA, R., McLAUGHLIN, J.L. 1990. Cytotoxic constituents of *Exostema mexicanum*. Planta Medica. 56:241.
38. PARIS, R.R., BASTIEN, M. 1960. On the hypoglycemic action of two plants called copalchi: *Coutarea latiflora* and *Croton niveus*. Ann. Pharm. Franc. 18:205-219.
39. BEVER, B.O., ZAHND, G.R. 1979. Plants with oral hypoglycemic action. Quarterly J. of Crude Drug Research. 17:139-196.
40. REGUERO, M.T., MATA, R., BYE, R., LINARES, E., DELGADO, G. 1987. Chemical studies of Mexican Plants used in traditional medicine. II. Cucurbitacins from *Hintonia latiflora*. J. of Natural Products. 50:315.

41. MATA, R., RIOS, L., CAMACHO, M. del R., REGUERO, M.T., LORENCE, D. 1988. Triterpenes fron *Cigarrilla mexicana.* Phytochem. 27:1887-1889.
42. MATA, R., CASTANEDA, P., CAMACHO, M del R., DELGADO, G. 1988. Chemical studies on Mexican plants used in traditional medicine, V. Cucurbitacin glucosides from *Cigarrilla mexicana.* J. of Natural Products. 51:836-839.
43. MATA, R., CAMACHO, M. del R., CERVERA, E., BYE, R., LINARES, E. 1990. Secondary metabolites from *Hintonia latiflora.* Phytochem. 29:2037-2040.
44. MATA, R., CAMACHO, M. del R., MENDOZA, S., CRUZ, M. del C. 1992. A phenylstyrene from *Hintonia latiflora* (Rubiaceae). Phytochem. 31. 3199-3201.
45. MANITO, P. 1981. Biosynthesis of Natural Products. Ellis Horwood Limited, New York, pp. 378.
46. CASTANEDA, P., ALBOR, C., MATA, R., BYE, R., LINARES, E. 1991. Alkaloids from *Simira mexicana.* Fitoterapia. 52:366.
47. MATA, R., ROJAS, A., SORIANO, M., VILLENA, R., BYE, R., LINARES, E. 1990. Ratibinolide, a new sesquiterpene lactone from *Ratibida latipalearis.* Heterocycles. 31:1111-1114.
48. HERZ, W., KULANTHAIVEL, P., GOEDKEN, V.L. 1985. Structures of the ratibidanolides, sesquiterpenes lactones with a new carbon skeleton, and unusual xanthanolides from *Ratibida columnifera.* J. of Organic Chemistry. 50:610- 617.
49. ELLMAUERER, E., PATHAK, V.P., JAKUPOVIC, J., BOHLMANN, F., DOMINGUEZ, X.A., KING, R.M., ROBINSON, H. 1987. 6-β-Lactonized xanthanolides from *Ratibida* species. Phytochem. 26:159-163.
50. ROJAS, A., VILLENA, R., JIMENEZ, A., MATA, R. 1991. Chemical studies on Mexican traditional medicine, XXI. Ratibinolide II, a new sesquiterpene lactone from *Ratibida latipalearis.* J. of Natural Products. 54:1279-1282.
51. SASTRY, K.N.S., NAYUDAMMA, Y. 1966. Leucocyanidin from *Dodonaea viscosa* bark. Leather Sciences. 13:174.
52. NAIR, R.A.G., SANKARA, S.S. 1973. Isorhamnetin and quercetin glycosides from *Dodonaea viscosa* and *Sapindus emarginatus.* Indian J. of Chemistry. 13:639-640.
53. DREYER, D.L. 1978. Kaempferol methyl ethers from flowers of *Dodonaea viscosa.* Revista Latinoamericana de Quimica. 9:97-98.

54. DOMINGUEZ, X.A., FRANCO, R., CANO, G., CHAVEZ, N. 1980. Aislamiento de 3,6,4'-trimethoxy-5,7-dihydroxy-flavona en el chapuliztle (*Dodonaea viscosa* var angustifolia Jacq., Sapindaceae). Revista Latinoamericana de Quimica. 11:150-151.

55. SACHDEV, K., KULSHRESHTHA, D.K. 1982. Aliarin, a new flavonoid from *Dodonaea viscosa* Linn. Indian J. of Chemistry. 21B:798-799.

56. SACHDEV, K., KULSHRESHTHA, D.K. 1983. Flavonoids from *Dodonaea viscosa*. Phytochem. 22:1253-1256.

57. SACHDEV, K., KULSHRESHTHA, D.K. 1986. Viscosol, a C-3' prenylated flavonoid from *Dodonaea viscosa*. Phytochem. 25:1967-1969.

58. HSU, Y.-H., CHEN, Y.P., KAKISAWA, H. 1971. Structure of hatriwaic acid. Phytochem. 10:28113-2814.

59. SACHDEV, K., KULSHRESHTHA, D.K. 1984. Dodonic acid, a new diterpenoid from *Dodonaea viscosa*. Planta Medica. 50:448-449.

60. DIMBI, M.Z., KAPUNDU, M., DARIMONT, E., WARIN, R., DELAUDE, C., HULS, R. 1985. Bull. Soc. Chim. Belg. 94:141-148.

61. MEKKAWA, A.G., MOSSA, J.S. 1981. Essential oil of *Dodonaea viscosa*. Die Pharmazie. 36:517.

62. WAGNER, H., LUDWIG, C., GROTJAHN, L., KHAN, M.S.Y. 1987. Biologically active saponins from *Dodonaea viscosa*. Phytochemistry 26:697-701.

63. MATA, R., CONTRERAS, J.L., CRISANTO, D., PEREDA-MIRANDA, R., CASTANEDA, P., DEL RIO, F. 1991. Chemical studies on Mexican plants used in traditional medicine. XVIII. New secondary metabolites from *Dodonaea viscosa*. J. of Natural Products. 54:913-917.

64. CALZADA, F., MATA, R., BYE, R., LINARES, E. 1990. A retrochalcone from *Anredera scandens*. Phytochemistry 29:2737-2738.

Chapter Three

CHEMICAL AND BIOLOGICAL EXTRACTIVES OF LAURACEAE SPECIES IN COSTA RICAN TROPICAL FORESTS

Oscar C. Castro *

National Institute of Biodiversity (INBio)
P.O. Box 22-3100, Santo Domingo
Heredia, Costa Rica

INTRODUCTION

The family Lauraceae is represented in the world by about 2500 species of trees. Only *Cassytha*, a genus of about 20 species, mostly Australian and with one unique pantropical species, contains herbaceous, parasitic species.[1] The family's greatest diversity is in Southeast Asia and South America, being common in the evergreen tropics and subtropics, with a few congeners in

*Dedicated to Prof. Luis. J. Poveda, Botanist of the Costa Rican National Museum herbarium and the National University of Heredia (UNA), for contributing to my understanding of the tropical forest of Costa Rica

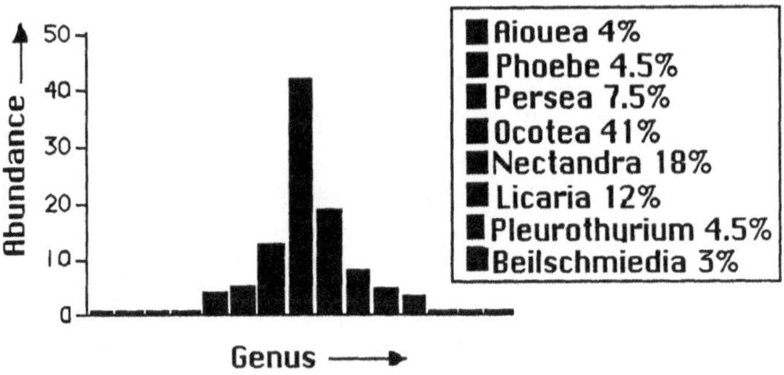

Fig. 1. Distribution of Lauraceae species in Costa Rica.

seasonally very dry regions. The major diversity of this family is found in the American tropics where approximately 40 genera containing about 1000 species are represented.[2] Because of its geographical tropical location, serving as a "bridge" between the continental masses and a barrier between the two oceans, and with a volcanic topography and geological and climate conditions, Costa Rica is one of the most diverse regions of the world in terms of its ecosystems. It contains a flora with approximately 1800 species of trees and shrubs, distributed in 124 families and 670 genera, constituting a major renewable resource with great potential.[3]

After the Leguminosae, the Lauraceae constitute the most important component of the Costa Rican tropical forest. In this country with less than 51,000 km^2 of surface, this family of plants is represented by 106 species distributed in 15 genera, 7 of which (*Aniba*, *Cariodaphnopsis*, *Cassyta*, *Endlicheria*, *Litsea*, *Povedsdaphne* and *Williamodendron*) contain only one unique species (Fig. 1).[4] The best represented genera are *Ocotea* and *Nectandra* with approximately 59% of all species, a composition similar to other tropical environments. Other genera represented by fewer species are *Licaria* (12%), *Persea* (7.5%) and *Phoebe* (4.5%), but these are important biomass components of the Costa Rican tropical forest. One genus, *Povedadaphne*, is known only in Costa Rica, and *Williamodendron* is represented by only by a single species collected from Colombia and the Amazon basin.

DISTRIBUTION AND ECOLOGY OF LAURACEAE IN COSTA RICA

Species of Costa Rican Lauraceae grow mostly in the mountain regions of the country between 500 and 2500 m above sea level, in places where the precipitation could reach 2000-3000 mm. This constitutes the humid forest, the very humid forest and the pluvial forest, using the classification system described by Holdridge.[5] In these regions one finds the greatest diversity of trees and shrubs, probably because the ecological surroundings favor such a diversification.[6] Only five species appear above 2500m; *Beilsmedia ovalis, Licaria excelsa, Ocotea calophylla, Ocotea pittieri, Ocotea austinii*, and six below 500m; *Aiouea obscura, Licaria cofudontisii, Licaria pergamentaceae, Cariodaphnopsis burgeri, Ocotea puberula* and *Ocotea helicterifolia* in the seasonally very dry lowlands of Guanacaste.

A close relationship has been found between the population density of birds and the abundance of Lauraceae fruits in our tropical forest, particularly in regard to annual fluctuations. It has been suggested that the fruits regulate breeding and also control migratory patterns.[7,8] Recent studies in the Monteverde National Reserve indicate that 4 species of *Nectandra* (*N. davidsoniana, N. membranaceae, N. hypoleuca, N. salicina*), together with species of *Ocotea, Persea* and *Phoebe* maintain a respectable population of frugivorous birds of *Procnias tricarunculata* and *Tityra semifasciata* (Cotingidae) for 5 months from June to October during which time they provide 59% of the diet. They also provide 63% of the diet for the quetzal *Pharomachrus mocinno*, (Trogonidae). Other birds like the tucans, *Aulacearhynchus presinus, Rhamphastos sulfuratus* (Ramphastidae), *Chamaespestes inucolor* (Crucidade), *Turdus plebejus* and *Myadestes melanops* (Turdidae) also feed on these plants.[9]

It is well known that most Lauraceae seeds are free of predation. The fruits are eaten by common rodents, but both birds and rodents reject the seeds. The larvae of the weevil *Heilipus draco* Farb (Curculionidae), however, develop almost exclusively in the seeds of species of this family, causing severe damage to fruit crops. In addition, small 'domains' are very common on the lower leaf surfaces of Lauraceous plants; they appear to be associated more with mites than with ants.[10]

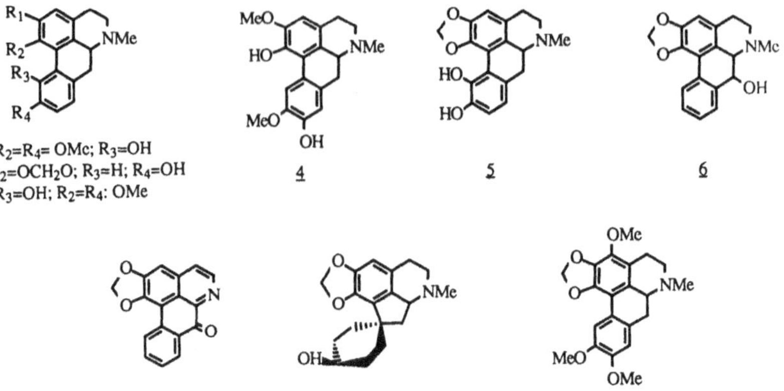

1: $R_1=R_2=R_4=$ OMe; $R_3=OH$
2: $R_1,R_2=OCH_2O$; $R_3=H$; $R_4=OH$
3: $R_1=R_3=OH$; $R_2=R_4$: OMe

4 5 6

7 8 9

Fig. 2. Alkaloids from *Phoebe clemenseii*, *P. formosana* and *P. porfiria*

THE GENUS *PHOEBE*

In the Americas the major center of speciation is in Mexico and Central America, with a secondary center in Tropical-Continental America. There are a few representatives in the Antilles. The *Phoebe* genus is not well defined. Many species described in Central America are better placed in *Ocotea*. Kostermans excludes all the American species, placing them in *Cinnamomun* .[11]

A short review of previous work in the genus indicated that three species *P. clemensii* (New Guinea), *P. formosana* (Taiwan) and *P. porfiria* (Argentina) elaborate aporphine alkaloids (12). The leaf alkaloids of *P. clemensii* include isocorydine (**1**), 10-hydroxy-1,2-methylenedioxy aporphine (**2**), and N-methyl-lindcarpine (**3**), while the bark contains largely laurolitsine (**4**). Nine alkaloids have been detected in *P. formosana*, but only four have been identified: the wood yielded roemerine (**5**) and laurolitsine, and the bark laurolitsine, ushinsunine (**6**) and liriodenine (**7**). In addition, an unusual proaporphine (**8**) was isolated from the same species.[13] Only ocoteine (**9**) was identified from *P. porfiria* bark (Fig. 2.), and flavan-3-ols and proanthocyanidins from *P. cinnamomifolia* (Colombia).[14]

None of the eleven *Phoebe* species represented in the herbarium of the Museum of Costa Rica were used in popular medicine. Four of these species contain alkaloids: *P. mollicella*, *P. pittieri*, *P. tonduzii* and *P. valeriana* (Table 1). Both known and new aporphines and their derivatives have been identified.

Table 1. Alkaloid distribution in *Phoebe* species for the Costa Rican rain forest

Plant	Part	Alk	Plant	Part	Alk
P. brenesii	(l,b,w,f)	-	*P. mollicella*	(l,b,w)	+
P. chavarriana	(l,b,w,f)	-	*P. neurophylla*	(l,b,w)	-
P. cinnamomifolia	(l,b,w)	-	*P. pittieri*	(l,b,w)	+
P. costarricana	(l,b,w)	-	*P. tonduzii*	(l,b,w)	+
P. hammeliana	(l,b,w,f)	-	*P. valeriana*	(l,b,w)	+
P. mexicana	(l,b,w,f)	-			

l= leaves; b= bark; w= wood; f= fruits.

Norlirioferine (**10**) and lirioferine (**11**) were identified from *P. pittieri* leaves, while lirioferine, norphoebine (**12**), norpurpureine (**13**) and the known benzylisoquinoline reticuline (**14**) were found in the bark and wood. Preliminary tests of alkaloid-containing crude extracts of this plant were shown to have some antibacterial and antifungal activities.[15,16] The known aporphine alkaloids, norpurpureine, purpureine and preocoteine (**15**), as well as a new alkaloid, norpreocoteine (**16**) that was isolated from *P. mollicella* bark.[13] Leaves of *P. tonduzii* (now *O. holdridgeiana*) yielded four aporphines with different substitution patterns on ring D: isocorydine (**1**), *o,o*-dimethylcorytuberine (**17**), 3-hydroxynuciferine (**18**) and 3-methoxynuciferine (**19**), together with the common flavonoids quercetin and catechin in large amounts (approximately 3%) (Fig 3).[17]

P. valeriana was the most exhaustively studied species. Three new penta-substituted aporphine alkaloids were isolated from leaves: phoebine (**20**), 3-hydroxynorphoebine (**22**), nordelporphine (**24**), along with one dehydro-aporphine (**26**), one new phenantrene alkaloid (**27**) and one new oxoaporphine (**28**). In addition, five known aporphine alkaloids (nantenine (**29**), thaliporphine (**30**), 3-hydroxy-glaucine (**25**), norphoebine (**21**), 3-hydroxyphoebina (**23**) and the oxoaporphine *o*-methylmoschatoline (**31**) were found (Fig 4).[18]

The presence of aporphines is consistent with the previous results reported for other species of this genus. It has been observed, however, that Costa Rican species frequently produce penta-oxygenated aporphines in the aromatic ring, and these, compared with the tetra-oxygenated aporphines, are relatively rare in nature .

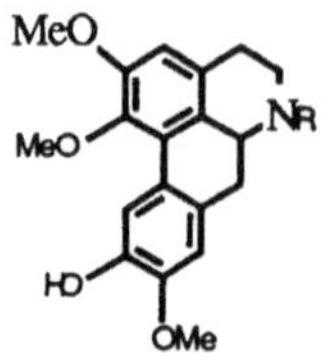

10: R=H
11: R = Me

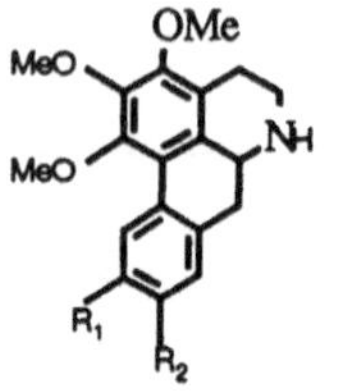

12: $R_1, R_2 = OCH_2O$
13 : $R_1 = R_2 = OMe$

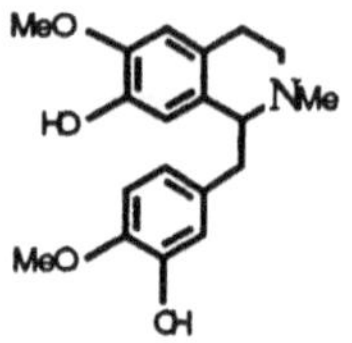

14

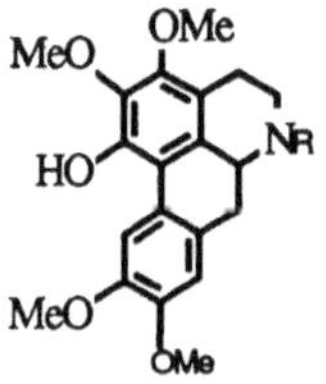

15: R = Me
16: R = H

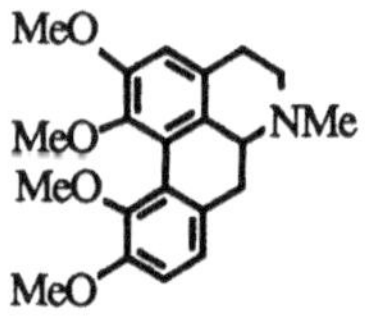

17

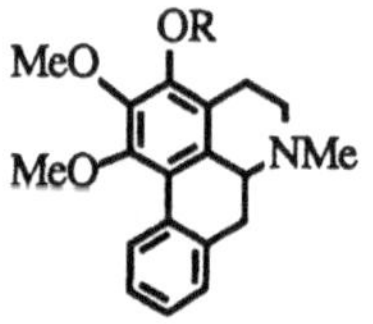

18: R =H
19: R = Me

Fig. 3. Aporphine alkaloids from *Phoebe pittier* (now *Ocotea pittieri*), *P. mollicella* (now *O. mollicella*) and *P. tonduzii* (now *O. holdridgeiana*).

All these alkaloids, when exposed to iodine vapors over silica gel, produce characteristic colors that go from the yellow to green, passing through brown and sky-blue colors. This can be easily seen after simple chromatographic methods. With ultraviolet light, phenanthrenic and oxoaporphinic structures give orange/yelow fluorescent colors. Generally these unsaturated alkaloids are usually present in small amounts.

We have found that those species of *Phoebe* that do not accumulate alkaloids elaborate mainly arylpropanoid derivatives, lignans and flavonoids. Stem bark of *P. mexicana* yielded the lignan (+)-sesamin (**32**) and the flavan-3-ol (-)-epicatechin (**39**), while *P. brenesii* stem bark produced benzyl *trans*-cinnamate (**34**) in appreciable quantities (0.3%), and the unusual methylsulfonylallyl *trans*-cinnamate (**35**) (0.22%), previously identified from *Cinammomun triplinervis*. (Fig. 5).[19,20] From leaves of *P. cinnamomifolia*, we identified two known lignans, epiashantin (**36**) and diayangambin (**37**). Previous work on the bark of this plant had indicated the presence of flavan-3-ols and proanthocyanidins (Fig. 5).[21] Recently, we found two common flavan-3-ols, (-)-epiafzelechin (**38**) and (+)-catechin (**33**), in *P. hammeliana*, as well as

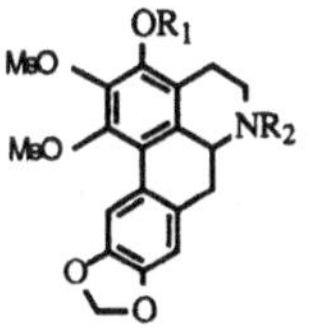

20: $R_1 = R_2 = Me$
21: $R_1 = Me; R_2=H$
22: $R_1 = R_2 = H$
23: $R_1 = H; R_2=Me$

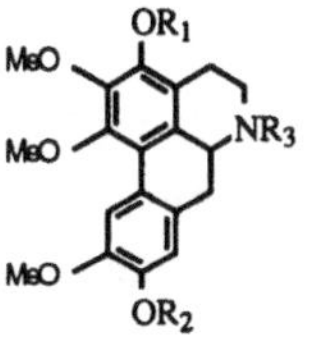

24: $R_1 = R_2 = R_3 = H$
25 :$R_1 = H; R_2=R_3=Me$

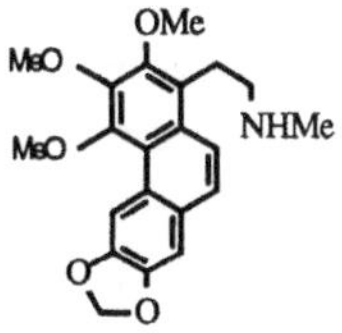

26

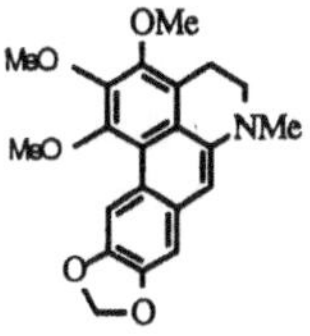

27

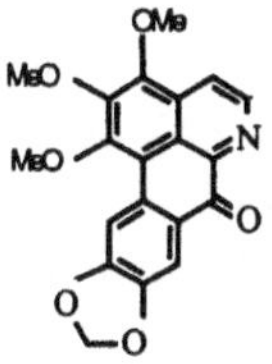

28

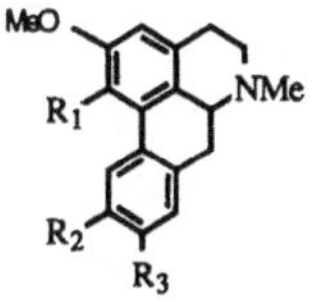

29: R1 = OMe; R2=R3= OCH_2O
30: R1 = OH; R2 = R3 = OMe

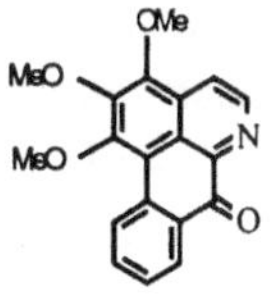

31

Fig. 4. Aporphine alkaloids from *Phoebe valeriana* (Now *Ocotea valerianoides*).

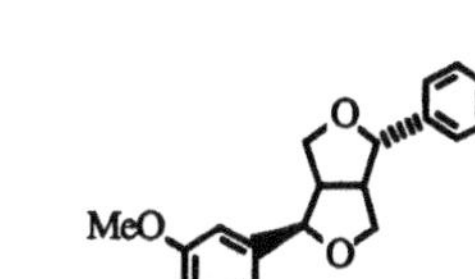

32

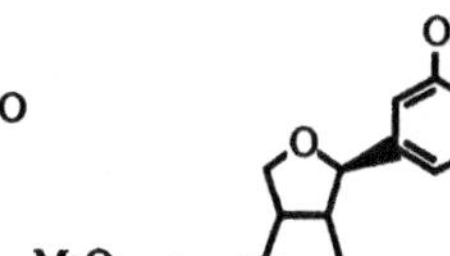

33: R=OH

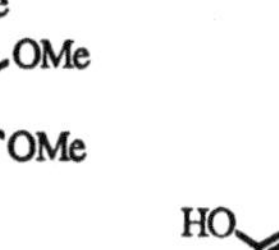

34

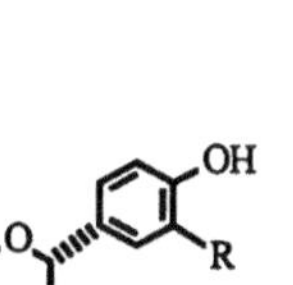

35

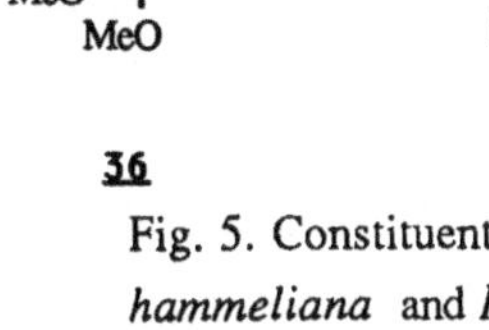

36

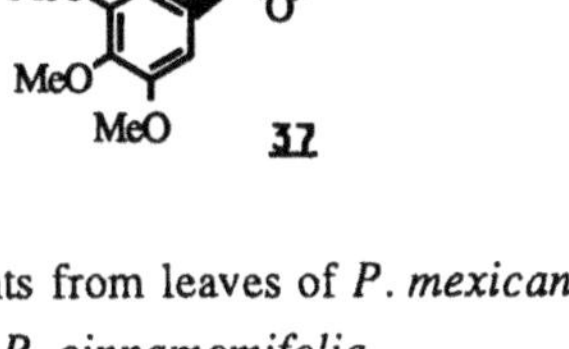

37

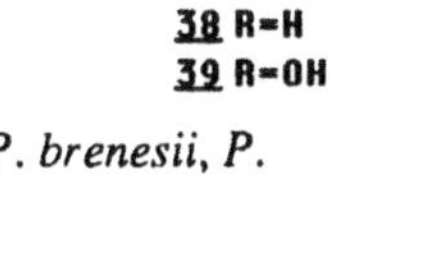

38 R=H
39 R=OH

Fig. 5. Constituents from leaves of *P. mexicana, P. brenesii, P. hammeliana* and *P. cinnamomifolia*.

the same sulfonylallyl derivative (**35**), which was detected in leaves and bark of *P. chavarriana.*[22]

According to a recent review of the Costa Rican Lauraceae, *Phoebe* species are characterized by three large staminodes with cordate-sagitate apices in each flower, nine fertile 4-thecous stamens, and fruits subtended by an expanded receptacle.[5] According to these criteria, *P. mollicella*, *P. pittieri*, *P. smithii*, *P. tonduzii*, *P. valeriana* have been transfered to the genus *Ocotea* and *P. costarricana* and *P. mexicana* have been combined as *P. cinnamomifolia*. Under this treatment, *Phoebe* in Costa Rica consists of only 5 species (*P. brenesii* , *P. chavarriana*, *P. cinnamomifolia*, *P. hamelianan* and *P. neurophylla*), three of which (*P. brenesii* , *P. chava* and *P. nevrophyllaare*) are highly restricted in distribution: *P. brenesii*, *P. chavarriana* (closely related to *P. hamelianan*), *P. neurophylla* (closely related to *P. cinnamomifolia*). *P. cinnamomifolia* has a wide ranging distribution from 0 to 1900 m and *P. hammeliana* is found only below 1500 m.

The chemical data are consistent with this new classification of *Phoebe* species in Costa Rica. Absence of alkaloids characterizes *Phoebe* species and the sulfonylallyl derivative **35** is a probable taxonomic marker for the genus. Considering that compound **35** is a major compound of *Cinnamomun triplinervis* [20] and is found in most species of this alkaloid-lacking genus, the suggestion of Kosterman that some Costa Rican species of *Phoebe* should be eliminated from this genus is chemically supported. Based on these data, we think that *P. clemensii*, *P. formosana* and *P. porfiria* should be reclassified within *Ocotea* or *Nectandra* .

OCOTEA AND *NECTANDRA* GENERA

Ocotea is a genus of 300-400 species centered in the American tropics; there are a few species in Africa and a group of species in Madagascar. *Nectandra* is a new genus of 100-150 species ranging from Florida to Argentina, well represented in Mesoamerica and the West Indies, but the majority of species are in South America. In Costa Rica a few species appear to be intermediate between *Ocotea* and *Nectandra*. The lack of the usual three well-developed staminodes distinguishes *Ocotea* from Costa Rican species of *Phoebe*. The position of the anthers may be the only means of distinguishing between *Ocotea* and *Nectandra*. According to these criteria, *Ocotea* is represented in Costa Rica by 44 species, constituting the main biomass component of Lauraceae family (41%) and *Nectandra* by 19 species (18%).[4]

Table 2. Distribution of Secondary Metabolites in *Ocotea* and *Nectandra*

	Ocotea	*Nectandra*
Aporphine alkaloids	53%	39%
Bisbenzylisoquinolinic alkaloids		
Morphinan alkaloids	6%	—
Indolic alkaloids	—	7%
Neolignan	35%	45%
Terpenes	9%	9%

A distribution of the main secondary metabolites isolated from the 34 species of *Ocotea* and 14 of *Nectandra* that have been reported in the literature are indicated in Table 2. From this table it is clear that aporphine alkaloids and neolignans characterize species of both genera. Only *N. megapotamica* synthesizes indole alkaloids,[23] *O. acutangula* and *O. brachybotra*,[24] accumulate morphine alkaloids, while *Ocotea rodiei*[25] and *O. venenosa*[26] produce bis-benzylisoquinoline alkaloids. Fruits of *O. venenosa* have been traditionally used for arrow poisoning by the natives of Kofan from Ecuador and Colombia .

We isolated from leaves of the 4 species of Phoebe, now transferred to *Ocotea*, and from *O. insularis* laudanine (**40**), thalictuberine (**41**), and a new phenanthrene alkaloid, 3-*o*-dimethylthalictuberine (**42**) (Fig 6).[27]

We reported aporphine alkaloids from two *Nectandra* species. Three known alkaloids, nordemesticine (**43**) norphoebine (**21**), phoebine (**20**) and a new 3-methoxy-nordomesticine (**44**) were found in *N. sinuata* bark (Fig. 7).[28] From *N. membranaceae* leaves, we isolated apoglaziovine (**45**), isoboldine (**46**) and glaziovine (**47**) (Fig. 8) [29,30]

PHARMACOLOGY OF APORPHINES

The main biological interest of aporphines is due to the fact that these compounds show antagonist effects to dopamine (DA). Such DA antagonists in current clinical use block DA receptors non-selectively and are associated with a continues to be useful in *in vitro* paradigms as ligands for the elucidation of the dopaminergic mechanism as binding receptors. The ability of several aporphines

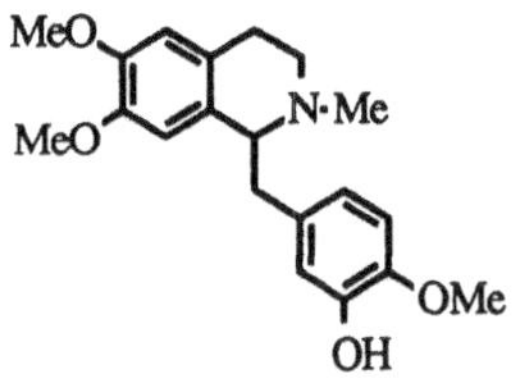

40

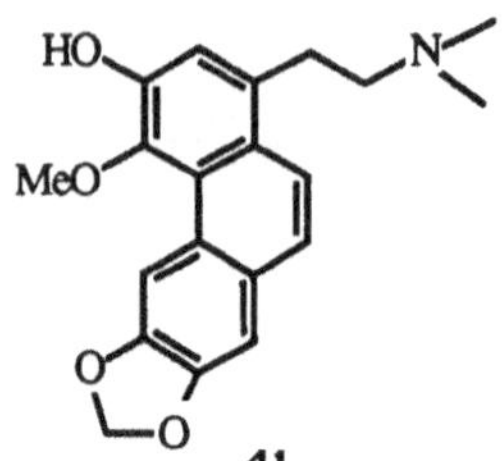

41

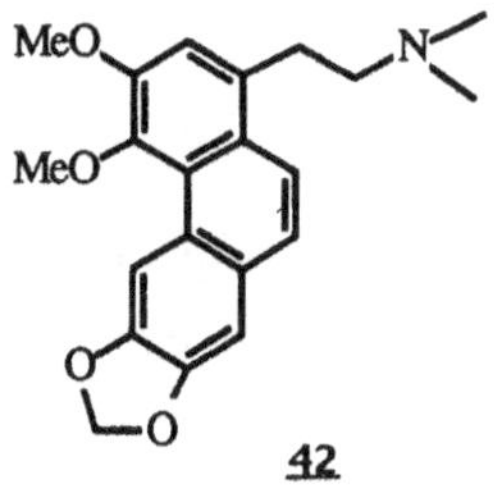

42

Fig. 6. Alkaloids from *Octoea insularis* (leaves).

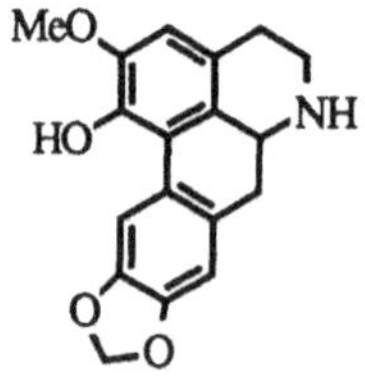

43

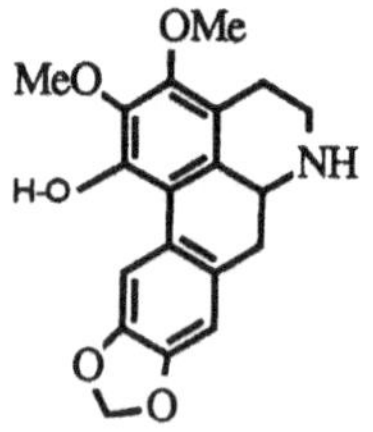

44

Fig. 7. Alkaloids from *Nectandra sinuate* (bark).

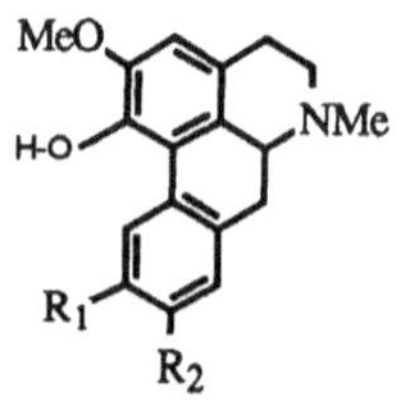

45: R1: OH; R2: H

46: R1: OMe; R2: OH

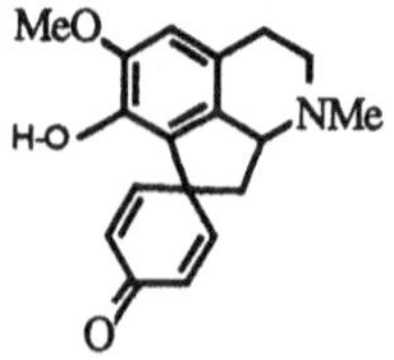

47

Fig. 8. Alkaloids from *Nectandra membranaceae* (leaves).

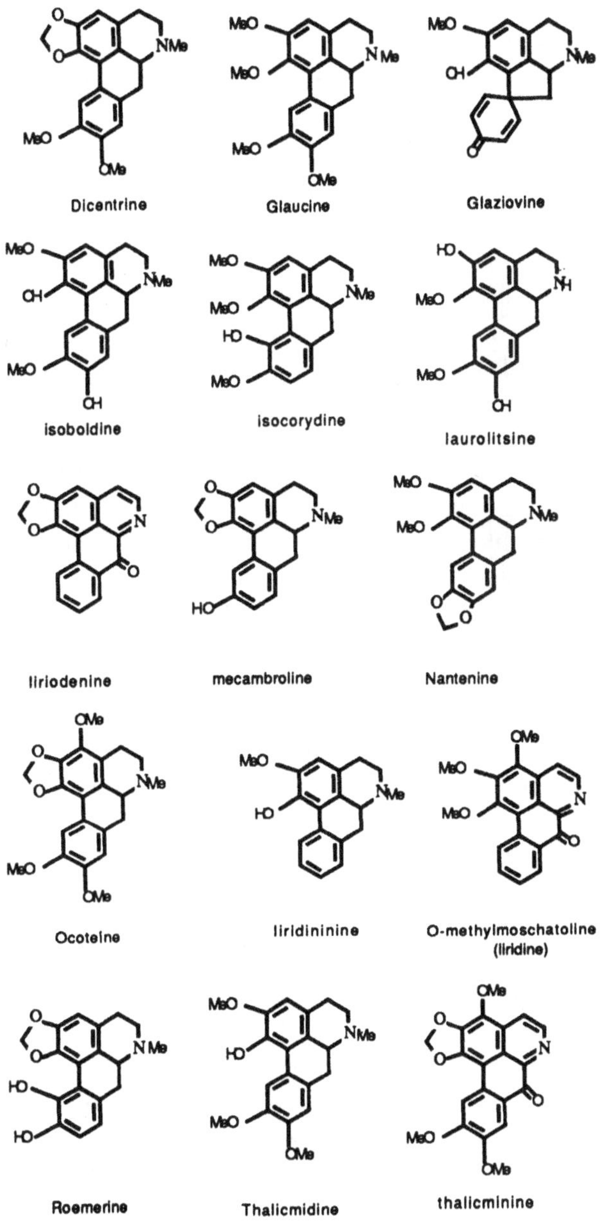

Fig. 9. Chemical structures of aporphinoid alkaloids with biological activity from *Ocotea*, *Nectandra* and *Phoebe*.

and benzylisoquinoline derivatives to inhibit mouse spontaneous locomotor activity at low doses and to reduce motor inhibitory effects at higher doses have been investigated.[31]

The anti-convulsant action of various aporphine derivatives acting as dopamine receptors has been well documented.[32] Studies of structure-activity with duodenal ulcerogens have implicated dopamine as a putative mediator and/or modulator in duodenal ulceration, and have showed that aporphines act as a dopamine agonist, decreasing the intensity of acute and chronic duodenal ulcers.[33]

Recently the broad spectrum of biological activities in aporphinoid alkaloids have been described.[34] Table 3 and Figure 9 show the pharmacological activities of some alkaloids isolated from *Nectandra*, *Ocotea* and *Phoebe* species. Therapeutic uses of glaziovine (**47**) as an anti-convulsant or muscle relaxant have been protected by patents.[35] Antitumoral, antifungal and antimicrobial properties are attributed to oxoaporphines (Table 3). The principal action mechanism must be due to the powerful antioxidant properties of these compounds. Very little information about toxicity and therapeutic doses is known for the compounds included in this table.

In contrast with *Ocotea*, the genus *Nectandra* has been less well studied. However, the following information reveals a need for intensified studies of species in this genus. From the bark of *N. megapotamica*, two alkaloids of tryptamine type were isolated. Plants like this are used by natives from South America, probably because of the hallucinogenic properties attributed to these alkaloids. These compounds also showed inhibitory properties to the growth *in vitro* of *Crithidia fasciculata* and *Trypanosoma cruzi*.[36] In *N. rigida* the co-occurrence of aporfinic alkaloids and neolignans were found to show promising cytotoxic activity.[37] Also, we have found similar biological functions of neolignans isolated from *O. veraguensis* and *Licaria triandra* fruits.

O. veraguensis is the only Lauraceae species that grows in the dry forest of Santa Rosa National Park (at 0-350 m) in Guanacaste Province, Costa Rica. According to D. Janzen, the fruits of this plant are the favorite food of many birds and rodents in this habitat.[38] The entire fruit is swallowed by frugivorous birds such as trogons (eg., *Trogon elegants*). The seeds, however, are regurgitated after the pulp is stripped off in the gizzard. Since such seeds or those that have fallen from the trees are not harvested by the spiny pocket mouse (*Liomys salvini*, Heteromyidae), agoutis (*Dasyprocta punctata*), pacas (*Agouti paca*) and peccaries (*Dicotyles tayassu*), common seed predators in this same

Table 3. Aporphine alkaloids isolated from *Ocotea, Nectandra* and *Phoebe* species and their biological properties [34]

ALKALOID	GENUS AND SPECIES	PHARMACOLOGICAL ACTIVITY
Bracteoline	*Licaria armenicana*	Hypotensive
Dicentrine	*O. macropoda* *O. brachybotra* *O. leucoxylum*	Antiparasitic Cytotoxic
Glaucine	*O. macrophylla*	Adrenalytic, anti-inflammatory, antimicrobial, antioedema, antithrombotic, antitussive, CNS-depressant, curare-like cytotoxic, expectorant, hypoglycemic, hypotensive, hypothermia, increase action of general anesthesics, increase cerebral levels of cAMP, increase cerebral levels of dopamine, inhibition of cAMP phosphodiesterase, inhibition of dopamine-sensitive adenylate cyclase, Inhibition of β-sestitive adenylate cyclase, parasympathomimetic peripheral adrenalytic, sedative, spasmolitic, sympatholitic, sympatomimetic, teratogenic.
Glaziovine	*O. glaziovii* *O. brachybotra* *O. variabilis* *N. membranaceae*	Ansiolytic, antidepressant, cytotoxic, hypotensive, psycotropic, tranquilizer.

Table 3, continued

Isoboldine	*O. glaziovii*	Hypotensive, hypothermia, inhibition of aldose-reductase, inhibition of dopamine-sensitive adenylate cyclase, insecticide, spasmolytic.
	N. pichurim	
	P. porfiria	
Isocorydine	*O. macrophylla*	
	P. clemensii	Adrenalytic, antiarrhytmic, antimicrobial, antitussive, catalepsic, CNS depressant, curare-like, decrease vascular strength, hypotensive, increase blood flow, increase corazole convulsant effects, increase lachrymal secretion, increase salivary secretion, inhibition of conditioned reflexes, parasympatholytic, parasympathomimetic, respiratory stimulant, spasmolitic.
Laurolitsine	*P. clemensii*	Hypotensive, hypothermia, spasmolitic.
Lirinidine	*O. macrophylla*	Antiserotoninic, hypotensive, spasmolitic.
Liriodenine	*O. formosana*	Analgesic, anti-leukemic, anti-microbial, antitumoral, cytotoxic, granulomatous, hypotensive, respiratory stimulant, sedative
Mecambroline (isofugapavine)	*P. clemensii*	Hypotensive
Nantenine (o-methyldomesticine)	*O. variabilis*	Antiserotoninic, hypotensive, inhibition of aldose reductase
	O. macrophylla	
	P. valeriana	

Table 3, continued

Ocoteine (Thalicmine)	*O. puberula* *O. minarum* *N. saligna* *P. porfiri*	Antimicrobial, antiparasitic, antitussive, catalepsic, cytotoxic, hypotensive, increase action of local anesthesics, increase salivary secretion, inhibition of conditioned reflexes, sedative, spasmogenic, spasmolytic.
o-methylmoschatoline (Liridine)	*P. valeriana*	Antimicrobial, antitumoral, cytotoxic.
Roemerine	*P. formosana*	Analeptic, antimicrobial, catalepsic, curare-like, hypotentive, increase lachrymal secretion, increase salivary secretion, inhibition of alcohol dehydrogenase, inhibition of conditioned reflexes, inhibition of glutamate dehydrogenase, serotonergic.
Thalicminine	*O. puberula* *O. minarua* *O. macropoda*	Hypnotic, hypotensive, respiratory stimulant.
Thalicmidine (Thaliporphine) (o-methylisoboldine)	P. valeriana	Cataleptic, CNS depressant, curare-like, decrease motility, expand respiratory movements, hypotensive, increase action of general anesthesics, increase lachrymal secretion, increase salivary secretion, inhibition of conditional reflexes.

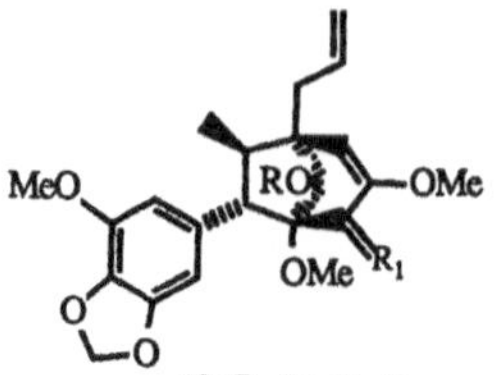

48: R=Ac; R_1=O
49: R=Ac; R_1=OH
50: R=OH; R_1=O
51: R=Ac; R_1=O
52: R=Ac; R_1=O

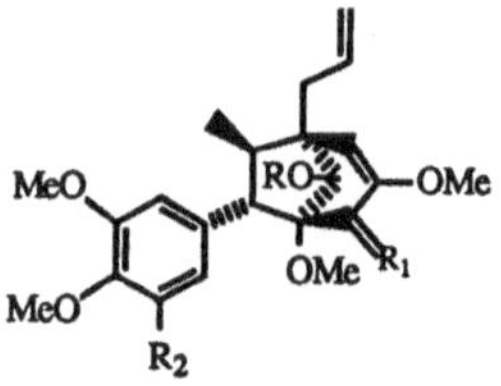

53: R=Ac; R_1=OH; R_2=H
54: R=Ac; R_1=O; R_2=OMe

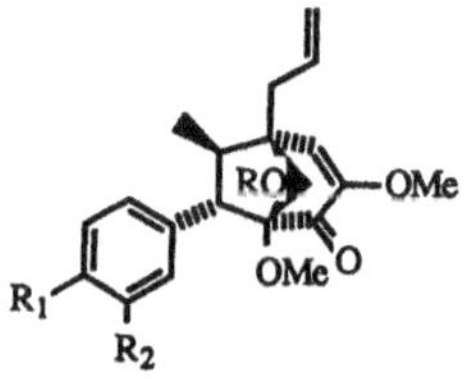

55: R=Ac; R_1=R_2=OCH_2O
56: R=OH; R_1=R_2=OCH_2O
57: R=Ac; R_1=R_2=OMe
58: R=OH; R_1=OH; R_2=OMe

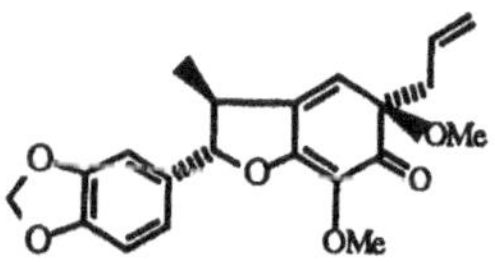

59

Fig. 10. Neolignans from *Ocotea veraguensis* fruits.

ecosystem, the seeds presumably contain chemicals that are toxic or repellent to such animals. However, in Santa Rosa, larval development of the weevil (*Heilipus draco*, Curculionidae) occurs only within individual seeds of *O. veraguensis*. Apparently the restriction of this insect to *O. veraguensis* as a host plant is based on metabolic changes that permit the larvae to overcome the toxicity of the chemical defenses of the seed.

Investigations were carried out on different parts of the fruit of *O. veraguensis* to understand some of these ecological interactions.Twelve neolignans compounds were found, 11 were bycyclo [3.2.1] octanoids and one was dihydrobenzofuranoid (Fig. 10). The distribution of the neolignans isolated from the fruit pulp, seed coat and seed are summarized in Table 4.[39] It is clear from this table that neolignans with methylenedioxy groups are concentrated in the seed coat and to a lesser extent in the inner parts of the seed, while with di-

Table 4 . Distribution of neolignans in *O. veraguensis* fruits (as % dry weight)

Compound	Seeds*	Seed Coat	Fruit Pulp
48	0.004	0.024	0.005
49	0.005	0.020	None
50	0.002	0.001	0.001
51	0.002	0.002	0.010
52	0.001	Traces	None
53	0.001	0.004	0.009
54	Traces	0.001	0.001
55	Traces	None	None
56	None	Traces	None
57	None	Traces	None
58	0.01	None	None
59	0.01	None	None

*Minus seed coat

and tri-methoxy- aromatic groups are predominant in the fruit pulp.[39] Herbivore theory predicts that compounds deterring seed predators should be at the highest concentrations in the seed coat; thus neolignans **48** and **49** are probably responsible for the deterrence described above. The level of **48** and **49** in the seed coat are four times that in the inner seed parts. In contrast, the level of **51** and **53** is greatest in the pulp fruit, but at approximately one-half the value for **47** and **48** in the seed coat.

Similar biodynamic interactions occur with the fruits of *Licaria triandra*, a tree of the evergreen, wet forest formation from near sea level to 1400 m on the caribbean side of Costa Rica and in the Central highlands. In this case, the weevil was identified as *Heilipus lauri* (Cucurbitaceae). The seed accumulates unusually large amounts of two neolignans of the megaphone type; triandrine A (aprox 1.8%) (**60**) and triandrine B (1.5%) (**61**) and a tetrahydrobenzofuranic neolignan, burcheline (**62**) (Fig 11).[40,41]

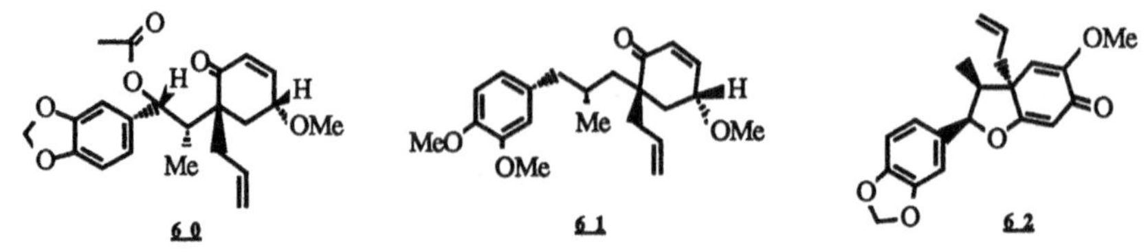

Fig. 11. Neolignans from *Licaria triandra* seeds.

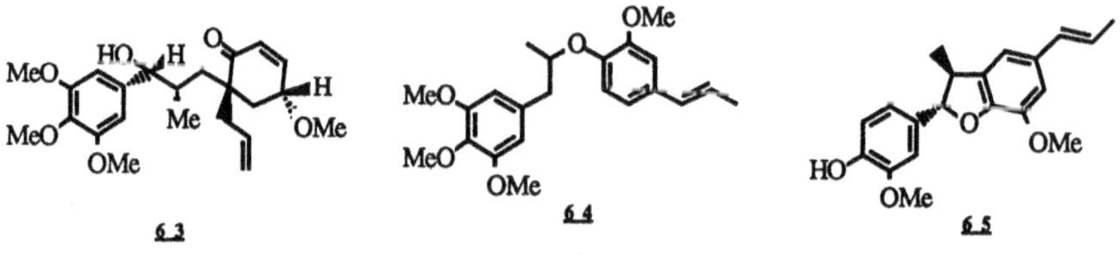

Fig. 12. Neolignans with cytotoxic activity.

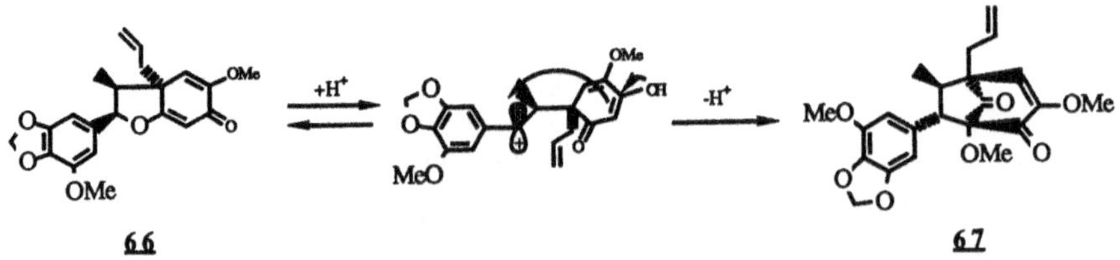

Fig. 13. Biomimetic rearrangement in acid media to convert benzofuranoid neolignans to bicyclo octanoids.

PHARMACOLOGY OF NEOLIGNANS

The neolignans megaphone (**63**) and surinamensine (**64**), isolated from *Aniba megaphylla* and *Virola surinamensis* respectively,[42,43] and the dehydrodiisoeugenol (**65**), identified from *Nectandra rigida*, [37] have showed cytotoxic activity (Fig. 12).[41,42] This cytotoxic characteristic has been attributed to the ease with which such neolignans and the benzofuranoid type like burcheline form benzylic cations in acidic media. They are very reactive as alkylating agents for the determination of molecular structures involved in cell division. This property was effectively used for the first time to transform benzofuranoid neolignans into bicyclo-[3,2,1]-octanoid derivatives. This rearrangement in acidic media of benzofuranoid (**66**) to give the octanoid bicycle (**67**) led to the determination of the absolute configuration of this neolignan (Fig 13).[44,45] No information is available regarding the biological activity of bicyclo-[3.2.1]-octanoid type of neolignans .

CONCLUSION

Species of *Phoebe* and *Cinnamomun* synthesize mainly cinnamic acid derivatives and furofuranoid lignans; the latter formed by a coupling between cinnamic acid derivatives and/or coniferyl alcohol. Genera such as *Ocotea*, and *Nectandra*, accumulate aporphine alkaloids and neolignan derivatives, while species of *Licaria* and *Aniba* contain mainly neolignans, metabolites formed from a coupling between propenyl benzene and/or allylbenzene.[46]

In contrast to the structural variety of neolignans, alkaloids produced in Lauraceae present a less complex biosynthetic origin. The defensive action of alkaloids against common forest predators has been shown to enhance the survival of plants that contain them.[47] This could explain the explosive distribution of Lauraceae in tropical forests. Nevertheless, the increase of alkaloids, although beneficial to the plants defense mechanism, is uneconomical in terms of the nitrogen utilization. If this economic distinction between alkaloids and neolignans is correct, species that produce mainly neolignans like *Licaria* and *Aniba* should have evolved from ancestors that accumulated alkaloids.

Aside from this evolutionary digression, there is no doubt that a complex biodynamic network exists within a tropical forest. Arboreal plants such as the Lauraceae produce a variety of oxygenated metabolites as defense agents with clear antioxidant properties We are just beginning to understand the meaning of these properties as a possibility to improve the welfare of men.

ACKNOWLEDGMENTS

This work was supported by the School of Chemistry CIPRONA and the Vice Director of Investigation of the University of Costa Rica, project Nº 115-90059. Special thanks to Dr. Frank R. Stermitz for the use of the NMR Spectroscopy Center of Colorado State University, NSF (INT-8919086), Ian

Castro for the valuable suggestions in the elaboration of this review and the support of the National Institute of Biodiversity (INBio), project "Chemical and Biological Prospectives", using funds of the MacArthur Foundation.

REFERENCES

1. KOSTERMANS, A.J. 1962. Taxonomic changes in Lauraceae, Leguminosae and Sterculiaceae. Reinwardtia. 6:115–187.
2. STANDLEY, P.C, STEYERMARK, J.A. 1946. Flora of Guatemala. Fieldiana. Botany 24: 302–344.
3. FOURNIER, L.A., SALAS, S., JIMENEZ,A., MARIN.F. (1966). Las Familias y Generos de Arboles y Arbustos de Costa Rica. Rev. Biol. Trop 14:317–328.
4. BURGER, W.B. 1990. Flora Costaricensis. Fieldiana. Botany New Series, Nº23, 138 pp.
5. HOLDRIDGE, R.L. 1982. Ecologia Basada en Zonas de Vida. Instituto Interamericano de Coorporación para la agricultura. San José, Costa Rica. 216 pp.
6. FOURNIER, L.A. 1969. Estudio Preliminar Sobre La Variacion Altitudinal en el Numero de las Familias de Arboles y de Arbustos en la Vertiente del Pacifico de Costa Rica. Turrialba 19:548-552.
7. CROME, F.H. 1975. The Ecology of Fruit Pigeon in Tropical Nothern Queensland. Aut. Wilde Res. 2:155-185 pp.
8. SNOW, D.W. 1977. Territorial Behavior and Courtship of the Male Three–Wattled Bellbird (Costa Rica). AUK 94 (4):623-645 pp.
9. CHACON, I.M. 1985. Tesis de Maestría Universidad de Costa Rica. Estudio Taxonomico y Ecologico del Genero *Nectandra* (Lauraceae) en Costa Rica. San José, Costa Rica. 108 pp.
10. PEMBERTON, R.W, TURNER, C.E. 1989. Occurrence of Predatory and Fungivorous Mites in Leaf Domantia. Am. J. Bot. 76:105–112.
11. KOSTERMANS, A.J.G.H. 1957. Lauraceae. Reinwardtia 4:193-256.
12. STERMITZ, F.R., CASTRO, O. 1983. Pentasubstituted Aporphine Alkaloids from *Phoebe molicella*. J.Nat.Prod. 46: 913–916.
13. LU, S.T., TSAI, J.L. 1984. Studies on the Alkaloids of Formosan Lauraceous Plants XXV. Two Hexahydro proaporphine Alkaloids, Lauroformine, and N–methyl–laroformine from *Phoebe formosana*. Heterocycles 22:1031–1033.
14. MARTINEZ, O.E., DELLE MONACHE, R., MARINI BETTOLO, G.B.

1983. Study of Colombian Lauraceae ll. Chemical Study of *Phoebe cinnamomifolia*. Rev. Latinoamer. Quim 14:76-77.

15. CASTRO, O.C., MUÑOZ, L.C. 1982. Alcaloides Aporfinicos en Hojas de *Phoebe pittieri*. Revista de Ingeniería y Ciencia Química 6:198-199.
16. CASTRO, O.C., LOPEZ, J., VERGARA, A. 1985. Aporphine Alkaloids from *Phoebe pittieri*. Phytochemistry 24:203–204.
17. RUIZ FLORES, I.E. 1990. Tesis Maestría Química. Estudio Fitoquimico de las Hojas de Ocotea holdridgeiana Familia Lauraceae. Universidad de Costa Rica. San José, Costa Rica, América Central. 140pp.
18. CASTRO, O., LOPEZ, J., STERMITZ, F.R. 1986. New Aporphine Alkaloids from *Phoebe valeriana* (Fam. Lauraceae). J. Nat. Prod. 49:1037–1040.
19. CASTRO, O., LOPEZ, J., VERGARA, A., STERMITZ, F.R. 1985. Phenylpropanoids in Alkaloid Free Species of Phoebe. J. Nat. Prod. 48:640-41.
20. RIPPERGER, H., DIAZ, M, SCHREIBER, K. 1981. Two Cinnamoyl Derivatives from *Cinnamomun triplinervis*. Phytochemistry 20:1453–1454.
21. CASTRO, O., SOTO, R.M. 1991. Tetrahydrofurofuran Lignans from *Phoebe cinnamomifolia* Leaves. Fitoterapia LXII, Nº5:457–458.
22. PÉREZ, C. G. 1993 Tesis de Lie Química. Evalvación fitoquímica del follage de *Phoebe hammeliana* W. Burger (Lauraceae). Universidad de Costa Rica, Escuels de Química, San Josée, Costa Rica. 80 pp.
23. DOS SANTOS, F.D., GILBERT, B. 1975. Alkaloids from *Nectandra megapotamica*. Phytochemistry 14:821–822.
24. VECCHIATTI, V., CASSAGRANDE, C., FERRARI, G. 1981. Alkaloids of *Ocotea acutangula*. J. Chem. Soc. Perkins. Trans I:578–581.
25. GUHA, K., MULHERJEE, B. 1979, Bis-benzilisoquiinoline alkaloids—a Review, J. Nat. Prod 42:1-84.
26. KOSTERMANS, A.J.G.H., PINKLEY, H.A. STERN, W.L. 1969. New Amazonian Arrow Poison, *Ocotea venenosa*. Bot. Mus. Leafl. Harvard Univ. 22:241–252.
27. HASBUN, C., CASTRO, O. 1992. A Nobel Phenanthrene Alkaloid from *Ocotea insularis*. Fitoterapia (in press).
28. CASTRO, O., HASBUN, C., CALDERON, M. 1991. New Aporphine alkaloid from *Nectandra sinuata*. Fitoterapia, Vol LXII Nº1, 72–73.
29. CASTRO, O., RODRIGUEZ, G., POVEDA, L. 1989. Proaporphine and

Aporphine Alkaloids of *Nectandra menbranaceae*. Fitoterapia, Vol LX 5:474.

30. HASBUN, C., CALDERON, M., CASTRO, O. 1991. *Nectandra menbranaceae* Alcaloides de su Corteza y Madera. Ing. Cienci. Quim. Vol 13:19–20.
31. BRADBURY, A.J., COSTALL, B., NAYLOR, R.J., NEUMEYER, J.L 1983. Motor Inhibition Induced by Aporphine Derivatives in the Mouse. J. Pharm. Pharmacol. 35:494-499.
32. ANLEZARK, G.M., BLACKWOOD, D.H.R., MELDRUM, B.S., RAM, V.J., NEUMEYER, J.L. 1983. Comparative Assessment of Dopamine Agonist as Anticonvulsants in Two Models of Epilepsy. Psychopharmacology 81:135-139
33. SZABO, S., NEUMEYER, J.L. 1983. Dopamine agonists and antagonists in duodenal ulcers disease. In: Dopamine Receptors. (C. Kaiser, J.W Kebatian, eds). Am. Chem. Soc. Washington, pp. 175-196.
34. RIOS, J.L., SIMEON, S., VILLAR, A. 1989. Pharmacology Activity of Aporphinoid Alkaloids. A Review. Fitoterapia, vol LX, 5:387-412.
35. FERRARI, G., CASSAGRANDE, C. US.Patent 3900564 (1975).
36. DOS SANTOS, F^{o}., SARTI, S.J. 1980. Alkaloides de Nectandra megapotamica Sprg. Chodat et Hassler– Actividade Inibitora do Desenvolvimento de *Crithidia fasciculata* y *Tripanosoma cruzi*. Rev. Fac. Farm. Odon. Rib. Preto 17:21–24.
37. LE QUESNE, P.W., LARRAHONDO, J.E., RAFFAUF, R.F. 1980. Antitumor Plants – Constituents of *Nectandra rigida*. J. Nat. Prod 43:353–359.
38. DODSON, G.D., STERMITZ, F.R., CASTRO, O.C., JANZEN, D.H. 1987. Neolignans from Fruit Parts of *Ocotea veraguensis*. Phytochemistry 26:2037–2040.
39. DODSON,G.D. Tesis Ph.D Chemistry. Neolignans of *Ocotea veraguensis* Mez. and Toxic Secundary Metabolites in the Human Diet. Colorado State University, Chemistry Department. 88 pp.
40. CASTRO, O., ULATE, L. 1988. Nuevos Bis–arilpropanoides en Semillas de *Licaria triandra*. Lauraceae. Rev. Latinoam. Quim. 1912:60–62.
41. ULATE, L Tesis Licenciatura en Química, Estudio Fitoquímico de los Frutos de *Licaria limbosa* kosterman ; (1986) Universidad de Costa Rica, Escuela de Química, 82 pp.
42. KUPCHAN, S.M., STEVEN, K.L., ROHLFING, E.A., SICKLES,

B.R., SNEDEN, A.T., MILLER, R.W., BRYAN, R.F. 1978, New cytotoxic neolignans from *Aniba megaphylla* , mez, J. Org. Chem 43:586-590.

43. BARATA, L.E.S., BAKER, P.M., GOTTLIEB, O.R. RUVEDA, E.A. 1978. Neolignans of *Virola surinamensis*. Phytochemistry 17:783–786.
44. CASTRO, O. 1977. Neolignanas de *Aniba simulans*. Tesis Ph.D Chemistry, University of São Paulo, Brasil. 133 pp.
45. AIBA, C.J., ALVARENGA, M.D., CASTRO, O., FERNANDEZ, J.B., GOTTLIEB, O.R., PAGLIOSA, F.M., YOSHIDA, M. 1978. Rearranjos Biomimeticos em Neolignanas. Química Nova 1:4-6.
46. GOTTLIEB, O.R. 1974. Lignans and Neolignans. Rev. Latinoamericana de Química 5:1-11.
47. SWAIN, T. 1974. Comprehensive Biochemistry (M. Florkin, E.H. Stotz, eds), Elsevier, Amsterdam Vol 29:125-127

Chapter Four

ZOOPHARMACOGNOSY: THE USE OF MEDICINAL PLANTS BY ANIMALS

Eloy Rodriguez and Richard Wrangham

Phytochemical & Toxicology Laboratory
University of California
Irvine, California 92717

Department of Anthropology
Harvard University
Cambridge, Massachusetts 01238

INTRODUCTION

The last years of the twentieth century offer both challenges and opportunities for phytochemists. The AIDS epidemic is still expanding and to date medical science has not been able to find effective treatments. Resistance to antibiotics, antimalarials, and other drugs occurs with increasing frequency. Crisis like these have created a new demand for phytochemical screening

*** Dedicated to the memory of the late Dr. Xorge A. Dominguez.**

programs to detect antiviral plant products. Consequently the attention of phytochemists has turned to the tropical rain forests that contain most of the world's species of plants and animals.

Unfortunately these tropical rain forests that contain the vast majority of the unidentified natural products are disappearing. This rapid disappearance of our principal source of new natural products makes our task more urgent and makes it incumbent upon natural products chemists to use innovative approaches to identify those compounds that have significant biological activity. Future phytochemists will not have access to the full resources of the tropical rain forests and will therefore have a much smaller array of species to screen for useful natural products.

This search for new techniques to identify prospective pharmacological products has increased interest in ethnobotanical studies of the tropical rainforest regions. In keeping with this trend we propose that phytochemists searching for novel natural products should investigate primates and other animals that have co-existed and co-evolved with the plants of the tropical rainforests for millions of years.

Pharmacognosy is the scientific study of the interaction of chemicals being investigated as potential drugs with the biological system of organisms that consume the drug, and requires knowledge of chemistry, chemical techniques, and a familiarity with anatomy and physiology of the animal being administered the drug. We propose the term "zoopharmacognosy" to describe the process by which wild animals select and use specific plants with medicinal properties for the treatment and prevention of disease.

The deliberate treatment of disease with medicines has commonly been assumed to be a unique human trait. However, anecdotal evidence from many naturalists who have observed animals in undisturbed habitats has suggested that animals often do consume plants containing high levels of chemicals that could have medicinal significance.[1] More recently investigators[2-5] have observed that European starlings select nest materials from plants with higher concentration of volatile terpenes. These authors speculate that these natural chemicals act as repellents, contact toxicants or natural fumigants to control parasite and/or pathogen populations in starling nests.

The most extensive work on zoopharmacognosy of mammals has centered in two areas, plant consumption by chimpanzees and fur rubbing by primates and carnivores. The investigations of the primatology groups at Harvard and Kyoto have identified a member of plants consumed by chimpanzees for medicinal properties. Several laboratories, most notably, the Phytochemistry and Toxicology at Irvine, have extracted and identified compounds with potent

Table 1. Chimpanzee Plant Medicine

Plant	Plant Part	Presumed Medical Compound
Ficus exasperata (Moraceae)	young leaves	furanocoumarins[25]
Rubia cordifolia (Rubiaceae)	young leaves	triterpenes, cyclic[26] oligopeptides
Aspilia mossambicensis (Asteraceae)	young leaves	thiarubrines[13]
Vernonia amygdalina (Asteraceae)	pith	steroidal glycosides, sesquiterpene lactones[28]

antibiotic activity. The results of the investigations of the plants consumed bythe chimpanzees are summarized in Table 1, and are discussed in the main portion of this article. Preliminary results of studies of fur-rubbing behaviour by other animals are discusssed in the last part of the paper.

ASPILIA X CHIMPANZEE INTERACTION

Interest in medicinal usage of plants by apes was initially stimulated by observations of chimpanzees.[6] The feeding behaviour of chimpanzees selecting *Aspilia* spp. leaves is unusual in several respects.[7] First, individual leaves are selected more slowly and carefully than normal; only young leaves are eaten, between about 2 cm and 10 cm long, and up to about 4 cm broad. This selection is carried out both visually and by touch or taste: chimpanzees sometimes close their lips over a leaf, remain still for a few seconds, and then abandon the leaf without detaching it from its stem (Fig. 1). Each leaf that is selected is held in the mouth for several seconds while it is rolled around by the tongue. Leaves are difficult to swallow since they are covered with bristly hair and have rough surfaces. The median rate of leaf consumption was 5 per min (range 2.4-15.4, n=7 bouts), compared to 37.3 ±6.5 leaves per min. (n=12 bouts) when chimpanzees ate leaves of *Mellera lobulata* (S. Moore) Acanthaceae, a commonly selected shrub with leaves of similar size which are collected and chewed in the

Fig. 1. A chimpanzee inspecting a leaf.

typical manner for most leaves.[7] Second, the leaves are not chewed and can be recovered whole from the feces. Third, Gombe chimpanzees tend to select *Aspilia rudis* (Oliv.) and *A. pluriseta* (Schweinf.) leaves within an hour of leaving their sleeping-nests, before their first big meal; presumably when their stomachs are empty and the postulated medicinal compounds are not diluted by large volumes of food. Fourth, although all individuals at least two years old occasionally select *Aspilia* spp. leaves, the sexes do differ in frequency. Females selected *Aspilia* spp. leaves significantly more often than males (median 11.4% of days (females), 2.6% days (males); Man Whitney U=11.5, n=11,7; $P < 0.02$). These data come from all times of year, with sexes sampled approximately equally at different times.

Other unquantified observation supports the fact that *Aspilia* spp. leaves are not eaten for nutritional purposes. First, on some days, chimpanzees make special journeys, which may be short or long, only to obtain *Aspilia spp.*

leaves. Yet on other days they do not forage for *Aspilia* even when fresh leaf is abundant close to the nest site. Medicinal properties of *Aspilia* spp. are indicated by the wide spread usage of *Aspilia* spp. by indigenous African societies. The number of reports of medicinal use of *Aspilia mossambicensis* (Oliv.) Wild, cited is greater than for 99% of all the species listed in Kokwaro's compilation of African medicinal plants.[8]

The supposition that chimpanzees might be consuming *Aspilia* leaves for medicinal purposes leads to the question: what disease or illness are the chimpanzees treating? Fecal examinations have shown that wild chimpanzees are affected by a variety of helminthic infestations of the digestive tract,[9-11] as illustrated in Table 2 by data from Kibale.[12]

These observations of chimpanzee behaviour prompted Wrangham and Nishida[6] to collect leaves of these plants, and send them to the research facilities of the senior author, E. Rodriguez, for chemical and pharmacological analysis. Chemical extraction from the plant material (leaves) yielded numerous natural

Table 2. Parasitic infection rate in chimpanzees at Kibale detected by Reid and Wrangham (N = 135) [12]

	Percentage of feces infested
Nematoda	
Ascaris	1
Strongles* (*Necator* and *Oesophagostomum*)	62
Strongyloides	23
Trichostrongyle	1
Trichuris	30
Percent infested with more than 1 species	83.7
Percent infested with more than 2 species	26.7
Protozoa	
Blastocystis	4
Coccidia	5
Unidentified Ciliate	7

* Only adult forms of *Necator and Oesophagostomum* can be distinguished by microscope; feces contained only eggs.

Table 3. *In vitro* toxicity[a] of thiarubrine A to second-stage larvae of *Meloidogyne javanica* [14]

Exposure Condition	D.W. (CK)[b] Control	1% EtoH Control	5	10 (ppm)	15	20
			% Live Nematodes[c]			
Light, 48 h	97.5	93.3	10.9	2.2	0	0
Dark, 48 h	99.5	93.7	1.7	0	0	0

[a] In 1% ethanol solution
[b] D.W. (CK) = distilled water (check)
[c] Mean of 8 replicates

products, but a major compound was identified as thiarubrine A, (Fig. 2) a polyacetylenic substituent that is intensely red, and unstable in the presence of light.[13]

In vitro experiments have established that a solution of 10 mg/ml of thiarubrine A is toxic to 100 percent of the plant parasitic nematodes *Meloidogyne javanica*[14](see Table 3). Similar effects have been noted *in vitro* in studies with the free-living, *Caenorhabitis elegans*[15] and the livestock parasite *Trichostrongyloides columbiformis.*[16]

Intraperitoneal injections of thiarubrine A into mice established that doses below 15 mg/kg were tolerated and caused no significant pathological symptoms to the major organs of mice.[14] Doses greater than this caused significant mortality, but the mechanism of toxicity remains unknown. One leaf of *Aspilia mossambicensis* contains an estimated 100 mg of thiarubrine; consequently a chimpanzee which weighs 40 kg could presumable consume 3,000 leaves of *Aspilia* without any significant toxicity problems from the thiarubrine. The published reports estimate that 30-100 leaves are consumed in one sitting; no toxicity from thiarubrine should be expected at this dosage.[6,7]

A simple calculation based on the estimate of 100-200 mg of thiarubrine per leaf leads to the conclusion that one leaf of *Aspilia* mossambi-

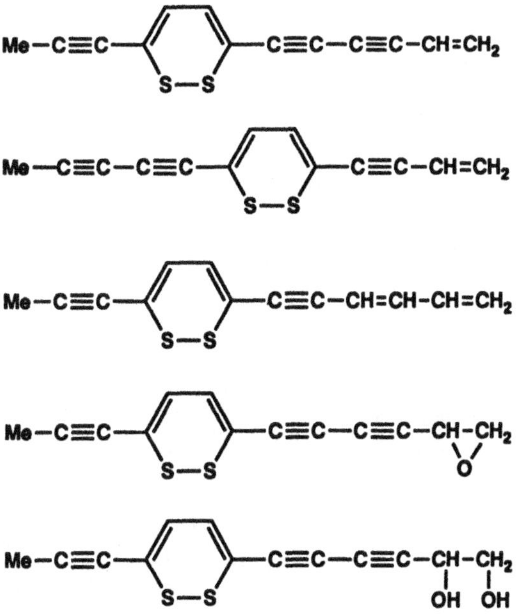

Fig. 2. Thiarubrine compounds from *Aspilia*, *Ambrosia*, and *Rudbeckia*.[13,17]

censis contains sufficient thiarubrine A to form 10-20 mls of a solution toxic enough to kill 100 percent of the nematodes. Therefore, 50 leaves should contain more than enough thiarubrine A to kill all the intestinal nematodes, and 100 leaves should contain sufficient thiarubrine for a therapeutic cocktail of one liter.

The fact that chimpanzee carefully swallow the leaves of *Aspilia* without chewing is highly significant. Any chewing would destroy the leaf structures that contain the thiarubrine and release them into the highly acidic digestive fluids of the stomach. We suggest that the intact leaf protects the thiarubrine and acts as a delivery vehicle that enables sufficient thiarubrine to survive passage through the stomach. The greatest number of parasitic helminths will be found in the small intestine and it is critical that thiarubrine be released in the small intestine and not in the stomach.

Laboratory tests have also been conducted with common bacteria and these results indicate that a concentration of 5 mg/ml of thiarubrine is not toxic

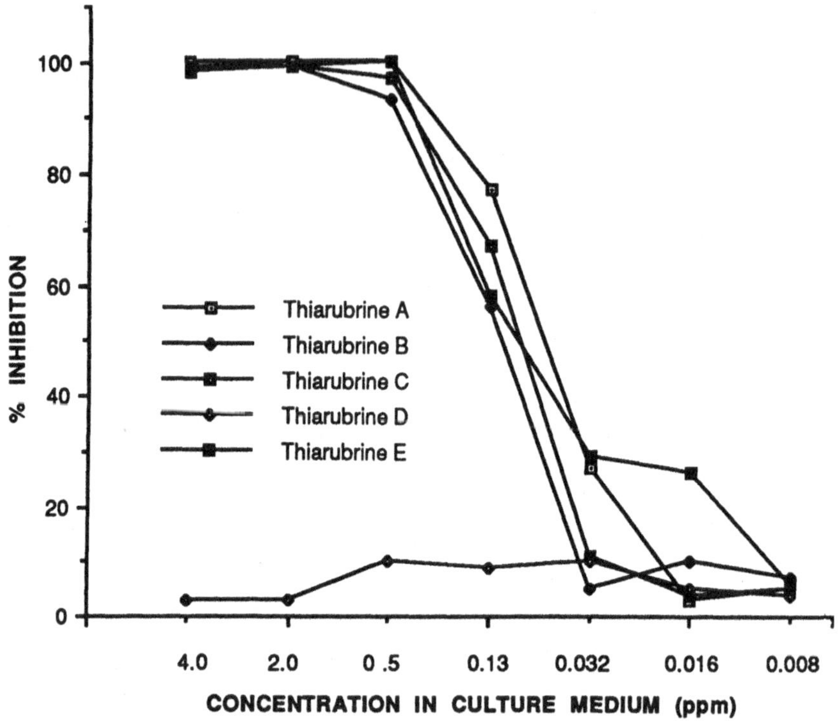

Fig. 3. Inhibition of growth of yeast, *Saccharomyces cerevisiae*, by thiarubrines, measured by change in optical density.[14]

to coliform bacteria such as *Escherichia coli* that are essential for the normal digestive process in apes. Similar concentrations of thiarubrine A would kill 70-80% of parasitic nematodes.

Subsequent work established that there are a series of thiarubine type compounds; preliminary research at the University of California, Irvine has been conducted with five of these thiarubrine compounds (Fig. 2). Thiarubrines A, B, D, and E were extracted from *Ambrosia chamissonis* (Less.),[17] and thiarubrine C was extracted from *Rudbeckia hirta* L plants raised in the greenhouse. Previous experiments indicate that compounds that lack dithiin moiety have only weak biological activity.

When tested with *Saccharomyces cerevisiae* there appears to be little difference between thiarubrine A, B, C, and E (Fig. 3). However, the epoxide form, thiarubrine D, appears to be inactive.[14] Results of the test with thiarubrine A are in general agreement with previous tests with *Candida albicans*.[14] Thiarubrine D (the epoxide form) is apparently devoid of significant nematicidal activity. This lack of activity in thiarubrine D was unexpected since epoxides are often potent antibiotics. A possible explanation for this anomaly would be that the epoxide moiety is labile and could form an adduct with a protein prior to exerting its toxic effects. Screening experiments performed previously indicate that the thiophenic derivatives of the dithiin compounds lack any significant activity to *Saccharomyces cerevisiae* or mammalian cell lines.[18]

Aspilia also contains a variety of other compounds that have been isolated and identified;[19] these include three ent-kaurenoic diterpenes from *Aspilia pluriseta* that have moderate activity against a variety of bacteria. The authors speculate that the widespread usage of *Aspilia* by Africans to treat surface wounds could be ascribed to the antibacterial activity of these compounds. Other researchers have reported the structures of three triterpene saponins that are important constituents of the leaves of *Aspilia kotschyi* (Oliv.).[20] Although the biological activity of these compounds is not reported, it is not unusual for this class of compounds to have antibacterial activity. Other researchers have reported that *Aspilia africana* contains powerful anticoagulant proteins that are not fully characterized to date.[21] There is also a report that alcoholic extracts of *A. africana* are vasoconstrictive.[22] Investigators in Canada have not found thiarubrine in *Aspilia* leaves and have proposed that the diterpenes present in *Aspilia* leaves could have effects on the reproductive system of female chimpanzees.[23]

FICUS X CHIMPANZEE INTERACTION

Four species of *Ficus* have been investigated for biological activities. Chimpanzees have been observed to consume leaves of *Ficus exasperata* (Vahl.) in a medicinal manner; the young leaves were carefully selected and then swallowed early in the morning without chewing.[24] Bioassay guided separation of the constituents in the laboratory led to the purification and isolation of a 5-methoxypsoralen; a well known compound with potent biological activity (Fig. 4).[25]

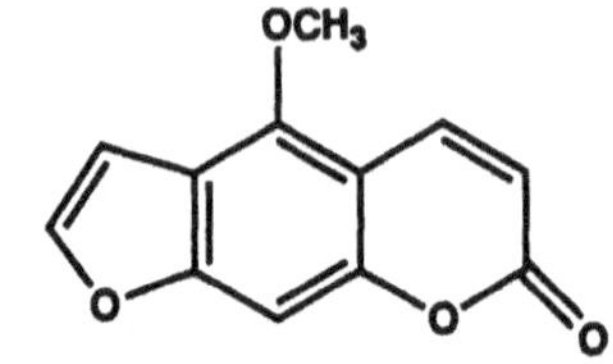

5-methoxypsoralen

Fig. 4. Antibiotic compound in *Ficus exasperata*.[25]

Subsequent studies of purified 5-methoxypsoralen with replicated trials established that a concentration of 130 mg/ml is toxic to 60 percent of the free living nematode *Caenorhabditis elegans*. The principal chimpanzee intestinal parasites are nematodes, and extensive drug screening trials have encountered few differences between nematode species in their susceptibility to potential nematicides.

Extrapolation from these numbers suggest 50-100 young leaves of *F. exasperata* should contain a sufficient quantity of 5-methoxypsoralen for significant nematicidal activity. The fact that chimpanzees carefully select young leaves of *Ficus exasperata* stimulated a study to compare the 5-methoxypsoralen content of young and old leaves. High performance liquid chromatographic analysis of leaf samples extracted with methanol established that young leaves contain about six times as much of the active compound; .023 mg/ml in young leaves and .004 mg/ml in old leaves. Wrangham has also collected leaf and fruit samples of *Ficus natalensis* (Hochst.), *F. conraui* (Warb.), and *F. asperifolia* (Miq.) for analysis at UC Irvine. A series of bioassays of the crude methanolic extract of the leaves of *F. asperifolia* indicate the presence of potent antibiotics. Fruits of *Ficus natalensis*, and *F. conraui* were examined and exhibited strong bacteriacidal activity against the gram (+) bacteria, *Bacillus cereus* and very weak activity against the gram (-) bacteria, *Escherichia coli*.[18] The normal coliform bacteria of chimpanzee are closely related to the human commensal bacteria. This finding is consistent with the fact that an effective medicine should not disturb the normal gut flora of the chimpanzee. Phytochemical investigations are continuing to fully characterize the antibiotic compounds in *Ficus*; preliminary evidence indicates the presence of antibiotic compounds that were not reported in the initial study.

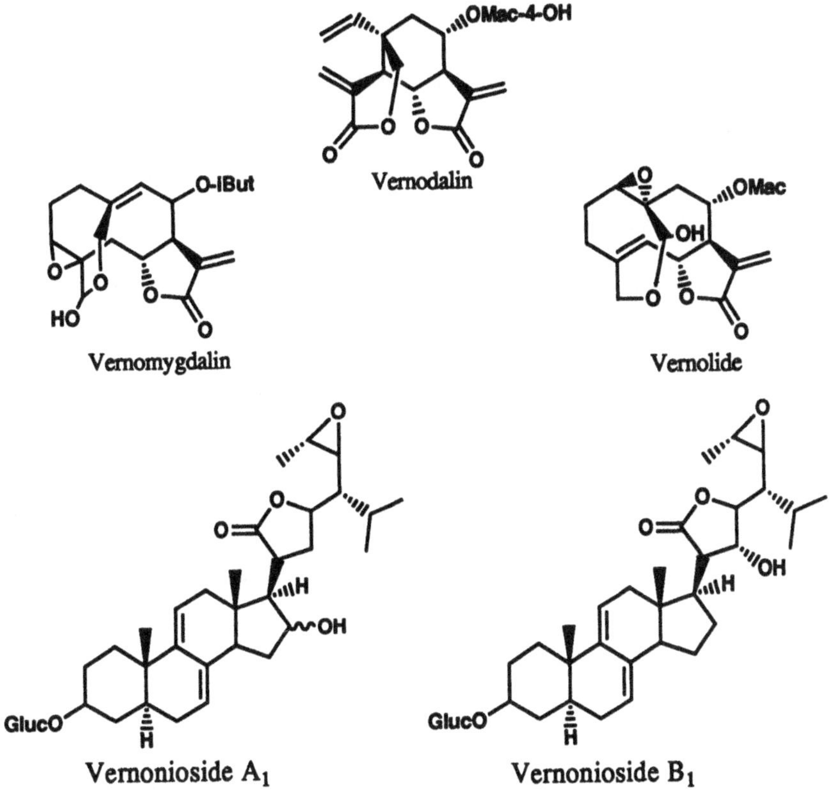

Fig. 5. Representative sesquiterpene lactones and steroidal glycosides from *Vernonia amydalina*.[28,29]

RUBIA CORDIFOLIA X CHIMPANZEE INTERACTION

Another plant swallowed whole by chimpanzees of the Kibale Forest that was suspected to be of medicinal value is *Rubia cordifolia* (L.).[24] The genus *Rubia* contains several species that are used extensively in the folk medicine of Africa and Asia. Plant material of this species was collected and examined. A bioassay guided fractionation determined that the most active fraction contained a previously reported triterpene, rubiatriol. Previous investigators had reported bioactive anthraquinones and a cyclic hexapeptide that is an extremely potent cytotoxic agent, and which is being investigated by the National Institute of Health as a therapeutic agent for cancer patients.[26]

VERNONIA AMYGDALINA X CHIMPANZEE INTERACTION

Huffman and Seifu observed that a sick chimpanzee deliberately split open the stems of *Vernonia amygdalina* (Del.) and swallowed the bitter juice from the pith.[27] Subsequent research identified three extremely bitter steroidal glycosides in the pith that appear to be toxic to a variety of parasites.[28] The leaves of *V. amygdalina* contain four well-known cytotoxic sesquiterpene lactones that have been extensively investigated (Fig. 5).[29] It is significant that the chimpanzee deliberately and carefully select the plant tissue (the pith) that contains the smallest amount of the toxic sesquiterpene lactones. *V. amygdalina* is a plant that is used extensively in many parts of Africa as an anthelmintic.[8]

OTHER PLANTS X CHIMPANZEE INTERACTION

Wrangham has observed several other plants being consumed by chimpanzees that are suspected of having medicinal properties. Preliminary studies with *Aneilema aequinoctiale*(Beauv.) Loudon (Commelinaceae) indicate the presence of an antibiotic agent that is being studied. Other species that are reported to be used as "medicines" by chimpanzees include *Lippia plicata* (Baker) and *Hibiscus aponeurus* (Sprague and Hutch).[30] Other medicinal plants are being recorded at new chimpanzee study sites such as Lope, Gabon[31] or through long term observation.[32] However, no phytochemical studies of these species have been conducted to date.

FUR-RUBBING BEHAVIOUR BY ANIMALS

Phytochemical research has recently begun to investigate the chemistry of plant products that are rubbed in the fur. This phenomena occurs in several mammals and is yet unexplained. The fact that nothing is ingested obviously excludes the possibility that the animals are consuming these plant products for nutritional purposes.

Table 4. Fur-Rubbing Plants

Animal	Plant	Plant Part
Coati	*Trattinickia aspera* (Burseraceae)	resin[36]
Bear	*Ligusticum wallichii* (Apiaceae)	root[35]
Capuchin monkey	*Hymenea coubaril* (Leguminosae)	trunk exudate[37]
	Piper sp. (Piperaceae)	
	Dieffenbachia longispatha (Araceae)	leaves[34*]
	Eugenia nesiotica (Myrtaceae)	fruit[34*]
	Laetia thamnia	fruit[34*]
	Tetrathylacium johansenii (Flacourtiaceae)	
	Protium sp. (Burseraceae)	fruit[34*]
	Virola surinamensis (Myristicaceae)	fruit[34*]

*Unpublished field observations.

One likely explanation of the fur-rubbing behaviour is that it repels or kills ectoparasites such as lice, ticks or mosquitoes; a similar procedure is practiced by birds and is termed "anting".[33] The chemicals rubbed into the fur could also stop or prevent topical fungal or bacterial infections of the animal engaging in the behaviour. However, at this point we cannot exclude the possibility that the peculiar odor of the plant product might serve as a repellent or attractant for members of the same species.

Table 4 lists some of the species of plants that have been reported to be used in this peculiar behaviour. The plants used by the capuchin monkey include groups that contain many volatile terpenes (*Eugenia*, *Protium*, *Virola*)[34] and (*Cebus* sp.) and bioactive alkaloids (*Piper* and *Virola*).[35] *Ligusticum* has been used by several human groups in North America,[35] and *Trattinickia* is in a family (Burseraceae) that has been used by many indigenous cultures.[36]

Recently it has been observed that capuchin monkeys in Costa Rica rub into their fur a mixture of rainwater and trunk exudate that collects in pools at the junctions of major limbs of the tree *Hymenea coubaril*.[37] Laboratory tests have given no indication that these mixtures have any significant antibiotic or antifungal activity. However, tests with laboratory ants indicate that these compounds are strongly repellent. More tests must be performed in the future, but at this point the preliminary data indicates that these natural products might act as a natural "OFF" or "6-12" that helps capuchin monkeys ward off biting dipterans and ectoparasites.

CONCLUSION

The studies of pharmacognosy conducted to date on the plants consumed by the chimpanzees and other animals indicate that they can be effective medicinal compounds. However, many additional studies will have to be conducted to fully characterize all the chemicals contained in the plants. Levels of the active compounds in the plants and their effects on the infectious parasites and their hosts will have to be determined independently. These studies will be conducted in the future to fully characterize the particular compounds in those plants and to determine their pharmacological potential.

ACKNOWLEDGEMENTS

We are grateful to National Institutes of Health, (AI18398 and AI24779) and the University of California, Irvine, for support; and to the Howard Hughes Medical Institute and the National Science Foundation (HRD9153663) for undergraduate research support. We also thank Dr. Susan Perry (Costa Rica) for an extract of *Hymenea coubaril* and Katherine MacKinon (Canada) for unpublished information on food plants of capuchins. We also thank Dr. Jan West and Dr. Manuel Aregullin (UCI) for their critical review of the manuscript.

REFERENCES

1. JANZEN, D.H. 1978. Complications in interpreting the chemical defenses of trees against tropical arboreal plant-eating vertebrates. In: The Ecology of Arboreal Folivores. (G.G. Montgomery, ed.). Washington D.C. Smithsonian Institution Press. pp. 73-84.
2. CLARK, L., MASON, J.R. 1985. Use of nest material as insecticidal and antipathogenic agents by the European starling. Oecologia 67:169-176.
3. CLARK, L., MASON, J.R. 1987. Olfactory discrimination of plant volatiles by the European starling. Animal Behavior 35:227-235.
4. CLARK, L., MASON, J.R. 1988. Effect of biolobically active plants used as nest material and the derived benefit to starling nestlings. Oecologica (Berl) 77:174-180.
5. CLARK, L., SMERASKI, C.A. 1990. Seasonal shifts in odor acuity by starlings. J. Exp. Zool 255: 22-29
6. WRANGHAM, R.W., NISHIDA, T. 1983. *Aspilia* spp. leaves: A puzzle in the feeding behaviour of wild chimpanzees. Primates 24:276-282.
7. WRANGHAM, R.W., GOODALL, J. 1989. Chimpanzee use of medicinal leaves. In: Understanding Chimpanzees. (P.G. Heltne, L.A. Marguart, eds.). Harvard Univ. Press, Cambridge, Massachusetts. pp. 22-37.
8. KOKWARO, J.O. 1976. Medicinal Plants of East Africa. East Africa Literature Bureau, Nairobi, Kenya.
9. FILE, S.K., McGREW, W.C., TUTIN, E.G. 1976. The intestinal parasites of a community of feral chimpanzees, *Pan troglodytes* (Schweinfurthii). J. of Parasitology 62 (2):259-26.
10. MCGREW, W.C., TUTIN, G.E.G., COLLINS, D.A., FILE, S.K. 1989. Intestinal parasites of sympatric *Pan troglodytes* and *Papio* spp. at two sites; Gombe (Tanzania) and Mt. Assirik (Senegal) Amer. J. Primatol. 17:147-158.
11. KAWABATA, M., NISHIDA, T. 1991. A preliminary note on the intestinal parasites of wild chimpanzees in the Mahale Mountains, Tanzania. Primates 32 (2):275-278.
12. REID, G., WRANGHAM, R.W. 1992. In preparation.
13. RODRIGUEZ, E., AREGULLIN, M., NISHIDA, T., UEHARA, S.,WRANGHAM, R.W., ABRAMOWSKI, Z., FINLAYSON, A., TOWERS, G.H.N. 1985. Thiarubrine-A, a bioactive constituent of

Aspilia (Asteraceae) consumed by wild chimpanzees. Experientia. 41:419-420.

14. WEST, J., WASSER, C., RODRIGUEZ, E. In preparation.
15. TOWERS, G.H.N., ABRAMOWSKI, Z., FINLAYSON, A.V., ZUCCONI, A. 1985. Antibiotic properties of thiarubrine A, a naturally occurring dithiahexadiene polyine. Planta Medica 51:225-227.
16. PZIFER PHARMACEUTICALS. 1991. Unpublished.
17. BALZA, F., LOPEZ, I., RODRIGUEZ, E., TOWERS, G.H.N. 1989. Dithiacyclohexadiene and thiophenes from *Ambrosia chamissonis*. Phytochemistry 28 (12): 3523-3524.
18. AREGULLIN, M., RODRIGUEZ, E. In preparation.
19. LWANDE, W., MACFOY, C., OKECJ, M., DELLE MONACHE, F., MARINI BETTOLO, G. 1985. Kaurenoic acids from *Aspilia pluriseta*. Fitoterapia 65 (2):126-128.
20. KAPUNDU, M., PENDERS, A. 1988. Structure du Kotschyioside, Saponside Novueau Isole de *Aspilia kotschyi* (Sch. Bip.) Oliv. Bull. Soc. Chim. Belg. Vol. 97 (5):329-342.
21. HANNA, M.M., NIEMITZ, J. 1987. Studies in anticoagulent action of *Aspilia africana*. Thrombosis Research 47:401-407.
22. AKABUE, P.I., MITTAL, G.C., AGUWA, C.N. 1983. Preliminary Pharmacological study of some African Medicinal Plants. Journal of Ethnopharmacol. 8:53-63.
23. PAGE, J., TOWERS, G.H.N. 1992. Personal Communication.
24. NISHIDA, T. 1990. A quarter century of research in the Mahale Mountains: an overview. In: The Chimpanzees of the Mahale Mountains: Sexual and Life History Strategies. (T. Nishida, ed.). University of Tokyo Press, Tokyo. pp. 3-35.
25. BARRERA, E., RODRIGUEZ, E. 1990. Studies on the African plant *Ficus exasperata* eaten by wild chimpanzees *Pan troglodytes*, (Schweinfurthii). Society for the Advancement of Chicanos and Native Americans in Science (SACNAS) Phoenix, Arizona.
26. ITOKAWA, H., TAKEYA, K. MIHARA, K., MORI, N., HAMANAKA, T., SONOBE, T., IITOKA, Y. 1983. Studies on the antitumor cyclic hexapeptides, obtained from *Rubia* radix. Chemical and Pharmcology Bulletin 37 (6):1670-1672.
27. HUFFMAN, M.A., SEIFU, M. 1989. Observations on the illness and consumption of a possibly medicinal plant *Vernonia amygdalina*

(Del.) by a wild chimpanzee in the Mahale Mountains National Park, Tanzania. Primates 30 (1):51-63.

28. JISAKA, M., OHIGASHI, H., TAKAGAKI, T., NOZAKI, H., TADA, T., HIROTA, M. I., RIE, R., HUFFMAN, M.A., NISHIDA, T., KAJI, M., KOSHIMIZU, K. 1992. Bitter steroid glucosides, vernoniosides, A., A2 and A3 and Related B1 from a possible medicinal plant, *Vernonia amygdalina*, used by wild chimpanzees. Tetrahedron 48 (4):625-532.
29. KUPCHAN, S.M., HEMINGWAY, R.J., KARIM, A., WERNER, D. 1969. Tumor Inhibitors. XLVII Vernodalin and vernomygdin, two new cytotoxic sesquiterpene lactones from *Vernonia amygdalina*. J. Org. Chemistry 34 (2):3908-3911.
30. TAKASAKI, H., HUNT, K. 1987. Further medicinal plant consumption in wild chimpanzees. Afr. Study Monographs 8 (2):125-128.
31. TUTIN, C. 1992. Personal Communication.
32. HUFFMAN, M. 1992. Personal Communication.
33. SIMMONS, KEL. 1964. Feather maintenance. In: A New Dictionary of Birds. (Sir A. Landsborough-Thomson, ed.). New York. McGraw-Hill. pp. 278-286.
34. MITCHELL, B. 1989. PH.D. Dissertation. University of California, Berkeley.
35. SIGSTEDT, S. 1992. Personal Communication.
36. GOMPPER, M.E., HOYLMAN, A.M. 1992. Grooming with *Trattinnickia* resin: Possible pharmaceutical plant used by coatis in Panama (In preparation).
37. PERRY, S. 1992. Unpublished field observations.

Chapter Five

INSECTICIDES IN TROPICAL PLANTS WITH NON-NEUROTOXIC MODES OF ACTION

J.T. Arnason,* S. MacKinnon,* A. Durst,* B.J.R. Philogène,* C. Hasbun,# P. Sanchez,# L. Poveda,# L. San Roman,# M.B. Isman,@ C. Satasook,@ G.H.N. Towers,@ P. Wiriyachitra,$ and J.L. McLaughlin^

*Ottawa Carleton Institutes of Biology and Chemistry
University of Ottawa, Ottawa, Canada, K1N 6N5

#Universidad Nacional, Heredia and Museo Nacional,
San Jose, Costa Rica

@Departments of Plant Science and Botany
University of British Columbia, Vancouver, Canada, V6T 1W5

$Research Center for Natural Products, Faculty of Pharmacy
Chiang Mai University, Thailand

^Department of Medicinal Chemistry and Pharmacognosy
Purdue University, W. Lafayette, U.S.A.

Phytochemical Potential of Tropical Plants
Edited by K.R. Downum et al., Plenum Press, New York, 1993

INTRODUCTION

Biotic interactions in the tropics are intense and have led to the evolution of a great diversity of phytochemicals which act as deterrents to phytophagous insects. While the diversity of tropical forests and the phytochemicals they produce is well known, certain other features of the chemical ecology of these areas have also been described. An early review of alkaloid-bearing plants suggested that the foliar concentration of alkaloids was higher in tropical environments at lower latitudes and altitudes than in subtropical and temperate environments at higher latitudes and altitudes.[1] Assignment of alkaloids to various toxicity classes suggested that tropical plants are also more biologically active as defense compounds than subtropical and temperate plants. Similar trends are seen even in crop plants such as maize, wherein resistance to borers and content of the insect resistance factor DIMBOA (2, 4-dihydroxy-7-methoxy-(2H)-1,4-benzoxazin-3(4H)-one) was inversely correlated to the altitude and latitude of origin of maize genotypes.[2] Janzen[3] suggested that chemical defenses are most important in areas of the tropics where soils are nutrient deficient (oligotrophic) and where loss of foliage to herbivory involves the loss of scarce plant macro- and micronutrients, which are more critical than replaceable carbohydrates. The forests of the Rio Negro were identified as an oligotrophic region which produces abundant amounts of secondary chemicals that probably contribute to the black waters. When the hypothesis that nutrient status of ecosystems results in different types of chemical defenses was investigated[4] in African forests, phenolic (quantitative[5]) defenses in leaves were found to be higher and leaf-feeding by Colobus monkeys found to be lower in an oligotrophic forest than a nutrient rich eutrophic forest. However, alkaloid (qualitative) defenses were higher in the eutrophic forest. These and other ecological studies of the tropics have been instrumental in defining general theories of antiherbivore defense[6] and suggest that tropical plants often have highly developed chemical defenses against insects.

Insect Control Agents From Tropical Plants

Besides shaping the natural interactions of plants and insects, phytochemicals from tropical plants include substances that are potentially useful as insect control agents or resistance factors in crop plants. In the past, many insect control agents have been developed from phytochemical leads, obtained largely from tropical plants.[7] Most of the materials developed have been neurotoxins, since these act rapidly and are high efficacious. However, recently published figures[8] estimate that 504 species of arthropods are now resistant to conventional insecticides/acaricides. Many display multiple resistance, i.e. to more than one insecticide, and at least 48 species are resistant to the synthetic pyrethroids.

While the neurotoxic mode of action of almost all conventional insecticides has promoted the rapid development of cross-resistance in insect populations, research into insecticidal phytochemicals has emphasized non-neurotoxic modes-of-action[9,10] such as antifeedant action, inhibition of molting, growth reduction, loss of fecundity, respiratory inhibition, phototoxicity, etc. These new modes-of-action could reduce the risk of cross-resistance in insect populations presently resistant to neurotoxins, as has been demonstrated with plant-derived phototoxins that are effective control agents for malathion- resistant *Culex* larvae.[11] Recent use of standardized plant extracts (botanicals) containing a mixture of active phytochemicals should reduce the rate of evolution of conventional resistance compared to the selection pressure exerted by a single pure toxin.

Investigations of phytochemicals as useful insect control agents now follow two distinct research and development paths: the synthetic route and the botanical route (see Figs. 1 and 2).

THE SYNTHETIC ROUTE: SYNTHESIS AND QSAR ANALYSIS OF THIOPHENES

The synthetic route (Fig. 1) generally involves the identification of an active "lead" molecule, with subsequent development of chemical analogues with enhanced efficacy, photostability, resistance to metabolism and selectivity (i.e.limited non-target impact). Synthetic pyrethroids[12] are the most successful example of this technology. Developments in molecular biology have allowed the cloning of receptors that permit rapid screening for potent biologically active agents aimed at specific biochemical targets. The application of quantitative structure-activity relationships (QSAR) is another development which raises the level of this technology from screening to prediction and effective molecular design.[12]

A recent example of the application of the synthetic route to phytochemicals involves the thiophenes, potent mosquito larvicides, of Asteraceae.[11] The thiophenes include alpha terthienyl (α-T) (**1a**) and related bithiophenes which are widely distributed in tropical and subtropical Asteraceae such as the genera *Flaveria*, *Tagetes*, *Porophyllum* and *Dyssodia*. *Tagetes* is traditionally

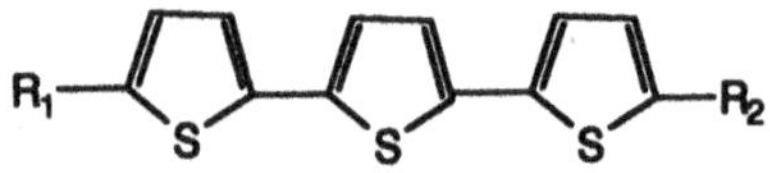

	R_1	R_2
1 a	-H	-H
1 b	-CN	-H
1 c	-S-CH_3	-H
1 d	-Si$(CH_3)_3$	-COOH
1 e	-CHO	-H
1 f	-$CONH_2$	-H
1 g	-CH_2OH	-H
1 h	-CH_2CH_2OH	-H
1 i	-C≡N	-C≡N
1 j	-CH_3	-H

1 k

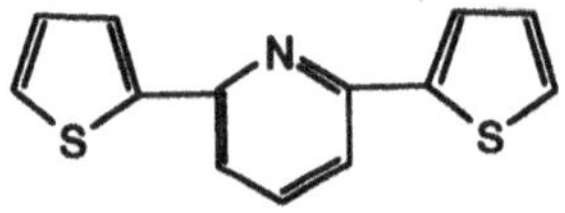

1 l

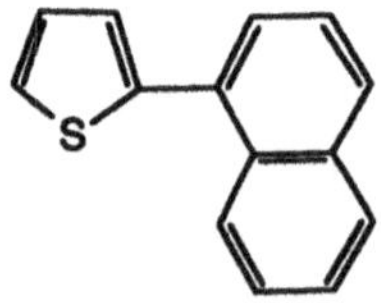

1 m

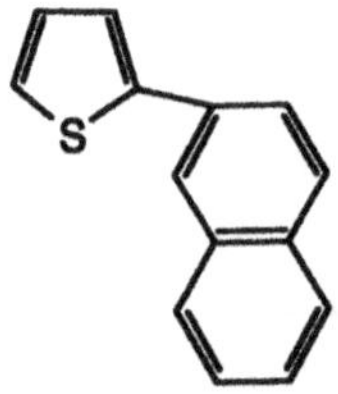

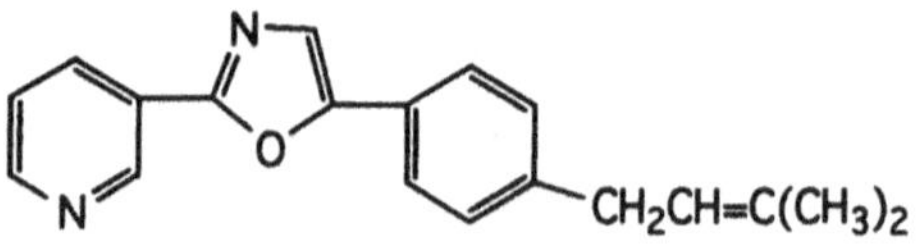

HALFORDINOL

2

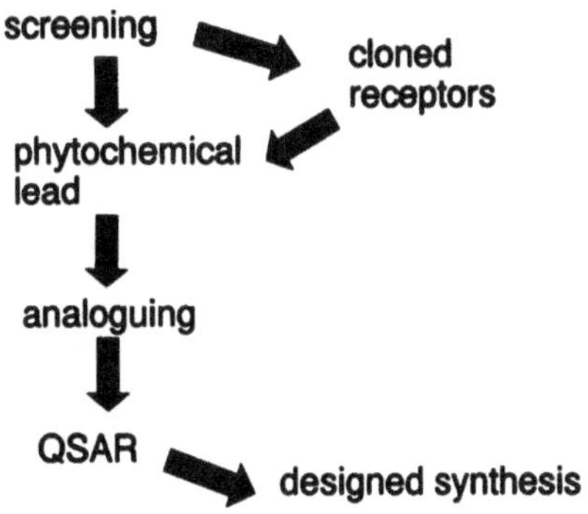

Fig. 1. Research and development scheme for synthetic insecticides from phytochemical "leads."

planted around maize fields (milpas) in Mexico as a protection against "mal viento" which can be interpreted as plagues of insects or disease. α-T is a well studied phototoxin which is acutely toxic and fast acting to the malaria mosquito *Anopheles* with an LC_{50} of 2 ppb.[13]

Over 50 derivatives and analogues of the "lead" molecule were prepared synthetically by a modified Gringard Wurz procedure.[14] Application of QSAR was envisaged as a way of improving efficacy and target vs non-target selectivity. QSAR for pesticides and drugs generally gives a good prediction of toxicity $[1/(LC_{50})]$ of analogues based on a parabolic or bilinear function of their calculated log octanol/water partition coefficients (CLOGP). Although this model predicts effectively neurotoxic activities of pyrethroids,[12] it gives a relatively poor prediction of phototoxicity mediated by UV (300-400nm) light $(1/(LC_{50}+UV))$. In order to develop a QSAR model for this new class of insecticide, we needed a new approach in which the photobiological characteristics of the molecule were considered. First, the quantum yield of singlet oxygen was determined by laser flash photolysis and the rate of singlet oxygen production (RATE) calculated for each compound based on the product of quantum yield and photons absorbed. A linear regression model developed by Marles[15] and based on the bilinear function of CLOGP and RATE gave a good prediction of the phototoxicity of thiophenes (**1a-1m**) to mosquito larvae:

$$\log(1/LC_{50}+UV) = 5.90\,(\pm 2.30)\,\text{CLOGP} - 6.57\,(\pm 2.47)\,\log(\beta\, 10^{\text{CLOGP}} + 1) + 0.11\,(\pm 0.05)\,\text{RATE} - 27.37\,(\pm 9.59)$$

where:

$$\log \beta = -4.31,\ n = 14,\ s = 0.23,\ r = 0.92,\ F_{1,9} = 17.78,\ \text{CLOGP}_o = 5.25$$

(β is the nonlinear coefficient, n is the number of data, s is the standard deviation, r is the correlation coefficient, F is the ratio of variances between the calculated and observed values and CLOGP_o is the optimum partition coefficient for toxicity, determined from the maximum of the bilinear curve)

The optimal partition coefficient (CLOGP_o) for phototoxicity to mosquito larvae was predicted to be 5.25, which is very close to the CLOGP (5.24) for the CN derivative (**1b**) of α-T. To test if this was indeed a better insecticide, synthesis of gram quantities of the cyano derivative was undertaken for field testing. These trials demonstrated that this material was a significantly better performing larvicide[16] than the parent molecule. The same model with different values for the parameters describes the phototoxicity of thiophenes to non-target organisms such as brine shrimp, and suggests analogues that will minimize toxicity to these organisms.[17]

The wider potential of tropical plants for phototoxicity by UV is largely uninvestigated but the Ottawa group maintains a screening program of tropical plant extracts and secondary chemicals. For example, Hasbun *et al.*[18] have recently isolated halfordinol (**2**) from *Amyris brenesii* (Rutaceae), collected in Rio Cuarto, Costa Rica. This 2,5-diaryloxazol represents a new class of phototoxic molecule.

THE BOTANICAL ROUTE

The second route to insect control agents, the botanicals route (Fig. 2), has recently received renewed interest with the development and registration of insecticides based on neem, *Azadirachta indica.*[19-22] Although some botanicals may produce "lead" phytochemicals similar to synthetic insecticides, with comparable overall effectiveness, botanical insecticides have distinctly different properties requiring different research and development paths. Environmental awareness and "green consumerism" has created a favorable environment for development of botanicals because they are generally non-persistent, and their natural origin and those with softer modes-of-action (e.g. deterrency versus

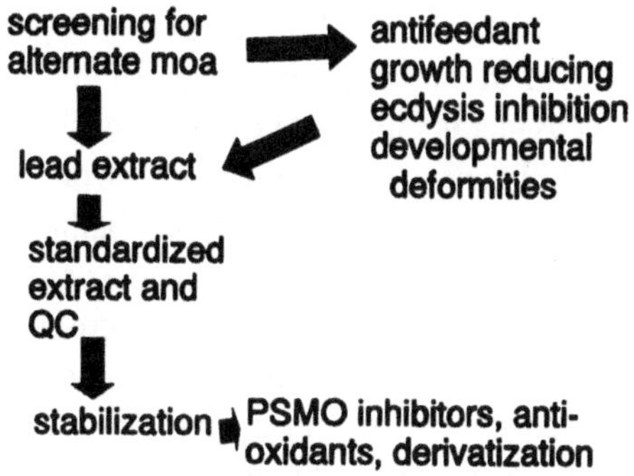

Fig. 2. Research and development scheme for botanical insecticides from plant extracts.

toxicity) are perceived favorably by the public. Although genuine safety in toxicological terms must be rigorously demonstrated on a case by case basis, examination of several advanced botanicals indicates there are several products (e.g. neem and endod) that are non-toxic in acute oral studies with rats at 3g/kg or greater.

With these advantages, however, come several major problems that must be addressed with botanicals. These include the best source of material, quality control of standardized extracts, the relatively high instability of botanicals compared to synthetics and unfavorable market-registration conditions.

Neem research has provided a model for addressing some of these problems. For example, the biological activity of commercial neem oils of different origin can be predicted on the basis of their azadirachtin content; this has provided a basis for chemical standardization.[19] Instability of neem products is one of the main reasons for failure and disappointment in the hands of the user community. Stabilization of neem products can be achieved through various means such as hydrogenation of extracts[21] which reduces azadirachtin degradation. We have found that the use of a naturally occurring polysubstrate monoxygenase inhibitor such as the neolignan dillapiol[23] synergizes the activity of azadirachtin.

New Sources of Botanicals

Because botanicals are produced by living plants, the source of supply is an important issue. For example, neem is very abundant in India but is far less so in the American tropics. Since it is desirable to produce botanicals from native species, scientists in our collaborative group in Costa Rica and Thailand have been attempting to identify alternatives to neem for the Americas and South East Asia. One potential source is the large amount of sawdust and bark waste containing insecticidal limonoids which can be obtained from tropical timber species of the Rutales (especially Meliaceae). Seed extracts of several Meliaceae species[24,25] have been reported to have potent growth-reducing activity to insects, but we have concentrated our efforts on the more accessible wood, bark and leaves and herein expand on a previous report.[26]

Collection and preparation of lyophilized ethanolic extracts of over 25 species of Meliaceae and Simaroubaceae, including the common indigenous tropical American genera *Cedrela*, *Guarea*, *Carapa*, *Swietenia* and *Trichilia*, were completed in the forest reserves of Costa Rica (by C. Hasbun, L. Poveda and P. Sanchez), with additional material from Belize (by J.T. Arnason and J.D.H. Lambert), and Fairchild Botanical Garden, Miami (by L. Swain and K. Downum). Insect bioassays were completed with two robust economic insects, second instar European corn borer, *Ostrinia nubilalis*, (in Ottawa) and with neonate variegated cutworm, *Peridroma saucia* (in Vancouver). It has been demonstrated[27] that cornborer larvae display little feeding aversion to plant chemicals in their diet, when fed soft artificial diets, because their chemosensillae become plugged by the diet. Thus, the results for corn borer represent primarily physiological growth reduction only, while the data for the variegated cutworm result from feeding deterrency and/or physiological growth reduction. The previously unpublished results (Table 1) suggest that a majority of the extracts from plants studied in the order Rutales inhibit growth significantly; some are more active than neem leaf extracts. *Swietenia mahogoni* bark, *Trichilia glabra* bark, *T. hirta* leaves, *T. americana* bark, *T. trifolia* wood, *T. pleana* wood and *Azadirachta indica* wood showed potent activity against the cutworm. Cedro corocolito bark, *Cedrela odorata* leaves, *Aphanamixis polystachys* wood, *T. glabra* wood and bark, and *T. pleana* bark showed good activity against corn borer larvae.

Comparison of the activity of the extracts suggests that wood and bark are generally more chemically defended than leaves. Loss of foliage is less expensive than replacement of structural materials such as wood. Of the indigenous Central American genera of Meliaceae, foliar extracts of *Cedrela odorata*, *C. tonduzii*, *Trichilia americana*, *T. glabra*, *Swietenia macrophylla* and *S. mahogoni* showed significant growth reducing activity to cornborer, but none of the leaf extracts of the three *Guarea* species had significant activity. Our

Table 1. Effect of Rutales extracts administered in diet on the growth and development of second instar European corn borer, *Ostrinia nubilalis* and neonate variegated cutworm, *Peridroma saucia*.

Plant	Tissues	*Ostrinia nubilalis* (Concentration in diet=0.4% fresh weight, n=15) RGR[a] (% of control)	*Peridroma saucia* (Concentration in diet=0.2% fresh weight, n=20) Larval Weight (% of control)	Collection Site
Aphanamixis polystachya	wood	35.52 h-l*	89.4 i-l	Fairchild Tropical Garden
Wall & Parker.	leaves	82.62 a-f	80.5 k-o	Miami, Florida
	bark	70.18 a-h*		
Azadirachta indica A. Juss.	bark	73.78 a-h*	10.4 x-z*	Fairchild Tropical Garden
(neem)	leaves	85.11 a-f	#*	Miami, Florida
Carapa guianensis Aubl.	bark	70.76 a-h*	NT	Puerto Viego, Limon,
	leaves	62.17 a-i*	NT	Costa Rica
	wood	93.71 a-c	NT	
"Cedro caracolito"	bark	19.54 l*	51.5 r-u*	Peninsula de Osa,
(*Reptilocarpon caracolito*)				Costa Rica

Table 1, continued

Plant	Part	*Ostrinia nubilalis* RGR (% of control)	*Peridroma saucia* Larval Weight (% of control)	Collection Site
Cedrela fissilis Vell.	bark	68.81 a-h	35.4 t-w*	Fairchild Tropical Garden
	leaves	90.78 a-c	32.5 u-w*	Miami Florida
	wood	78.99 a-g	85.5 j-n	
Cedrela odorata L.	bark	63.56 a-i*	96.0 j-k	Carara, Costa Rica
	leaves	74.58 a-h*	89.7 i-l	
	wood	43.76 f-l*	63.6 o-r*	
Cedrela salvadorensis Standl.	bark	49.14 d-l*	75.5 l-p*	Ciudad Colon San Jose,
	fruits	68.59 a-h*	75.8 a*	Costa Rica
	leaves	83.31 a-f	192.0 l-p*	
Cedrela tonduzii	leaves	90.89 a-i	18.3 w-z*	outside Heredia, Costa Rica
(younger) C.DC.	wood	83.12 a-f	#	
Cedrela tonduzii(older) C.DC.	leaves	69.47 a-h*	99.4 f-k	outside Heredia, Costa Rica

Table 1, continued

Plant	Part	*Ostrinia nubilalis* RGR (% of control)	*Peridroma saucia* Larval Weight (% of control)	Collection Site
Cedrela toona (Endlicher)	bark	79.11 a-g*	150.8 b*	Fairchild Tropical Garden,
	leaves	84.42 a-d	109.6 e-h*	Miami Florida
Chukrasia tabularis	bark	58.48 b-j*	124.2c-l*	Fairchild Tropical Garden,
A. Juss.	leaves	65.38 a-h*	73.6 l-q*	Miami Florida
	wood	55.42 c-j*	117.3 d-f	
Dysoxylum fraseranum	bark	83.91 a-f	153.4 b*	Fairchild Tropical Garden,
(A. Juss.) Benth.	leaves	89.09 a-d	130.3 c-d*	Miami, Florida
Guarea glabra Vahl.	bark	70.71 a-h	89.0 i-l	La Pacifica Conos,
	leaves	78.3 a-g	67.14 m-r*	Guanacosta, Costa Rica
Guarea pterorachis Harms.	leaves	82.18 a-f	NT	Costa Rica
	twigs	77.12 a-d	NT	
	wood	82.97 a-f	NT	

Table 1, continued

Plant	Part	*Ostrinia nubilalis* RGR (% of control)	*Peridroma saucia* Larval Weight (% of control)	Collection Site
Guarea pyriformis Pennington	bark	82.43 a-f	125.3 c-l*	Reserva Biologica, Carara, Costa Rica
	leaves	81.36 a-f	151.5 b*	
	wood	23.4 j-l*	63.3 o-r*	
Swietenia macrophylla . King	bark	79.42 a-g	155.5 b*	La Pacifica Conos, Guanacosta, Costa Rica
	leaves	76.47 a-h*	52.7 r-t*	
	wood	55.35 b-j*	113.8 d-g	
Swietenia mahagoni (L.) Jacq.	bark	63.33 a-i*	12.1 x-z*	Fairchild Tropical Garden, Miami Florida
	leaves	69.45 a-h*	75.1 l-p*	
	wood	61.72 a-i*	74.7 l-p*	
Simarouba glauca D.C.	bark	63.32 a-i*	65.8 n-r*	San Jose Succotz, Belize
	leaves	58.33 c-k*	41.5 s-v*	
Simarouba peapansis D.C.	leaves	83.00a-e	84.8 j-n	San Ramon, Costa Rica

Table 1, continued

Plant	Part	*Ostrinia nubilalis* RGR (% of control)	*Peridroma saucia* Larval Weight (% of control)	Collection Site
Trichilia americana (Sesse &	bark	85.07 a-f	24.8 v-y*	Costa Rica
Mocino) Pennington	leaves	75.69 a-h*	43.2 s-v*	
Trichilia glabra L.	bark	29.04 i-l*	9.4 x-z*	La Pacifica,
	leaves	55.78 c-l*	54.1 q-t*	Guanacosta, Costa Rica
	wood	46.11 e-l*	#*	
Trichilia havanensis Jacq.	leaves	107.52 a	64.30 o-r*	outside Heredia, Costa Rica
	wood	102.13 ab	104.00 f-j	
Trichilia hirta L.	leaves	84.57 a-d	5.30 y-z*	Fairchild Tropical Garden,
	wood	72.50 a-h*	#*	Miami Florida
Trichilia martiana C.DC.	bark	81.19 a-f	NT	Ciudad Colon San Jose,
	leaves	94.21 a-c	86.20 j-m	Costa Rica

Table 1, continued

Plant	Part	*Ostrinia nubilalis* RGR (% of control)	*Peridroma saucia* Larval Weight (% of control)	Collection Site
Trichilia pleeana . (A. Juss.) C.DC	bark	55.71 b-j*	8.30 x-z*	Reserva Biologica Carara, Costa Rica
	leaves	89.47 a-c	63.60 o-r*	
	wood	67.85 a-h*	66.70 m-r*	
Trichilia quadrijuga subsp. cinerascens (C.DC) Pennington	leaves	93.75 a-c	27.80 v-x*	Reserva Biologica Carara, Costa Rica
Trichilia trifolia L.	wood	71.93 a-g	6.40 y-z*	Costa Rica

Note: Previously unpublished data. Means followed by the same letter are not significantly different in Tukey's multiple range test (P=0.05). For *Ostrinia nubilalis* the mean (s.d) of the RGR of the control = 1.486 (0.226). Identifications of all Costa Rican material was undertaken by L. Proveda and P. Sanchez, Fairchild Topical Gasrden collections by garden staff, and Belize material by J. Arnason and J.D.H. Lambert. @ - RGR = relative growth of larvae; * = significantly different from control; # = all insects died before assessment; NT = not tested .

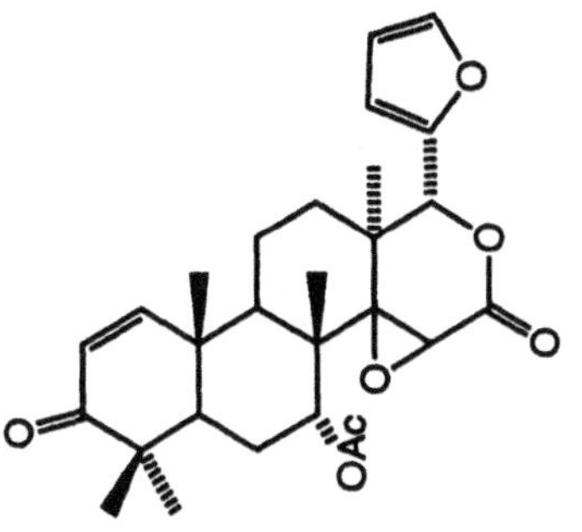

GEDUNIN

3a

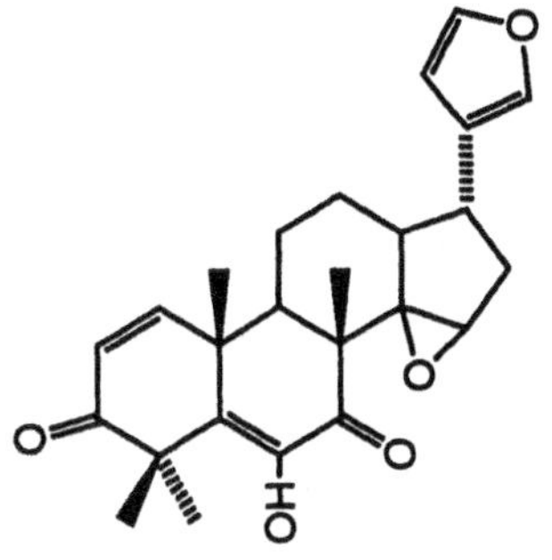

CEDRELONE

3b

bioassay results are somewhat comparable to observations in the field, where *Cedrela* spp. and *Trichilia* spp. generally have leaves with little insect damage while the *Guarea*'s are often found with considerable feeding damage. *Guarea* leaves are much thicker and tougher than the other genera. As suggested by Champagne, *Guarea*may use structural defenses because its leaves are long lived,while the other genera rely on chemical defenses. In the literature,[28] A,D seco limonoids such as obacunone found in *Guarea* spp. are considered low in biological activity to many insects. Apo-euphol limonoids such as trichilins found in *Trichilia* spp. are known to be among the most active limonoids to insects, next to the C-seco limonoids which include azadirachtin.[28]

The genus *Cedrela* is another important source of active materials. The Ottawa group has isolated gedunin (**3A**), a D-seco limonoid in 0.1% yield from tropical cedar, *Cedrela odorata* wood collected in Belize. At a dietary concentration of 50 ppm, gedunin causes growth reduction and mortality to neonates (Fig. 3) and possesses significant feeding deterency at 500 ppm on leaves.[31] One approach to exploiting and improving the activity of this substance has been the synthesis of fragments related to the compound.[29,30] We are currently investigating ways to improve the efficacy and stability of extracts containing gedunin by derivatization, as has been achieved with neem extracts containing azadirachtin. *Cedrela* spp. are some of the most abundant timber trees in the Americas with industrial production of large amounts of waste sawdust and bark which could be exploited for the production of a botanical.

Related Asian species (also grown in parts of the Americas) such as *Cedrela toona* (synonym, *Toona ciliata*) are a source of cedrelone (**3B**).

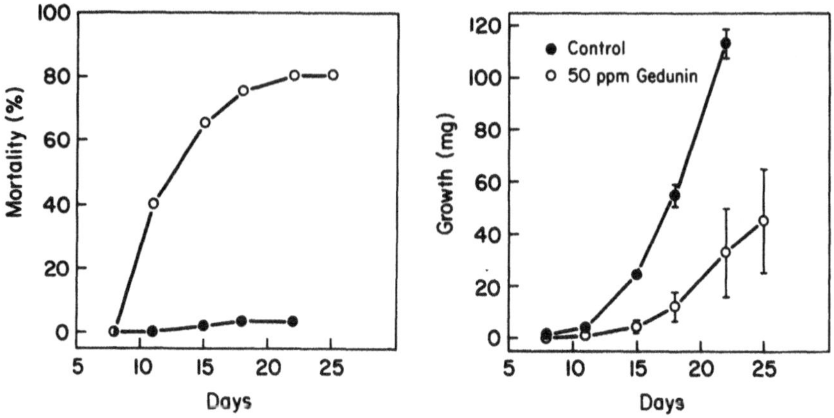

Fig. 3. Effect of dietary administration of 50 μg/g gedunin (from *Cedrela odorata*) on growth and mortality of corn borer larvae.

Table 2. Larval growth inhibition of lepidopterans by cedrelone

Plants[a]	Dietary EC_{50} (PPM)[b]
Spodoptera frugiperda[36]	2
Pectinophora gossypiella[36]	3
*Heliothis zea*l[36]	8
Mamestra configurata[33]	14
Peridroma saucia[33]	53

[a]References
[b]EC^{50} = effective concentration for 50% inhibition of growth

Cedrelone is an apo-euphol limonoid which is a more active antifeedant and growth reducer than gedunin to corn borer,[31] and reduces larval growth of variegated cutworms and bertha armyworms (*Mamestra configurata*) by 50% at dietary concentrations of 53 and 14 ppm respectively.[33] It is even more active to three other species of lepidopteran pests[36] (Table 2). Although cedrelone has

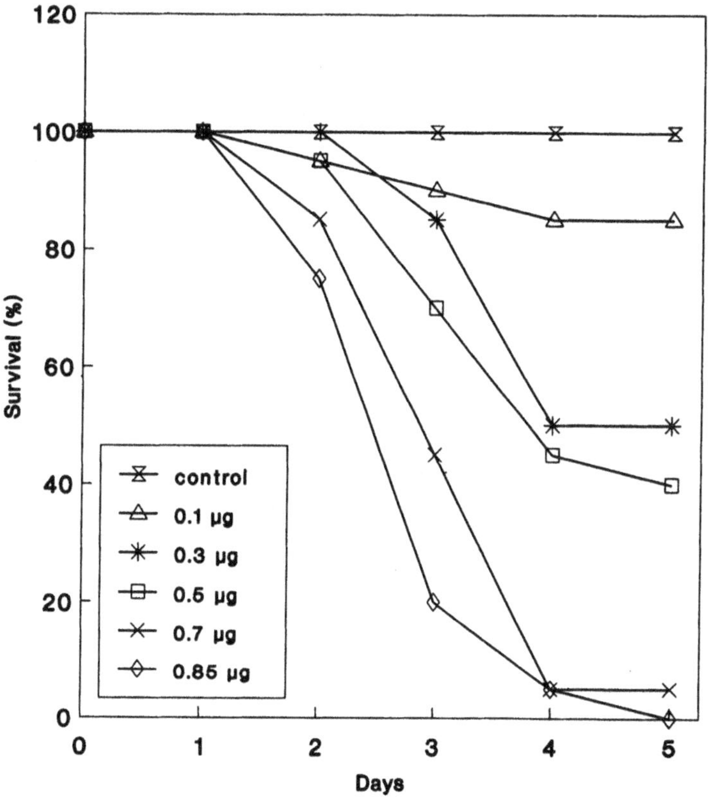

Fig. 4. Effect of dietary administration of "stopfeed" (*Aglaia odorata*) extract on growth of corn borer larvae (unpublished data).

some direct antifeedant action, topical application to fourth instar cutworms indicated that growth inhibition was caused by postdigestive toxicity rather than a chemosensory effect.[34]

Mackinnon, Hasbum and Towers are currently isolating active compounds from a number of the other species listed, including "cedro caracolito," whose botanical assignment is uncertain, but may be closely aligned to the Meliaceae (Table 1). It was the most active extract against corn borer. The two species of *Simarouba* (Simaroubaceae) were also tested because of their bitter taste, reputed in folk medicine, and were found to have activity.

An Advanced Botanical: "Stopfeed" from *Aglaia odorata*

The most active Meliaceae species from South East Asia in the present screening and in previous screening trials at Chiang Mai and U.B.C. is *Aglaia odorata*, a reputed medicinal plant of the Indo-Malayan area. An advanced standardized extract developed in Chiang Mai and given the tradename "Stopfeed," produces substantial growth reduction to corn borers (Fig. 4). It was a slow acting toxin to borers also causing remarkable delays in development that we have ever observed with any extract (Table 3). The active principles have been identified at U.B.C. and Chiang Mai to be the benzofuran, rocaglamide (**4**) and several related analogues (desmethyl rocaglamide, methyl rocaglamide and rocaglaol), rather than a limonoid.[30-32]

In topical applications to fourth instar cutworm larvae, rocaglamide results in an almost immediate cessation of feeding and terminal anorexia; mortality does not become apparent until 48 hours after treatment and is only fully expressed on the fourth day after application (Fig. 5). Similar results are seen with cornborer. In our experience, rocaglamide is the second most potent insecticidal phytochemical next to azadirachtin isolated to date, being about one fifth as active. However, rocaglamide is at least twenty times more active than the simple limonoid cedrelone (Table 4).

Botanicals from the Annonaceae

Another group of slow-acting insecticidal toxins from plants are acetogenins, such as asimicin and bulatacin, isolated from the Annonaceae at Purdue.[35] Asimicin (**5**) and a standardized botanical called "F020" prepared from

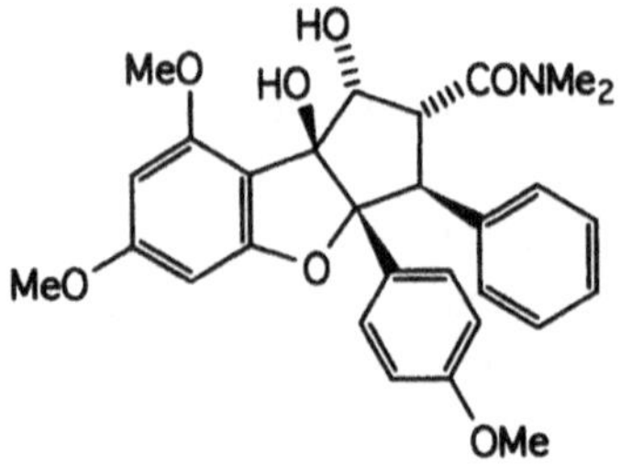

ROCAGLAMIDE

4

Table 3. Effect of "Stopfeed" (*Aglaia odorata*) extract on *Ostrinia nubilalis* growth parameters [a]

Concentration (µg/g)	Time to pupation (days)	Survival to pupation (%)	Time to adult emergence (days)	Survival to emergence (%)	Sex Ratio (♀ / total)
0	24.1	90.	28.1	85.	0.70
12.5	34.7*	70.	43.2*	50.	0.40
25	51.0*	25.	60.4*	25.	0.20
50	83.8*	20.	86.3*	10.	-
100	-	-	-	-	-

[a]Mean time to pupation and adult emergence were subjected to Kruskal-Wallis test: Means with asterisks are significantly different from controls ($P < 0.05$) (unpublished data).

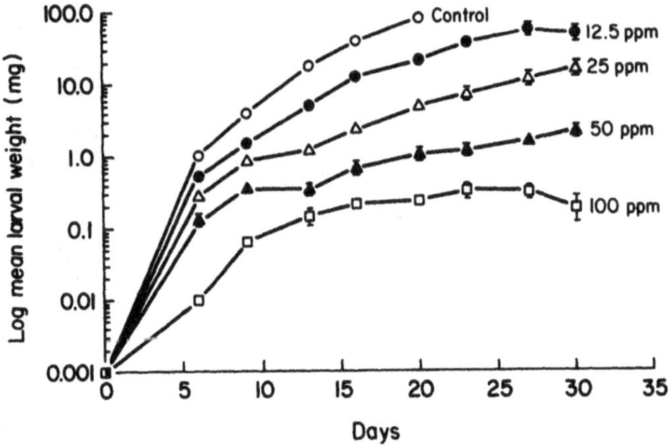

Fig. 5. Survival of fourth instar variegated cutworms following topical application of rocaglamide. Microgram quantities represent the acute dose per larva (unpublished data)

Table 4. Antifeedant and toxic effects of selected compounds on *Peridroma saucia* larvae [a]

Phytochemical	Dietary EC_{50} (ppm)	Dietary DC_{50} (ppm)	Leaf disk IC_{50} ($\mu g/cm^2$)
Azadirachtin	0.3	2.7	0.008
Rocaglamide	1.4	11.2	3.5
Cedrelone	53.1	220	27.2

[a]Data from references [30,33,37] The EC_{50} is is the effective concentration for reduction of first instar larval growth rates by 50% in 'no choice' tests. The DC_{50} is the deterrent concentration for 50% reduction of feeding on treated diets in a 'choice' test with second instar larvae, and the IC_{50} is the concentration for 50% reduction in feeding on treated leaf disks with fifth instar larvae in 'choice' tests.

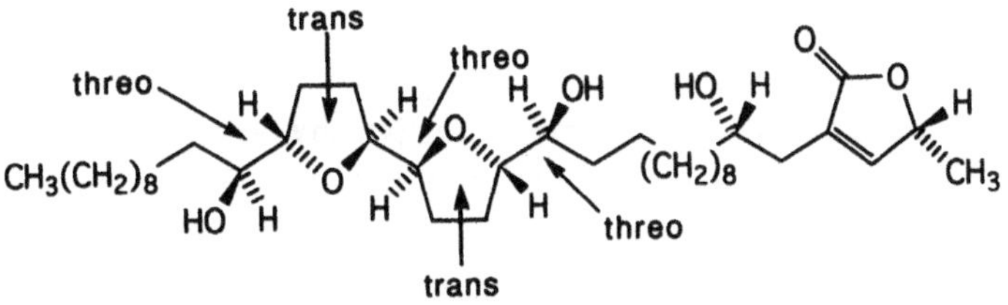

ASIMICIN

5

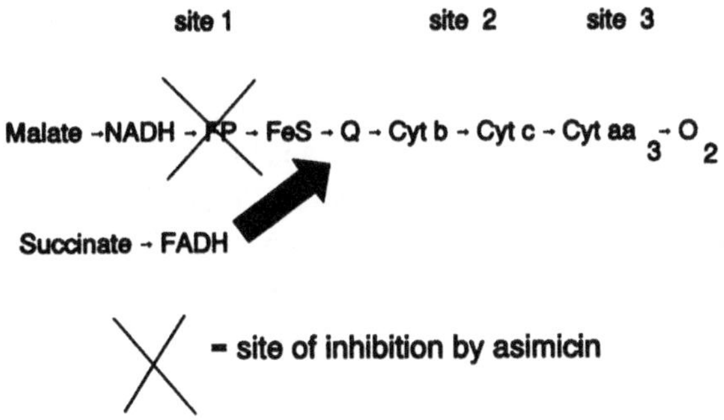

Fig. 6. The electron transport scheme of respiration and the site of asimicin inhibition.

Asimina triloba containing this and other acetogenins have apparent feeding deterrency and toxic effects on a variety of insects.[35] However, in this case, feeding deterrency is mainly a consequence of toxicity as in the case of cedrelone.

Topical application of bullatacin to fourth instar borers at up to 1 mg/g produced no mortality in 7 days but the insects failed to feed and grow. By day 9 an ED_{50} of 6.8 μg/g for growth inhibition was obtained. Investigations of the mode-of-action of asimicin and bullatacin at the physiological and biochemical level has shown that these materials inhibit insect respiration *in vivo* and mitochondrial respiration *in vitro*. In particular, respiration supported by malate and pyruvate was inhibited by asimicin but could be restored when succinate was the respiratory substrate.[34] These results clearly point to site 1 in cellular respiration as the site of action of this new botanical (Fig. 6).

CONCLUSION

A number of promising new alternative phytochemicals and botanicals are now becoming available from tropical plants. Use of these materials should eventually provide effective alternative control agents for insect pests that address problems of both resistance and local production. Their use enhances the conservation value of tropical forest but care must be taken to avoid over-exploitation of this renewable resource by uncontrolled market demand.

ACKNOWLEDGMENTS

This research was supported by grants from the Natural Science Engineering and Research Council of Canada (Strategic program) and the International Development Research Centre. We thank Lee Swain and Kelsey Downum, Florida International University for collections of material from the Fairchild Tropical Garden and J.D.H. Lambert for assistance with collections in Central America. The expert technical assistance of F. Duval, Y Zheng and A. Sciano is acknowledged.

REFERENCES

1. LEVIN, D.A., YORK, B.M. 1978. The toxicity of plant alkaloids: an ecogeographic perspective. Biochem. Syst. & Ecol. 6:61-76.
2. REID, L., ARNASON, J.T., NOZZOLILLO, C., HAMILTON, R. 1990. Resistance of maize to the European corn borer, *Ostrinia nubilalis*, as related to the geographical origin of maize. Can. J. Bot. 68:311-316.
3. JANZEN, D. 1974. Blackwater rivers. Biotropica 6:69.
4. McKEY, D., WATERMAN, P.G., MBI, C.N., GARTLAN, J.S., STRUTTSAKER, T. 1978. Phenolic content in two African rain forests: ecological implications. Sci. 202:61-64.
5. FEENY, P. 1976. Plant Apparency and Chemical Defense. Rec. Adv. Phytochemistry 10:1-40.
6. COLEY, P., BRYANT, J., CHAPIN, F.S. 1983. Resource availability and plant antiherbivore defense. Science 230:895-899.
7. KLOCKE, J.A. 1987. Natural plant compounds useful in insect control. In: Allelochemicals in Forestry and Agriculture. (P. Hedin, ed.). A.C.S. Wash. D.C.

8. GEORGHIOU, G.P. 1990. Resistance potential of biopesticides and consideration of countermeasures. In: Pesticides and Alternatives: Innovative Chemical and Biological Approaches to Pest Control. (J.E. Casida, ed.). Elsevier, N.Y. pp 409-420.
9. SCHOONHOVEN, L.M. 1982. Biological aspects of antifeedants. Entomol. Exp. Appl. 31:57-69.
10. VAN BEEK, T.A., DE GROOT, A. 1986. Terpenoid antifeedants. Recl. Trav. Chim. Pays Bas. 105: 513-527.
11. ARNASON, J.T., PHILOGÉNE, B.J.R., MORAND, P., IMRIE, K., HASSPIELER, B., DOWNE, A.E.R. 1989. Naturally occurring and synthetic thiophenes as insecticides. Amer. Chem. Soc. Symp. Ser. 387:164-172.
12. NISHIMURA, K., KUROGOCHI, N., FUJITA, T. 1990. Quantitative structure activity relationships of pyrethroids. Pestic. Biochem. Physiol. 37:41-52.
13. HASSPIELER, B.M., ARNASON, J.T., DOWNE, A.E.R. 1988. Toxicity, toxicokinetics and localization of the phototoxin α-terthienyl in mosquito larvae. J. Amer. Mosq. Control Assoc. 4:79-84.
14. SOUCY-BREAU, C., MacEACHERN, A., LEITCH, L.C., ARNASON, T., MORAND, P. 1992. Synthesis and characterization of alkyl, alkenyl, acyl, and azo substituted derivatives of the potent phototoxin α-terthienyl. Heterocycles (in press).
15. MARLES, R.J., COMPADRE, R.L., COMPADRE, C.M., SOUCY-BREAU, C, REDMOND, R.W., DUVAL, F., MEHTA, B., MORAND, P., SCAIANO, J.C., ARNASON, J.T. 1991. Thiophenes as mosquito larvicides: structure-toxicity relationship analysis. Pesticide Biochemistry and Physiology 41:(in press).
16. FIELDS, P.G., ARNASON, J.T., PHILOGENE, B.J.R., AUCOIN, R., MORAND, P., SOUCY-BREAU, C. 1991. Phototoxins as insecticides and natural plant defenses. Mem. Entomol. Soc. Canada 159:29-38.
17. MARLES, R.J., ARNASON, J.T., COMPADRE, R.L., COMPADRE, C.M., SOUCY-BREAU, C., MORAND, P., MEHTA, B., REDMOND, R.W., SCAIANO, J.C. 1991. Quantitative structure-activity relationship analysis of natural products: phototoxic thiophenes. In: Modern Phytochemical Methods. (Nikolaus H. Fischer, Murray B. Isman, H.A. Stafford, eds.). Plenum Press, New York. Rec. Adv. Phytochemistry 25:371-395.
18. HASBUN, C., CALDERON, M., ROMERO, I. 1990. 2,5-Diaryloxazols

and coumarins from roots of *Amyris brenesii*. Fitotherapia 61:88.

19. ISMAN, M., KOUL, O., LOWREY, T., GAGNON, D., STEWART, J.G., SALLOUM, G.S. 1990. Development of a neem based insecticide for Canada, Proc. "Neem's potential in US pest management programs." USDA ARS-86, Beltsville pp. 32-39.
20. ISMAN, M., KOUL, O., STEWART, J.G., ARNASON, J.T., SALLOUM, G.S. 1991. Developing a neem insecticide for Canada, Mem. Ent. Soc. Can. 159:39-47.
21. WOOD, T. 1990. Efficacy of neem extracts and neem derivatives against several agricultural insect pests, "Neem's potential in US pest management programs." USDA ARS-86, Beltsville pp. 76-83.
22. SCHMUTTERER, H. 1990. Properties and potential of natural pesticides from the neem tree and other tropical plants. Ann. Rev. Entomol. 35:271-297.
23. BERNARD, C., ARNASON, J.T., PHILOGÉNE, B.J.R., WADDELL, T., LAM., J. 1989. Effect of lignans and other representative secondary metabolites of the Asteraceae on PMSO activity of the European corn borer. Phytochemistry 28:1373-77.
24. MIKOLAJCZAK, K.L., ZILKOWSKI, B.W., BARTELT, R.J. 1989. Effect of Meliaceous seed extracts on growth and survival of *Spodoptera frugiperda*. J. Chem. Ecol. 15:121-128.
25. MIKOLAJCZAK, K.L., REED, D.K. 1987. Extractives of seeds of the Meliaceae: effects on *Spodoptera frugiperda*. J. Chem. Ecol. 13:99-111.
26. CHAMPAGNE, D., ISMAN, M.B., TOWERS, G.H.N. 1989. Insecticidal activity of phytochemicals and extracts of the Meliaceae. Amer. Chem. Soc. Symp. Ser. 387:195-109.
27. CANNEY, P., GARDNER, D. 1988. Effects of artificial and natural diets on success in tip recording and on the galeal chemosensillum morphology of European corn borer. Physiol. Entomol. 14:13-19.
28. CHAMPAGNE, D., KOUL, O., ISMAN, M.B., SCUDDER, J., TOWERS, G.H.N. 1992. Biological activity of limonoids from the Rutales. Phytochemistry 31:377-394.
29. MATEOS, A.F., de la FUENTE BLANCO, J.A. 1990. A simple synthesis of compounds related to insect antifeedants: azadiradion and gedunin. J. Org. Chem. 55:1349-54.
30. MATEOS, A.F., de la FUENTE BLANCO, J.A. 1991. Synthesis of limonoid insect antifeedants through sterioselective aldol addition reactions. J. Org. Chem. 55:1349-54.

31. ARNASON, J.T., PHILOGÉNE, B.J.R., DONSKOV, N., KUBO, I. 1987. Limonoids disrupt growth and development of *Ostrinia nubilalis*. Entomol. Exp. & Appl. 43:221-226.

30. SATASOOK, C., ISMAN, M., WIRIYACHITRA, P. 1992. Toxicity of rocaglamide, an insecticidal natural product to the variegated cutworm. Pestic. Sci. (in press).

31. JANPRASERT, J., SATASOOK, C., SUKUMALANAND, P., CHAMPAGNE, D. ISMAN, M., WIRIYACHITRA, P., TOWERS, G.H.N. 1992. Rocaglamide, a new insecticide from *Aglaia odorata*, Phytochemistry (in press).

32. ISHIBASHI, F., SATASOOK, C., ISMAN, M., TOWERS, G.H.N. 1992. Insecticidal 1*H*-Cyclopentatetrahydro[*b*]benzofurans from*Aglaia odorata*. Phytochemistry (in press).

33. KOUL, O., ISMAN, M.B. 1992. Toxicity of the limonoid cedrelone to noctuid larvae. Entomol. Exp. & Appl. 64:281-287.

34. LEWIS, M.A., ARNASON, J.T., PHILOGÉNE, B.J.R., RUPPRECHT, J.K., MCLAUGHLIN, J.L. 1993. Inhibition of respiration at site I by asimicin, an acetogenin of the pawpaw. Pest. Biochem. Physiol. (in press).

35. ALKOFAHI, A., RUPPRECHT, J.K., ANDERSON, J.E., McLAUGHLIN, J.L. MIKOLAJCZAK, K.L., SCOTT, A. 1988. Search for new pesticides from higher plants. Amer. Chem. Soc. Symp. Ser. 387:25-43.

36. KUBO, I., KLOCKE, J.A. 1986. Insect ecdysis inhibitors. Amer. Chem. Soc. Symp. Ser. 296:206-219.

37. ISMAN, M.B., KOUL, O., LUCZYNSKI, A., KAMINSKI, J. 1990. Insecticidal and antifeedant bioactivity of neem oils and their relationship to azadirachtin content. J. Agric. Food Chem. 38:1406-1411.

Chapter Six

INSECT CONTROL AGENTS FROM TROPICAL PLANTS

Isao Kubo

Division of Entomology and Parasitology
College of Natural Resources
University of California, Berkeley, California 94720

INTRODUCTION

Since tropical plants are exposed throughout the entire year to attack by various parasites such as bacteria, fungi and insects, they are confronted with harsh conditions for survival. This leads to efficient built-in defense mechanisms, and thus tropical plants offer a rich and intriguing source of secondary metabolites possessing attractive pesticidal properties. These phytochemicals are mainly biodegradable and, more importantly, they are renewable. The efficient use of such renewable natural resources is becoming increasingly important worldwide.

There is no doubt that many plant secondary metabolites affect insect behavior, development and reproduction. Identifying these substances is an important first step in understanding the effects of plants on insect life at the

Phytochemical Potential of Tropical Plants
Edited by K.R. Downum et al., Plenum Press, New York, 1993

molecular level. Besides benefits to basic science, accumulation of this knowledge may provide us with a more rational and scientific approach to insect pest control. Plant defenses to insect attack usually involve diverse secondary metabolites. In this chapter I will describe multichemical defenses in two plants used in our studies.

SELECTION OF PLANTS FOR INVESTIGATION

An important consideration to ensure the success of the project is the selection of plants to be studied. Wherever I go for plant collection, I first try to obtain suggestions of appropriate plants from naturalists such as botanists, entomologists and ecologists, as well as from local people. If no information is available during plant collection, I usually collect plants which are relatively free from insect attacks in nature. Interestingly, many plants reported to have an effect on insects taste bitter or hot to us. For example, *Azadirachta indica* (Meliaceae),[1,2] *Harrisonia abyssinica* (Simaroubaceae),[3] *Trichilia roka* (Meliaceae),[4] *Bersama abyssinica* (Melianthaceae),[5] *Ajuga remota* (Labiatae),[6] *Vernonia amygdalina* (Compositae),[7] *Schkuhria pinnata* (Compositae),[8] *Croton cajucara*,[9] *C. jatrophoides*[10] (Euphorbiaceae) and *Castela tortuosa* (Simaroubaceae),[11] *etc.* are extremely bitter. *Warburgia ugandensis*, *W. stuhlmannii* (Canellaceae)[12,13] and *Fagara macrophylla* (Rutaceae),[14] are exceptionally hot to us. Active principles isolated from these plants are mainly highly oxygenated terpenoids and steroids. Therefore, a bitter or hot taste may be considered to be a valuable clue. In fact, plants collected based on this kind of information have a much higher probability of containing active compounds than those collected for other purposes such as medicinal use. However, some plants such as *Podocarpus gracilior*,[15] *P. nagi*[16] (Podocarpaceae), *Vitex fisherii*[17] and *V. strickeri*[18] (Verbenaceae), which also have effects on insects, are tasteless. Nevertheless, tropical plants are good potential sources of insect control agents.

In general, plant materials can be air dried before the extraction is made, but in some cases the biological activity may be lost if plants are left for a long period of time. This should be taken into account especially when unstable compounds or glycosides such as saponins are the biologically active target molecules. For example, although several limonoids isolated from oil of the Indian neem, *A. indica*, such as azadirachtin,[19] exhibit potent insect growth inhibitory activity,[20] they are not stable in the oil squeezed out from the fresh fruit. Stabilization of these active limonoids is one of the important concerns for their commercial use. Nevertheless, most of the above mentioned tropical plants were received in a dried form from which we were able to isolate many active compounds.

In our investigation, the plants collected from tropical countries are assayed against mainly North American pest insects. The data can, therefore, document the physiological properties of phytochemicals, but they do not reflect the ecological implications of these properties in their native habitats.

BIOASSAYS

More important consideration should be given to the selection of appropriate bioassays as guides in order to isolate active compounds. It is difficult to generalize the bioassay, since plant secondary metabolites interfere with many biological activities occurring before and after ingestion, and also since each insect is different in many respects. There is still no simple bioassay that can fully explain resistance mechanisms in plants to insect attack. For example, one complication is that a chemical identified as a feeding stimulant to some insects can be a feeding deterrent to others, and that a plant contains a number of secondary metabolites that may act synergistically. The selection of insect species to be tested is, therefore, important. In our study, several lepidopteran herbivores such as the pink bollworm, *Pectinophora gossypiella*, the fall armyworm, *Spodoptera frugiperda* and the tobacco budworm, *Heliothis virescens* were selected in order to make our investigation beneficial to society, although additional insects were occasionally assayed if materials permitted. In other words, we have been primarily searching for insect control agents against these lepidopteran pests, unless otherwise specified.

Artificial Diet Assay

We usually test first the effects of an artificial diet feeding assay against the lepidopteran herbivores, *P. gossypiella*, *S. frugiperda* and *H. virescens*.[21] The general scheme of the feeding assay is shown in Figure 1. The plant extracts or purified substances are dissolved in solvents such as methanol or acetone. The sample solution is added to α-cellulose, a non-nutritive filler used as a carrier, and evaporated to dryness under reduced pressure. This mixture is combined in sequence with a vitamin solution and nutrient powder, and finally mixed with a gelling agent *via* vigorous stirring. Homogenous incorporation of the samples into the diet is the key to a successful assay. One day-old larvae are carefully transferred to diet in plastic cups. The size of cups depends on the test insects.

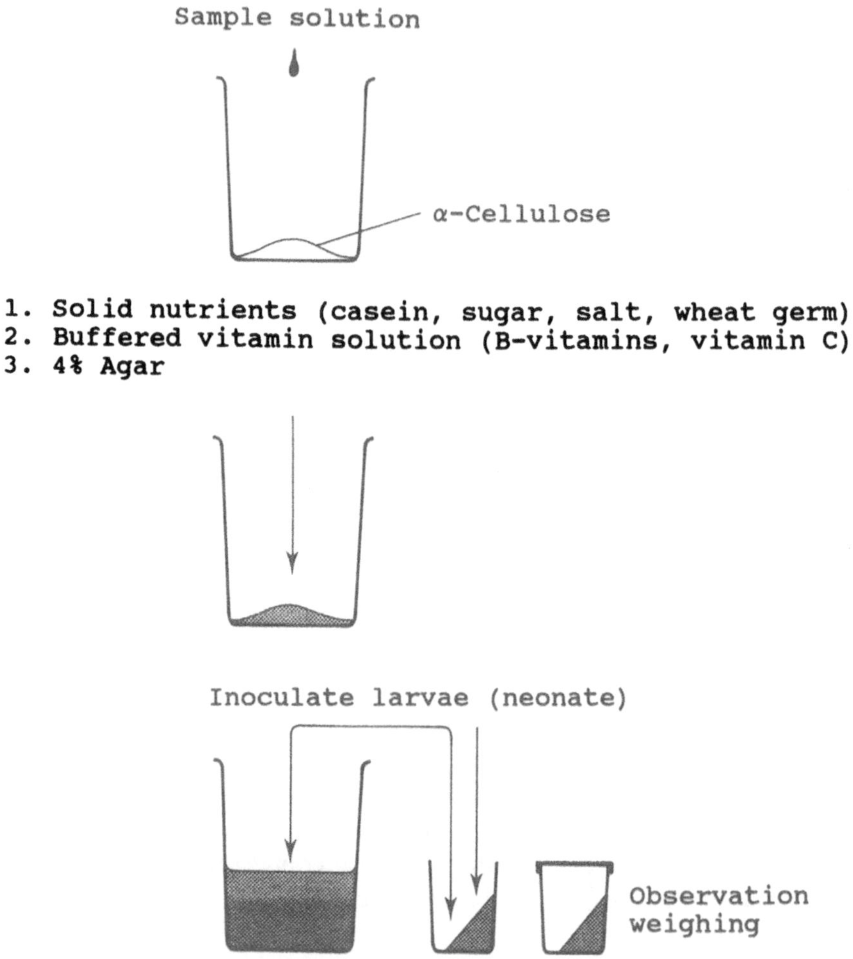

Fig. 1. Artificial diet assay for lepidopteran larvae.

Two larvae of each species are used per cup. The smaller of the two is removed after 3-4 days to reduce variability. Larvae are incubated for 8-10 days at 28°C in an incubator under 12-14 hrs day length. Larval weights are determined during the assay period and daily observations are made with the stereo dissecting microscope. Needless to say, any differences from the control should not be overlooked. Careful observation of insects throughout the assay is essential to make the project successful.[21,22]

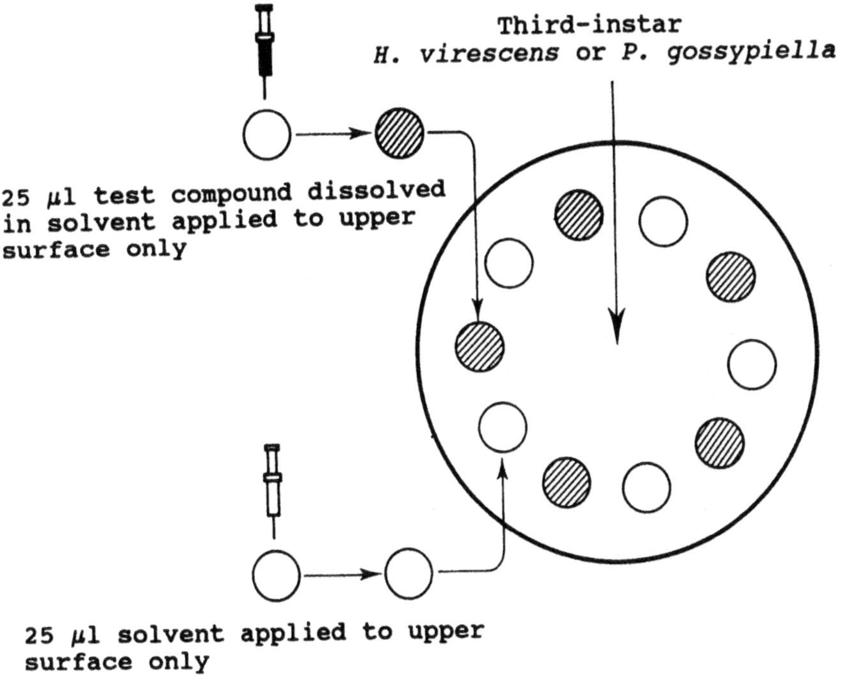

Fig. 2. Leaf disk "choice" antifeedant assay for lepidopteran larvae.

Antifeedant Activity Assay

In addition to the artificial diet feeding assay, a screening for antifeedant activity is also carried out with the same insects. Since all insects have varying feeding behaviors, the particular antifeedant assay differs according to the insect. Even though the target insects are narrowed to the three lepidopterans, *P. gossypiella*, *S. frugiperda* and *H. virescens*, there are still several methods to test antifeedant activity. We usually conduct a simple "leaf disk" assay. The plant extract is applied to the upper surface of host plant disks, then insects are placed on them for 2 days, and the disks are examined visually. The general scheme of the leaf disk assay is shown in Figure 2. Disks (1 cm diameter) are punched out from leaves of a glandless cotton cultivar (Pima S-4 of *Gossypium barbadanse*, normally a favored host of our test insects). The leaf disks are symmetrically arranged (12 disks per dish) in a circle on moistened filter papers in polyethylene which forms grids inside glass Petri dishes (100 X 15 mm). Alternating disks are

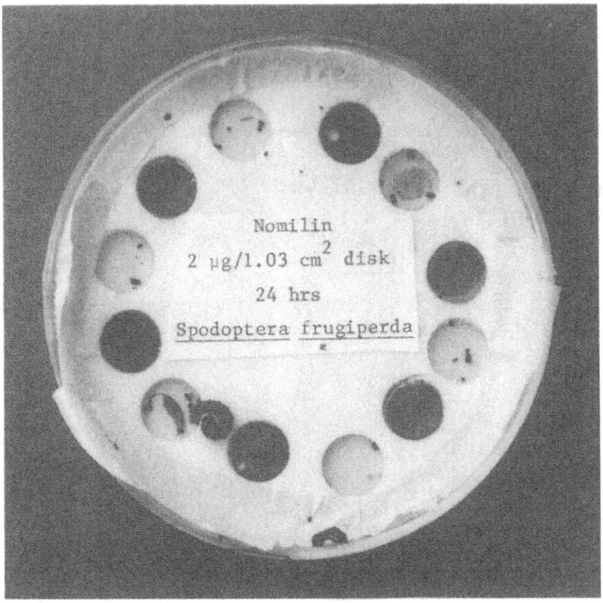

Fig. 3. Photograph of the leaf disk assay.

treated on their upper surface with either 25 μl acetone or with 1-100 μg of a chemical dissolved in 25 μl acetone (applied with a microsyringe). Newly-molted third-instar larvae, reared from eggs on a meridic artificial diet, are placed, three larvae per dish, at 25°C and 80% relative humidity in a dark incubator. After 48 hrs, the larvae are removed and the disks are examined visually. Activities of the antifeedants are compared in terms of 95% protective concentration (PC_{50}) values. Experiments determining PC_{50} values are repeated at least six times. Figure 3 is a photograph of an actual leaf disc assay.

The promising samples also are tested against other insects, if materials permitted. For example, a bioassay with a pest aphid species is occasionally conducted to test for feeding deterrent effects of the water soluble fractions and purified compounds. The samples are dissolved directly into an aqueous diet consisting of vitamins, sucrose, amino acids, trace metals, salts, and cholesterol, which are brought to pH 8.7 with K_3PO_4. The aqueous diet is placed into polyethylene vial caps and each of these caps is fitted with circular holes punched into the plastic snap-on lids for polystyrene catsup cups (1 oz). Between 50-100

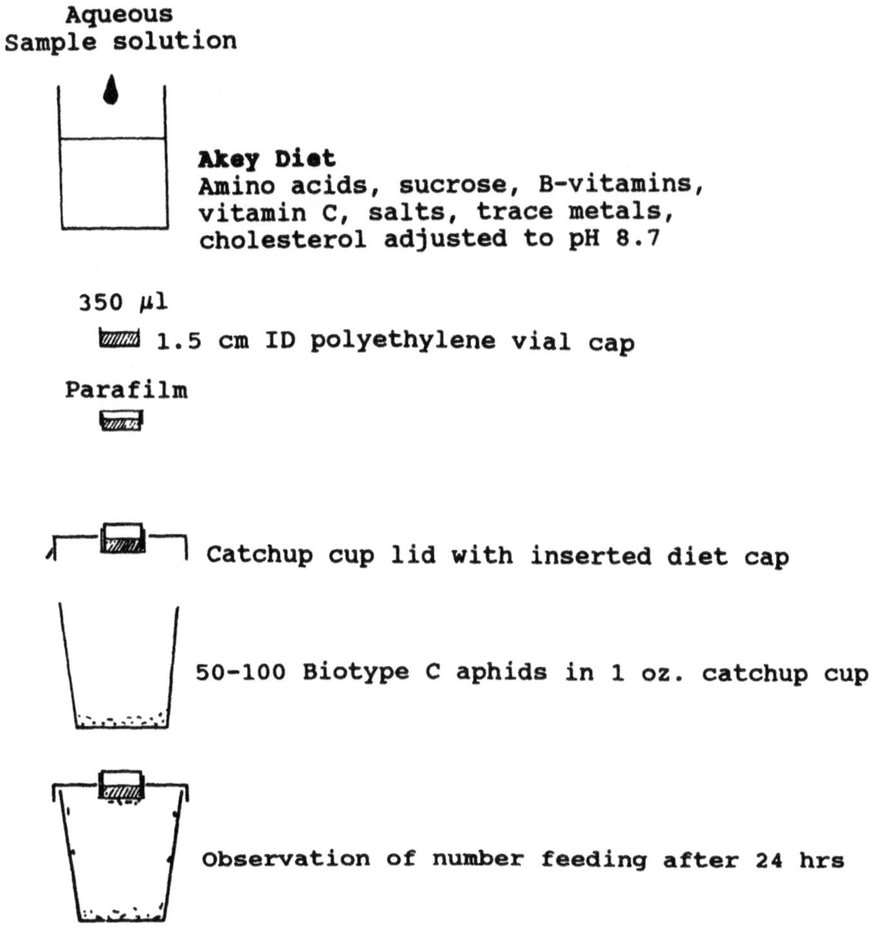

Fig. 4. Antifeedant bioassay for the greenbug, *S. graminum*.

Biotype C greenbugs, *Schizaphis graminum*, an important pest on sorghum, are transferred from sorghum plants into each of the 1 oz cups, which are immediately fitted with the diet cap-containing snap-on lids. Appropriate controls, which consistently resulted in >90% feeding, are then used to compare to each of the treatments in order to determine ED_{50} values for each test. After 24 hrs at room temperature, the number of aphids feeding/total number aphids is determined for each treatment. Figure 4 shows the general scheme of the artificial diet feeding assay against the greenbug, *S. graminum*.

INSECT CONTROL IN *AJUGA REMOTA*

A bitter tasting tropical medicinal plant, *Ajuga remota* (Labiatae), is relatively free from insect attack in nature. In order to test for chemical factors involved in this observed resistance, the methanol extract of *A. remota* foliage was first tested against the African armyworm, *Spodoptera exempta* by the leaf disk assay. Subsequently, three bitter tasting diterpenoids, ajugarin-I (**1**), -II (**2**) and -III (**3**) were isolated as antifeedants.[6,23] Interestingly, ajugarin-I was also found to exhibit potent antifeedant activity against the desert locust, *Schistocerca gregaria*, but not against the vagrant grasshopper, *S. vega*.

In addition to creating a behavioral barrier, the secondary metabolites may also interfere with metabolism after ingestion of the samples. Therefore, 0.1% of the methanol extract of *A. remota* foliage was incorporated into the artificial diet optimized for *P. gossypiella* and *H. virescens*. Visual analysis of the pink bollworm which were fed *A. remota* extract revealed a developmental disruption in which the insect died in the pharate condition following the initiation of molting (apolysis), but before completion (ecdysis). The newly-molted larva died while encased in the old cuticular skin and head capsule (pharate condition). The pharate condition prevents feeding because the mouthparts are masked by the head capsule as shown in Figure 5. This also prevents locomotory and excretory functions, since the whole body is entrapped in the old cuticle. Bioassay guided fractionation with a combination of various chromatographic methods led to the isolation of three active principles.[24] Spectroscopic identification revealed three phytoecdysteroids, namely 20-hydroxyecdysone (**4**), cyasterone (**5**) and ajugasterone C (**6**). Figure 6 is an electron micrograph of a pink bollworm after ingestion of 20-hydroxyecdysone. The insect possessed three head capsules because it underwent two failed molting cycles before death. That is, although feeding became impossible after the first inhibited ecdysis, because the adhering second head capsule covered the mouthparts, the larva still synthesized a third head capsule.[25]

An additional artificial diet feeding assay demonstrated that the silkworm, *B. mori* is particularly sensitive to the effects of phytoecdysteroids. Since its eggs and artificial diet are commercially available, *B. Mori* may be a good choice in searching for phytoecdysteroids. In addition to the three phyto-ecdysteroids, the bioassay guided fractionation with *B. mori* led to isolation of an insecticidal diterpene, ajugarin-IV (**7**), from the ether extract of *A. remota.*[26]

The same bioassay-guided fractionation using *P. gossypiella* led to the identification of various phytoecdysteroids in two other *Ajuga* species,*A. reptans* and *A. chameacistus*,[27] and four *Vitex* trees, *V. fisherii*,[17] *V. madiensis*,[28]

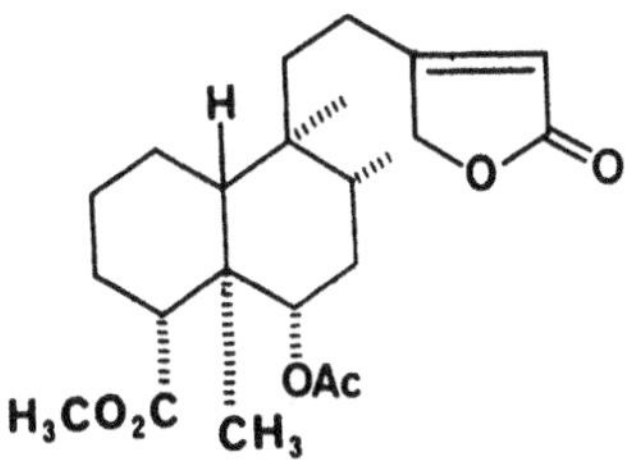

1

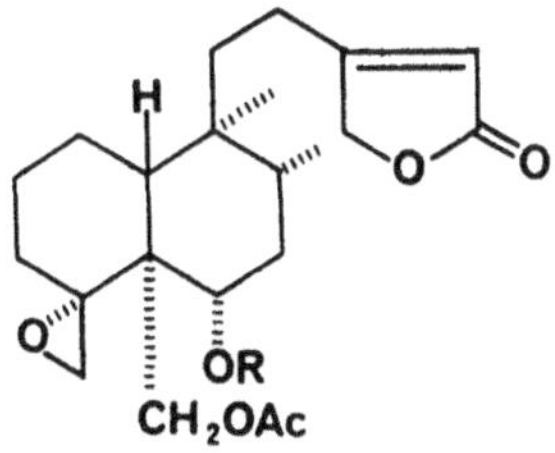

2 R=Ac

7 R=H

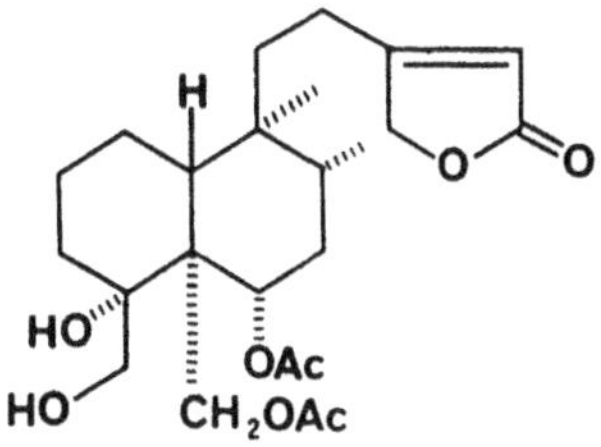

3

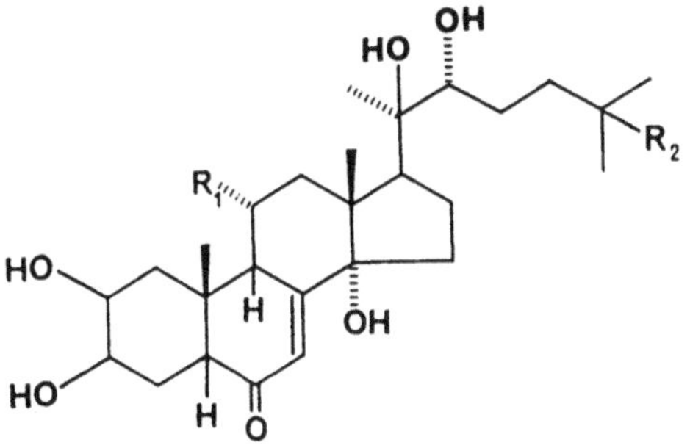

4 R_1=H, R_2=OH

6 R_1=OH, R_2=H

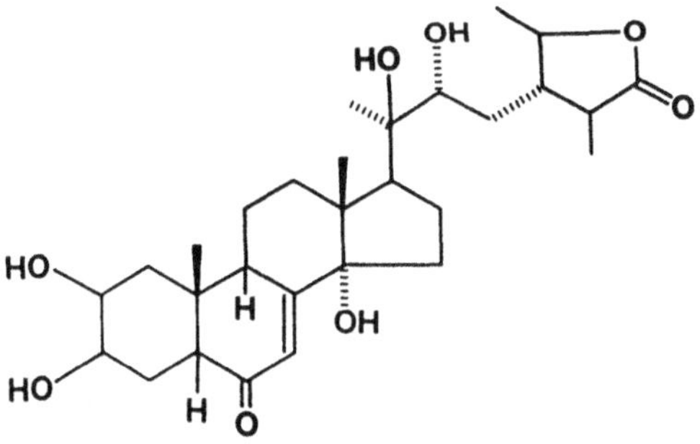

5

V. thyrsiflora[29] and *V. strickeri*.[18] In contrast to *Ajuga*, where phytoecdysteroids tend to be foliar, phytoecdysteroids are usually concentrated in the root bark in *Vitex*.

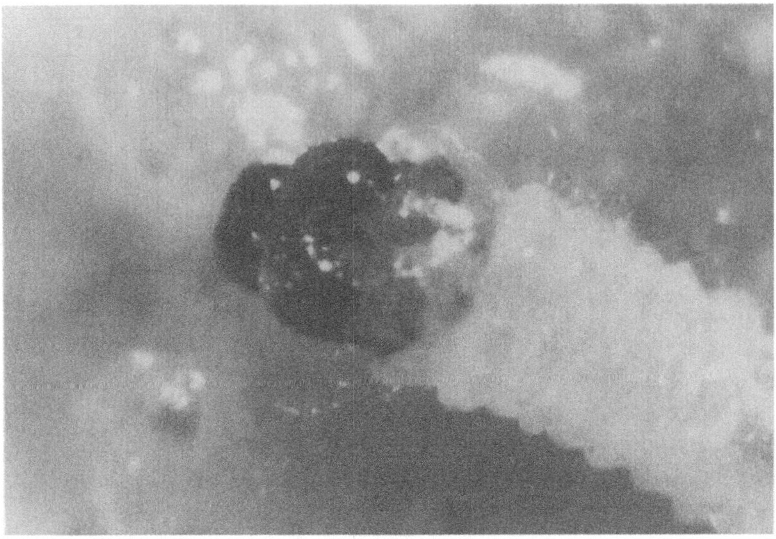

Fig. 5. A larva of the pink bollworm, *P. gossypiella*, becomes much smaller in size from the control caused by the ingestion of a crude methanol extract of *A. remota*foliage. Another remarkable observation is the molting cycle failure; the insect undergoes a normal apolysis, but fails to complete ecdysis. Larva has three head capsules which mask the functional mouthparts. The insect eventually starves to death.

Several other bioassays have also been developed for the detection of the phytoecdysteroids, but since these bioassays entail unnatural application of the compounds, they can be used only for detection and not for understanding the physiological phenomena involved in plant-insect interactions. Since the artificial diet feeding assay, on the other hand, is based on the actual ingestion of dietary phytoecdysteroids, it may be comparable to the effects of ingestion of living plants.

Although ingested phytoecdysteroids have a potent and unique hormonal activity against susceptible species like the pink bollworm and the silkworm, other insect species, such as *H. virescens* larvae, are unaffected by dietary phytoecdysteroids. When this polyphagous insect is fed more than 0.3% of 20-hydroxyecdysone in the artificial diet, it shows no obvious morphological or developmental changes. The closely related species, *H. zea*, is also unaffected. These herbivores may have developed mechanisms to detoxify phyto-

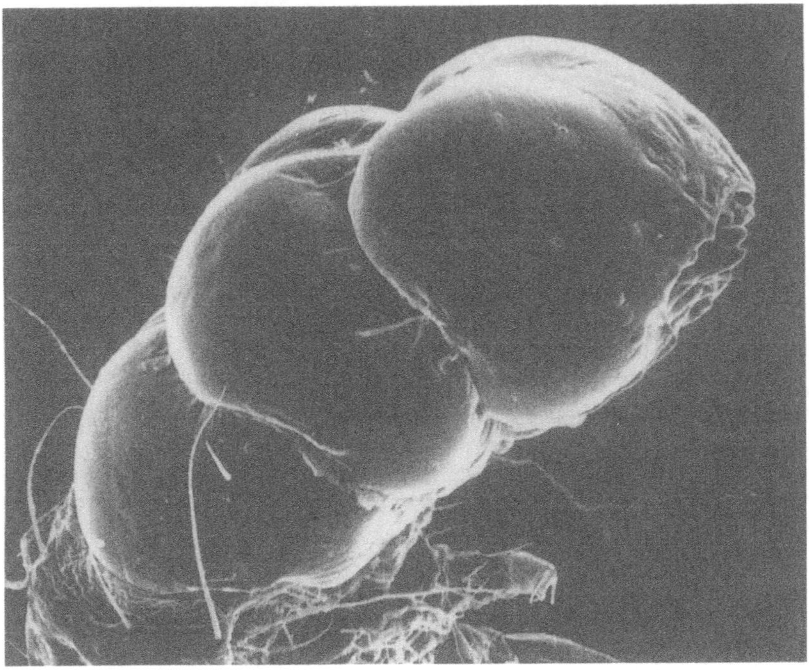

Fig. 6. Electron micrograph of the anterior end of a larva of the pink bollworm, *P. gossypiella*, after ingestion of 20-hydroxyecdysone. Larva has three head capsules which mask the functional mouthparts. The insect eventually starves to death. Magnification X 125.

ecdysteroids.[30] These results indicate that if we had assayed only against *H. virescens*, none of the phytoecdysteroids would have been isolated from *A. remota*. Further study of the detoxification mechanisms using *H. virescens* has shown that the ingested phytoecdysteroids are transformed to their 22-acyl esters in the larval gut by a novel enzyme, ecdysteroid-22-acyltransferase. This conversion is different from the normal metabolic pathway in which the ecdysteroids are usually transformed to more water-soluble products. Most ecdysteroid-22-acyltransferase activity was found in the midgut epithelial-cell membrane and was active only during feeding stages. The esterification was specific to the 22-hydroxy group.[31,32]

The additional artificial diet feeding assay with the greenbugs, *S. graminum*, was also conducted. Interestingly, the above mentioned

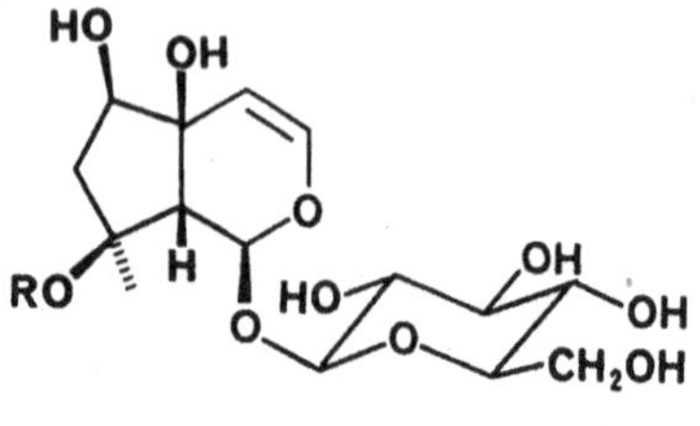

8 R=H

9 R=Ac

Table 1. Feeding deterrency of three phytoecdysteroids on the greenbug, *S. graminum* [a]

Compounds	ED_{50} (ppm in diet)[b]
20-Hydroxyecdysone	650
Cyasterone	2000
Ajugasterone C	62

[a] Biotype C of *S. graminum* from a mixed population in a 24 hr no-choice assay.
[b] ED_{50} is the effective dose for 50% feeding compared to control.

phytoecdysteroids were found to exhibit antifeedant activity. As summarized in Table 1, ajugasterone C (**6**) was found to be 10-fold more potent than 20-hydroxy-ecdysone (**4**), and 30-fold more potent than cyasterone (**5**). The same antifeedant bioassay also led to the isolation of two bitter tasting iridoid glycosides, harpagide (**8**) and its acetate (**9**). These iridoid glycosides were, however, inactive in the leaf disk and artificial diet feeding assays against the lepidopteran insects.

Phytoecdysteroids are hormonally active in several insects. However, since some species of insects are unaffected by dietary phytoecdysteroids, these compounds alone may not be responsible for the resistance of *A. remota* to insect attack in nature. As already pointed out, a behavioral barrier of the antifeedant may be important. Insects overcoming this behavioral barrier must then contend with at least one more chemical barrier—a developmental barrier due to three phytoecdysteroids.

INSECT CONTROL IN *PODOCARPUS GRACILIOR*

Podocarpus species in general are considered to be relatively immune to insect attack. At least some of this immunity may be attributed to ponasterones, compounds with molting hormone activity which occur predominantly in the Taxaceae and the Podocarpaceae. Recently, more than 50 nor- and bisnor-diterpene dilactones have been isolated from various tissues of *Podocarpus* species.[33] Some of these dilactones have a variety of biological effects, including insecticidal ones. The norditerpene dilactones, along with the phytoecdysteroids, may be important elements in the defense against insect attack. Besides the phytoecdysteroid ponasterone A (**10**), two additional classes of chemicals from *P. gracilior* foliage were isolated. Feeding deterrent activity was ultimately coupled to an insecticidal activity due to nagilactones C (**11**), D (**12**) and podolide (**13**), and to a growth inhibitory activity caused by podocarpusflavone A (**14**) and 7",4"'-dimethyletheramentoflavone (**15**)[15].

A concentrated methanol extract of dried leaves of *P. gracilior* was partitioned between ethyl acetate and water. The ethyl acetate fraction was dried and washed with ether to remove pigments. The remaining ether extract was subjected to column chromatography on silica gel 60 in 10% methanol in chloroform. When assayed in the artificial diet bioassay, the early eluted fractions were lethal; the later fractions caused growth inhibition. Still later fractions caused inhibition of ecdysis (completion of molting). Growth inhibitory podocarpusflavone A and 7",4"'-dimethyletheramentoflavone (Table 2), insecticidal nagilactones C and D and podolide (Table 3), and ecdysis inhibitory ponasterone A were purified by various chromatographic procedures, asayed by the artificial diet feeding assay using *P. gossypiella* and *H. virescens*. Incorporation of either biflavonoids, as high as 5% in the artificial diet, resulted in abnormally small larvae. The growth inhibitory activity might be due to some effect on larval metabolism, since no feeding deterrence response in the leaf disk assay was evoked. In the artificial diet feeding assay, the test insects are not given a choice of food source. Nagilactones C and D, and podolide increased mortality and these compounds were relatively potent growth inhibitors at lower concentrations. This might be due, at least in part, to a feeding deterrent effect of these compounds (Table 4). In *P. gracilior*, the feeding deterrent action of the dilactones represents a behavioral defense barrier in the leaf disk assay.

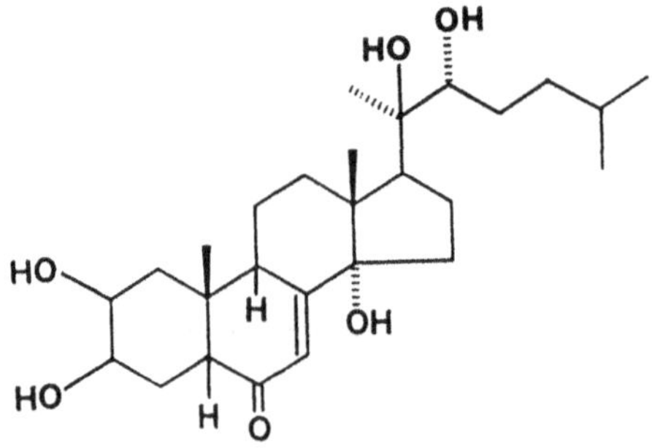

10

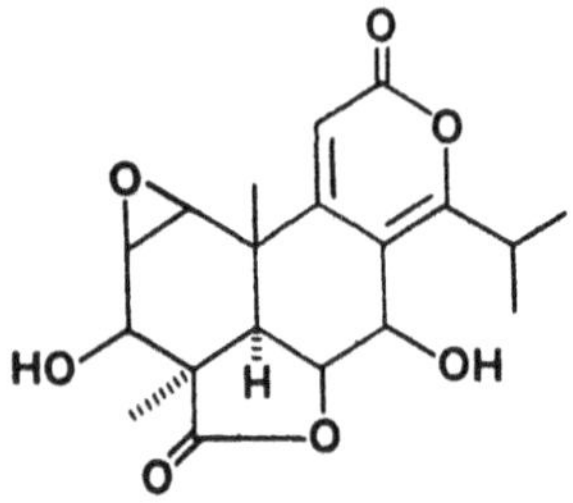

11

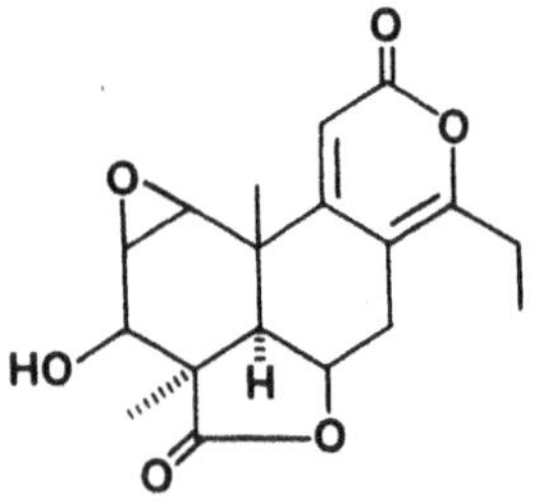

12

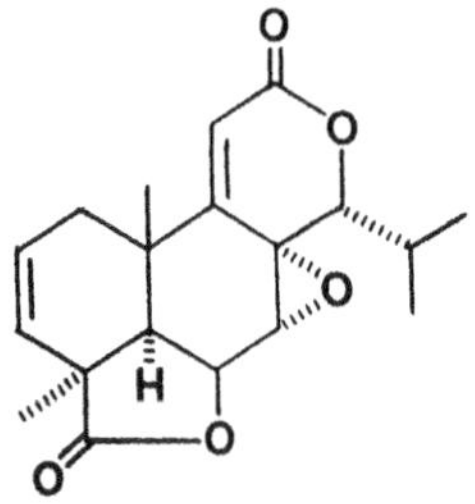

13

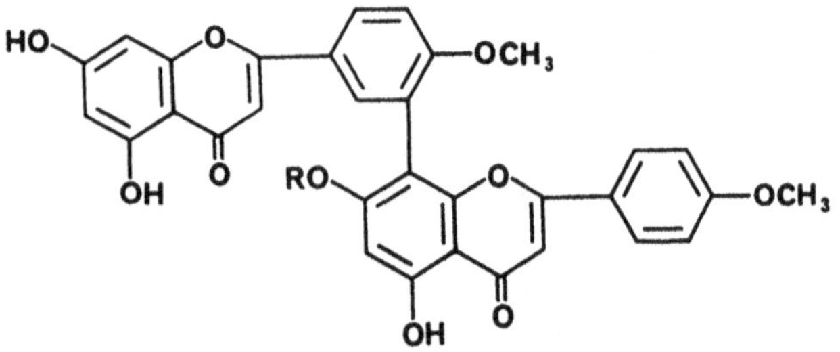

14 R=H

15 R=CH_3

Nagilactone D (**13**) was found to be the most potent component in the artificial diet feeding assay. A third line of defense in *P. gracilior* is a developmental barrier which must be overcome by attacking insects. Due to its molting hormone activity, ponasterone A exemplifies this developmental barrier.

Table 2. Growth inhibitory activity of two biflavones put in the artificial diet [a]

Insect species	Test compounds	ED_{50} (mmol)	ED_{50} (ppm)[b]
H. virescens	Podocarpusflavone A	1.1	625
	7",4'"-Dimethyletheramentoflavone	4.4	2500
P. gossypiella	Podocarpusflavone A	4.2	2300
	7",4'"-Dimethyletheramentoflavone	4.1	2300

[a] No deaths were recorded with either biflavone up to 5000 ppm.
[b] ED_{50} values are the effective doses for 50% growth inhibition and are the means of five determinations.

Table 3. Effects of three norditerpene dilactones on the growth and survival of neonate larvae treated in 12 day feeding assay

Insect species	Test compounds	LD_{90} (ppm)[a]	ED_{50} (ppm)[b]
H. virescens	Nagilactone C	1500	20
	Nagilactone D	800	4
	Podolide	—	12
P. gossypiella	Nagilactone C	1500	14
	Nagilactone D	200	4
	Podolide	300	9

[a] LD_{90} values are the lethal doses for 90% of deaths and are the means of five determinations.
[b] ED_{50} values are the effective doses for 50% growth inhibition and are the means of five determinations.

Table 4. A feeding deterrant assay: 24-hr leaf disk choice assay with two norditerpene dilactones[a] against third-instar larvae of *H. virescens*

Insect species	Test compounds	PC_{50} (μg/disk)[b]
H. virescens	Nagilactone C	38
	Nagilactone D	15

[a] Podolide was not tested in the leaf disk assay. Podocarpusflavone A and 7",4'"-dimethyletheramentoflavone were found inactive in this assay.
[b] PC_{50} values are the concentration of test compounds at which less than 5% of the treated leaves are eaten, while greater than 50% of the untreated leaves are eaten. The experiment was terminated when more than 50%, but less than 80% of the control leaf disks were consumed.

The diversity of protective strategies included in the chemical defense of *P. gracilior* is essential because of the wide variation in susceptibility of insect species to any one defense. For instance, *H. virescens* has overcome the developmental barrier of ponasterone A, but is still susceptible to behavioral and metabolic barriers. Any insect species can develop resistance to any one barrier, but cross-resistance to other barriers is unlikely, since both the structures of defensive chemicals and the activities that they induce differ widely. Thus, the use of more than one chemical in the defensive strategy confers a potential for resistance to insect attack in *P. gracilior* foliage.

CONCLUSION

In our continuing search for alternative insect control agents from tropical plants against several lepidopteran herbivores, a number of secondary metabolites have been isolated. Multichemical defensive strategies, such as those described in *A. remota* and in *P. gracilior*, may be more the rule than the exception in resistant plant cultivars. The elucidation of these strategies, particularly the chemical aspects of them, will generate a more rational and scientific approach to insect pest control. Presently, some of the compounds identified provide useful information for pesticide syntheses. In addition, some of the plant extracts or fractions can be considered for practical use to control insect pests. In the future, plant breeding programs may genetically enhance the chemical defenses of beneficial plant species. However, multichemical defense mechanisms in plants against insect attack should be taken into consideration.

ACKNOWLEDGEMENTS

I am greatly indebted to the following colleagues who have been engaged in studies: namely, Dr. J.A. Klocke, Dr. S. Asano, Dr. T. Mataumoto, Dr. A. Matsumoto, Dr. J.F. Hanke, Dr. G. de Bore, Dr. Y. Asaka, Dr. S. Komatsu, Dr. J.F. Ayafor, Mr. A. Chapya and Mr. M. Zhang.

REFERENCES

1. KUBO, I., MATSUMOTO, T., MATSUMOTO, A., SHOOLERY, J.N. 1984. Structure of deacetylazadirachtinol, application of 2D ^{1}H-^{1}H and ^{1}H-^{13}C shift correlation spectroscopy. Tetrahedron Lett. 4729-4732.
2. KUBO, I., KLOCKE, J.A. 1982. Azadirachtin, insect ecdysis inhibitor, Agric. Biol. Chem. 46:1951-1953.
3. LIU, H.W., KUBO, I., NAKANISHI, K. 1982. A Southern armyworm antifeedant, 12 β-acetoxy-harrisonin from an African shrub, *Harrisonia abyssinica.* Heterocycles 17:67-71.
4. KUBO, I., KLOCKE, J.A. 1982. An insect growth inhibitor from *Trichilia roka* (Meliaceae). Experientia 38: 639-640.
5. KUBO, I., MATSUMOTO, T. 1984. Abyssinin, a potent insect antifeedant from an African medicinal plant, *Bersama abyssinica.* Tetrahedron Lett. 4601-4604.
6. KUBO, I., LEE, Y.W., BALOGH-NAIR, V., NAKANISHI, K., CHAPYA, A. 1979. Structure of ajugarins. J.C.S. Chem. Commun. 949-950.
7. GANJIAN, I., KUBO, I., FLUDZINSKI, P. 1983. Insect antifeedant elemanolide lactones from *Vernonia amygdalina* (Compositae). Phytochemistry 22:2525-2526.
8. PETTEI, M.J., MIURA, I., KUBO, I., NAKANISHI, K. 1979. Insect antifeedant sesquiterpene lactones from *Schkuhria pinnata*; the direct obtaintion of pure compounds using reverse-phase preparation liquid chromatography. Heterocycles 11:471-480.
9. KUBO, I., ASAKA, Y., SHIBATA, K. 1991. Insect growth inhibitory

nor-diterpenes, dehydrocrotonin and *cis*-dehydrocrotonin from the Amazonian medicinal plant *Croton cajucara* Benth (Euphorbiaceae). Phytochemistry 30:2545-2456.

10. KUBO, I., HANKE, F.J., ASAKA, Y., MATSUMOTO, T., HENG, H.C. CLARDY, J. 1990. Insect antifeedants from tropical plants I. Structure of dumsin. Tetrahedron 46:1515-1552.
11. KUBO, I., MURAI, Y., CHAUDHURI, S.K. 1992. Structure of chaparramarin, a quassinoid from *Castela tortuosa*. Phytochemistry 31:3262-3264.
12. KUBO, I., LEE, Y.W., PETTEI, M., PILKIEWICZ, F., NAKANISHI, K. 1976. Potent army worm antifeedants from the East African *Warburgia* plants. J.C.S. Chem. Commun. 1013-1014.
13. KUBO, I., GANJIAN, I. 1981. Insect antifeedant terpenes, hot-tasting to humans. Experientia 37:1063-1064.
14. KUBO, I., MATSUMOTO, T., KLOCKE, J.A., KAMIKAWA, T. 1984. Molluscicidal and insecticidal activities of isobutylamides isolated from *Fagara macrophylla*. Experientia 40:340-341.
15. KUBO, I., MATSUMOTO, T., KLOCKE, J.A. 1984. Multichemical resistance of the conifer *Podocarpus gracilior* (Podocarpaceae) to insect attack. J. Chem. Ecol. 10: 547-559.
16. ZHANG, M., YING, B.P., KUBO, I. 1992. Nagilactones from *Podocarpus nagi* and their effects on the feeding and growth of tobacco budworm. J. Nat. Prod. 55:1057-1062.
17. KUBO, I., ASAKA, Y., STOUT, M.J., NAKATSU, T. 1990. Structure of a novel phytoecdysteroid, vitexirone, and efficient isolation of phytoecdysteroids from *Vitex fisherii*. J. Chem. Ecol. 16:2581-2588.
18. ZHANG, M., STOUT, M.J., KUBO, I. 1992. Isolation of ecdysteroids from *Vitex strickeri* using RLCC and recycling HPLC. Phytochemistry 31:247-250.
19. KUBO, I., NAKANISHI, K. 1979. Some terpenoid insect antifeedants from tropical plants. In: Advances in Pesticide Science (H. Geissbúhler, G.T. Brooks, P. Kearney, eds.). Pergamon Press, Oxford, pp. 284-294.
20. JACOBSON, M. 1986. The neem tree; natural resistance par excellence. In: Natural Resistance of Plants to Pests, ACS Symposium Series 296 (M.B. Green, P.A. Hedin, eds.). The American Chemical Society, Washington, D. C., pp. 220-232.
21. KUBO, I. 1991. Screening techniques for plant-insect interactions. In: Methods in Plant Biochemistry, Vol. 6 (K. Hostettmann, ed.). Academic Press, London, pp. 179-193.
22. KUBO, I. 1992. Screening methods for postingestive effects of phytochemicals on insects. In: Screening Techniques of Biologically active

Phytochemicals (I. Kubo, ed.). CRC Press, Boca Raton. In press.
23. KUBO, I., KIDO, M., FUKUYAMA, Y. 1980. X-ray crystal structure of 12-bromoajugarin-I and conclusion on the absolute configuration of ajugarins. J.C.S. Chem. Commun. 897-898.
24. KUBO, I., KLOCKE, J.A., ASANO, S. 1981. Insect ecdysis inhibitors from the East African medicinal plant, *Ajuga remota* (Labiatae). Agric. Biol, Chem. 45:1925-1927.
25. KUBO, I., KLOCKE, J.A., ASANO, S. 1983. Effect of ingested phytoecdysones on the growth and development of two lepidopterous larvae. J. Insect Physiol. 29:307-316.
26. KUBO, I., KLOCKE, J.A., MIURA, I., FUKUYAMA, Y. 1982. Structure of ajugarin-IV. J.C.S. Chem. Commun. 618-619.
27. KUBO, I., KLOCKE, J.A., GANJIAN, I., ICHIKAWA, N., MATSUMOTO, T. 1983. Efficient isolation of phytoecdysones from *Ajuga* plants by high-performance liquid chromatography and droplet countercurrent chromatography. J. Chromatogr. 257:157-161.
28. KUBO, I., MATSUMOTO, A., AYAFOR, J.F. 1984. Efficient isolation of a large amount of 20-hydroxyecdysone from *Vitex madiensis* (Verbenaceae) by droplet countercurrent chromatography. Agric. Biol. Chem. 48:1683-1684.
29. KUBO, I., MATSUMOTO, A., HANKE, F.J., AYAFOR, J.F. 1985. Analytical DCCC isolation of 20-hydroxyecdysone from *Vitex thyrsiflora* (Verbenaceae). J. Chromatogr. 321:246-248.
30. KUBO, I., KOMATSU, S., ASAKA, Y., DeBORE, G. 1987. Isolation and identification of apolar metabolites of ingested 20-hydroxyecdysone in frass of *Heliothis virescens* larvae. J. Chem. Ecol. 13:785-794.
31. ZHANG, M. KUBO, I. 1992. Possible function of ecdysteroid-22-*O*-acyltransferase in the larval gut of tobacco budworm. *Heliothis virescens*. J. Chem. Ecol. 18:1139-1149.
32. ZHANG, M. KUBO, I. 1992. Characterization of ecdysteroid-22-*O*-acyltransferase from tobacco budworm, *Heliothis virescens*. Insect Biochem. Molec. Biol. 22:599-603.
33. SASSE, J.M., KINGSLEY, S.R., GALBRAITH, M.N. 1981. Distribution of podolactone-type plant growth inhibitors in the Coniferae. Phytochemistry 20:2195-2204.

Chapter Seven

PHYTOCHEMISTRY AND BIOLOGICAL ACTIVITY OF METABOLITES FROM TROPICAL MELIACEAE

Heinz Rembold and Ingo Puhlmann

Insect Biochemistry Group
Max-Planck-Institute for Biochemistry
D-8033 Martinsried near Munich, Germany

INTRODUCTION

Extracts of Meliaceae, leaves, barks and especially seeds have long been used for medicinal and plant protection purposes. The Indian neem tree, *Azadirachta indica* A. Juss, is part of Indian folklore and one of the meliacean species which has interested several laboratories for a long time. Most of the results have been presented in three international neem conferences.[1-3] The chemistry of the most intensively studied secondary metabolites from Meliaceae, i.e. the limonoids or tetranortriterpenoids, has been reviewed regularly.[4-7] The primary intention of this overview is to concentrate primarily on one pharmacologically interesting aspect of the bioactive metabolites from tropical Meliaceae, their disrupting effect on growth, development and reproduction of insects.

Phytochemical Potential of Tropical Plants
Edited by K.R. Downum et al., Plenum Press, New York, 1993

Azadirachtin is one of the major insect antifeedants isolated from neem seeds. Apart from this effect, it is responsible for the many disturbances of insect metamorphosis, mortality and decrease in fecundity. These aspects have been reviewed in detail.[8-10] Feeding deterrence is usually based on an interaction of the respective plant metabolite with defined chemoreceptors of the feeding insect, whereas the effects on growth and development are based on disturbances of the hormonal regulation of metamorphosis and reproduction. An interference of a certain plant metabolite with the insects' endocrine system may be masked or even be prevented by an antifeedant effect which through starvation acts on the endocrine system. Consequently, the selection of the respective insect species for bioassay depends on the isolation of an antifeedant or of a growth inhibitor from the plant material. Our standard isolation procedure for insect growth inhibitors is based on *Epilachna varivestis*, the Mexican bean beetle, which is more tolerant against phagodeterrents than most lepidopteran larvae. The standard bioassay[11] will be described briefly.

EPILACHNA VARIVESTIS BIOASSAY

The standardized assay[11] (cage test) is based on the rearing of freshly moulted fourth (last) instar larvae of the Mexican bean beetle, *E. varivestis*, on sprayed bean (*Phaseolus vulgaris*) plants for two days. Rate of survival and weight gain of the larvae is recorded during this 48 hr period. Those larvae which survive are transferred onto unsprayed bean leaves for further development. The plant extracts often induce a delay in further development of the treated larvae and/or pupae or they interfere with larval and pupal moult. Such effects are recorded. The growth disrupting effect of a certain plant extract or metabolite is expressed as the concentration of the material (in ppm), dissolved in 4 ml methanol and sprayed onto 10 young bean plants, that permitted the survival of 50% healthy adults in comparison with the solvent treated control (ED_{50}). The test starts with 20 larvae, followed by a careful statistical treatment of 3 to 10 independent tests. The given ppm value for an ED_{50} usually reflects a total number of 60-200 test larvae (see Table 1). A Petri dish test[11] that makes use of individual last instar larvae may also be used.

A study of the whole period of pupal and adult development is a time-consuming procedure taking 25-30 days per assay. Sometimes the next generation originating from eggs of the pretreated adults was also followed to detect any possible transfer of damage not lethal for the parent generation.

FRACTIONS WITH GROWTH-INHIBITING EFFECTS

Crude and partially purified seed extracts from *Melia volkensii* have shown a high potential for mosquito[12] and for locust[13] control. The same holds true for neem,[14] *M. azedarach*,[15] *A. integrifolia*[16] and *A. excelsa*.[17] However, the chemical basis for the biological effects is only known for neem; in this case, a group of isomeric azadirachtins has been identified that is primarily responsible for the insecticidal activity of the crude extracts.[8-10] All data on insect growth disrupting effects discussed below are based on the growth inhibitory effects in the *Epilachna* bioassay.

Consideration of the high biological activity of these crude extracts makes one wonder why the isolation and structural elucidation of the active compounds has attracted so few chemists and why so little is known about the chemistry of these botanical insect growth inhibitors. One reason of course is the three week period required for one complete *Epilachna* bioassay. Another drawback for those interested in botanical insect growth inhibitors is that very often the crude starting material is not touched by the larvae due to its deterring effect on food uptake. The larvae, instead, survive and recover from their weight loss due to starvation after being transferred to the untreated bean plants, often without showing any negative effect on their further growth and development. In such cases, the starting material must be diluted so that feeding deterrence no longer masks a possible growth disrupting effect of other plant metabolites. Unfortunately, new botanical growth inhibitors may be missed because of this masking effect. Another sometimes critical problem in the isolation of a bioactive compound is possible due to synergistic effects of other plant metabolites in a mixture which are lost during the purification of the active metabolite. All these difficulties may, at least to some extent, explain the low number of potent botanical insect growth inhibitors which have been purified and structurally elucidated up to now.

Whereas an assay for feeding inhibitors usually takes only 1-2 days and purification of the plant metabolite responsible is achieved within weeks, one may have to invest years in the isolation of a new 'structural class' of insect growth inhibitors. Such a structural class' may increase the chance of finding a completely new chemical strategy in the field of plant protection. For example, a new type of chemical defence seems to have occurred during the long coevolution of the tropical Meliaceae and their phytophagous insect pests. Some insects like the desert locust, *Schistocerca gregaria*, will not even touch a neem tree. Others, such as mosquitoes are repelled by the plant's volatiles. The isomeric azadirachtins from the neem tree are specifically directed against the insect neuroendocrine control system and represent a new class of neurotoxic chemicals that directly attack an insect-specific target, but are non-toxic to mammals.

THE ISOMERIC AZADIRACHTINS—NATURAL PRODUCTS

Several insect growth disrupting metabolites have been isolated from neem seeds which, in the *Epilachna* bioassay, do not inhibit feeding in a series of insects.[8-10,19,20] All these compounds induce the same biological effects and are structural isomers. It was therefore proposed to classify them as a group of azadirachtins, marked initially with the letters A-D.[18] With the exception of azadirachtin C, the structure of which has only partially been elucidated, all of these metabolites can be derived from azadirachtin A (Structure I) which is the most abundant representative.

The C-ring of this basically tetracyclic ring structure is opened to a C-seco compound with carbon atoms 24-27 lacking and the A,B rings *trans*-connected. Consequently this plant metabolite belongs to the group of tetranor-triterpenoids which are also known as limonoids. The number of plant-derived azadirachtins has now reached the letter L (Table 1) and will increase, due to many more *Epilachna* active metabolites found in extracts of different Meliaceae.

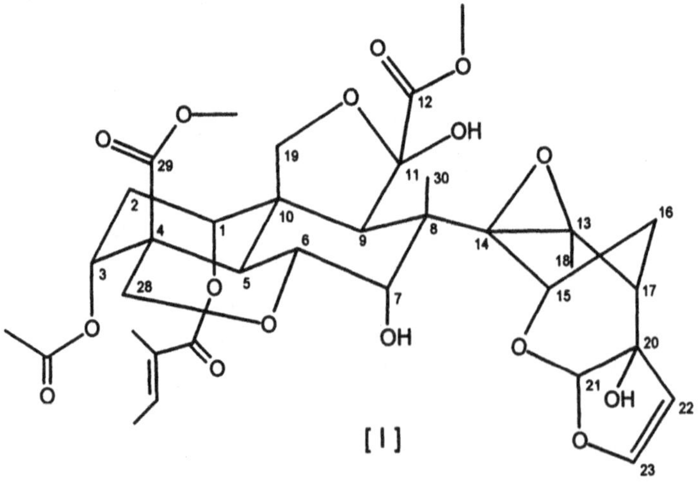

[I]

Table 1. Structures of the naturally occurring azadirachtins A-L. The ppm value gives a 50% rate of survival (ED_{50}) for the respective dose applied under standard *Epilachna* bioassay conditions. The Roman numbers refer to the respective azadirachtin structures.

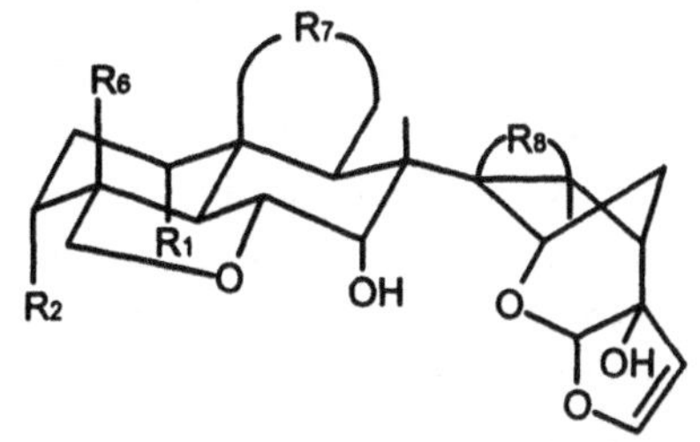

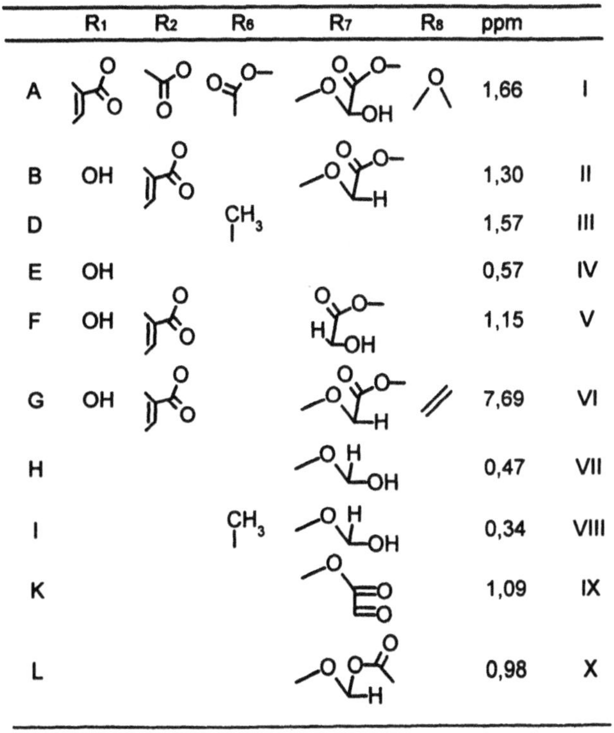

The structures of azadirachtins A-G have already been reviewed.[8-10] Recently azadirachtins H, I[19] and K[20] have been isolated from neem and their structures elucidated. Finally, azadirachtin L was isolated from two other Meliaceae, *A. excelsa* and *M. volkensii* (manuscripts in preparation). Whereas the former Melia also contains azadirachtin A in its seeds,[17] there is not a trace of this predominant neem metabolite present in the seeds of *M. volkensii*.

All ten of the azadirachtins listed in Table 1 are highly active in the *Epilachna* bioassay. There are some characteristic structural differences which become obvious when comparing the structures. Azadirachtin B (II) shares a free OH-group at C-1 with azadirachtins E [IV], F [V] and G [VI], and the tigloate at position C-3 with [V] and [VI]. Another interesting position is at C-11, where [II] has no hydroxyl group, and the stereochemistry of [VII] and [VIII] is opposite at C-11 to that of [X]. However, all four of these metabolites induce more or less the same biological activity. It can be concluded from a detailed structural comparison that free or esterified OH-groups at C-1 and C-3, presence or absence of an OH group at C-11, reduction of the carboxyl group at C-29, or removal of the 13,14 epoxy ring do not seriously affect the biological activity in the *Epilachna* bioassay. Also the carboxyl group at position C-12 can be removed without any negative effect [VII-X]. However, nothing is known about an involvement of the other hydroxyl groups (C-7 and C-20) or of the acetal at C-21 in the biological activity of the azadirachtins. In order to study their effect, we will have to modify the structures by chemical derivatization of the natural products.

AZADIRACHTINOIDS: CHEMICALLY DERIVED FROM AZADIRACHTINS

Azadirachtins are defined as plant metabolites that share the same structural elements and similar biological activity. This at present holds true for the ten compounds listed in Table 1. If neem seed is used as a source, azadirachtin A [I] makes up almost 60% and azadirachtin B [II] nearly 20% of the total azadirachtins. These two plant metabolites were therefore used for chemical modification of their structure. Six such azadirachtinoids are listed in Table 2 in decreasing order of biological activity. Compared to the natural products [I] and [II], a remarkable increase [XI-XIII] as well as a decrease [XIV-XVI] in growth disrupting activity is achieved. Methylation of the two hydroxyl groups at C-11 and C-20 [XIV] or acetylation at C-11 [XV] reduces, but does not abolish the biological activity, as happens with the trimethoxy derivative [XVI]. On the other hand, removal of tigloate [XIII] or of acetate [XII] clearly increases the biological activity if compared with the metabolite [I]. An even higher activation is achieved with [XI], which is the detigloyl-azadirachtin B.

STRUCTURE—ACTIVITY RELATIONSHIPS

With these sixteen structures in mind, we can begin to come to an understanding of the structural elements which are important for the growth

Table 2. Structures of some synthetic derivatives of azadirachtin A and B (azadirachtinoids) compared with their mother compounds. (See also Fig. 1)

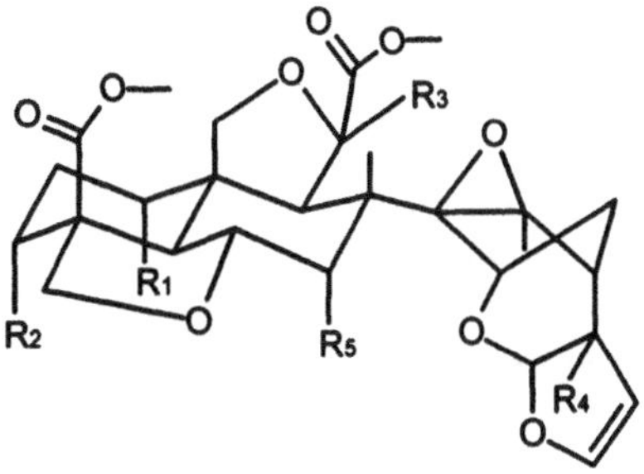

R_1	R_2	R_3	R_4	R_5	ppm	
OH	OH	H			0,08	XI
	OH				0,38	XII
OH					0,57	XIII
OH	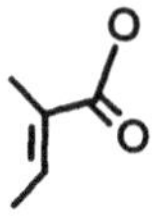	H			1,30	II
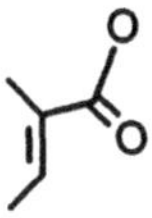	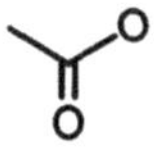	OH	OH	OH	1,66	I
		O-Me	O-Me		10	XIV
		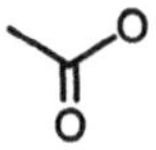			60	XV
		O-Me	O-Me	O-Me	>100	XVI

inhibitory activity of the azadirachtin isomers. Some of this has been discussedalready in view of the biological data collated in Tables 1 and 2. The esters at C-1 and C-2 seem to be superfluous. The same holds true for the carboxyl groups at C-12 and C-29. Also, neither of the ether bridges between C-11 and C-19 or the oxiran ring at C-13,14 seem to be involved in the biological effect of azadirachtin on neuroendocrine control of insect hormone pools. Due to the fact that all these considerations have been guided by the *Epilachna* bioassay, we can identify some biologically inactive metabolites that share structural similarities with a reduced azadirachtin structure.

One promising candidate is a cyclization product[21] [**XVIII**] of 22,23-dihydro-azadirachtin A [**XVII**]. Whereas the latter is a highly active compound, the cyclization product in which the epoxide is opened and forms an ether bridge with the former C-7 hydroxyl group, is devoid of any activity in the bioassay. This indicates an involvement of the C-7 hydroxyl group and of the acetal moiety, the latter due to a different sterochemistry of [**XVIII**], possibly through an abolished free rotation at the C-8,14 bond.

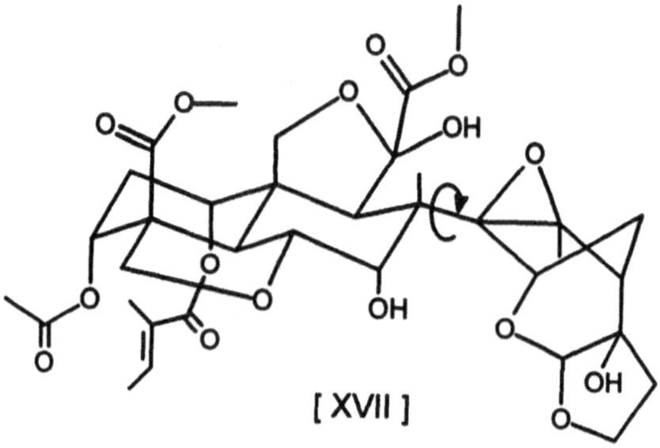

[XVII]

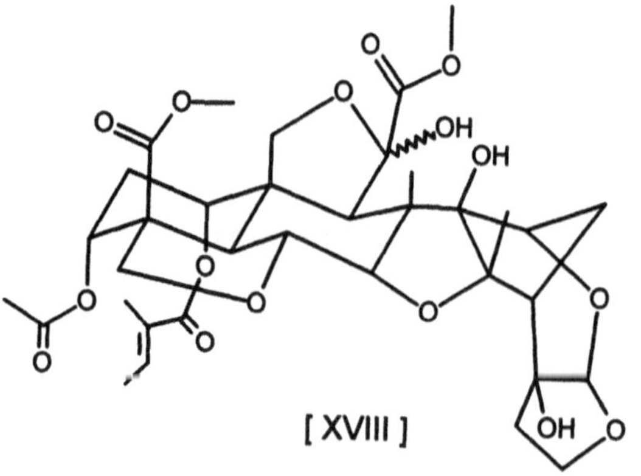

[XVIII]

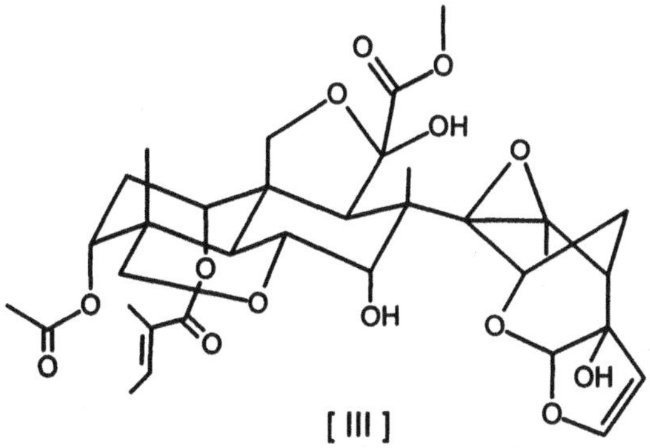

[III]

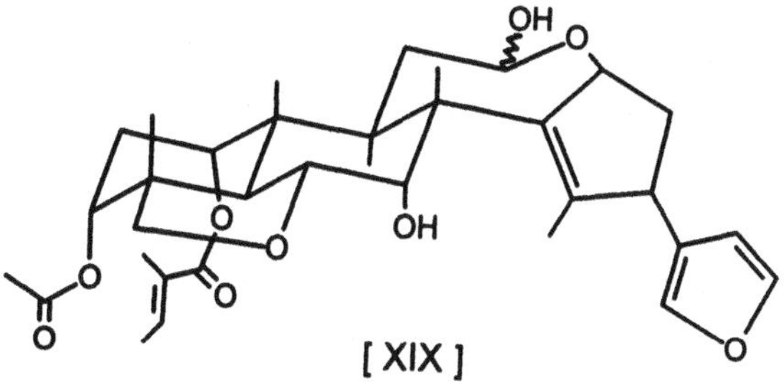

[XIX]

Another example is given by comparison of azadirachtin D **[III]** with volkensin[22] **[XIX]**, which also is completely inactive in the *Epilachna* bioassay. These compounds are almost identical in the chemistry of the left (decalin) part of the molecule, including the free hydroxyl group at C-7 and the methyl group at C-4. Azadirachtin D, however, has an ED_{50} of 1.57 ppm (Table 1), whereas that of volkensin is >100 ppm. This result also demonstrates the involvement of both parts of the azadirachtin molecule in insect growth inhibitory effects.

CONCLUSION

Based on present state of the art, we can make some conclusions concerning the effective structural elements of the azadirachtin molecules. There is obviously an upper part of the molecule which is not involved in binding to a hypothetical protein receptor. The sensitive part of the azadirachtin molecule,

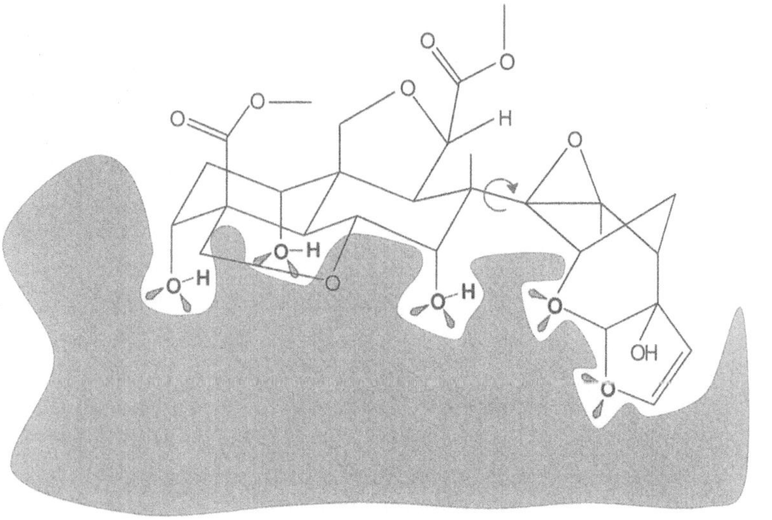

Figure 1. Schematic representation of the binding anchors in the azadirachtin molecule to a hypothetical receptor protein. For more details see text.

that binds with high affinity to the hypothetical receptor, seems to include the oxygens at C-1, 2, and 7 at the decalin moiety, and the acetal structure of the D-ring part of the triterpenoid. A significant reduction in biological activity after acetylation of the 11-hydroxyl group (from 1.66 to 60 ppm, Table 2) as well as the deshielding effect of azadirachtins B and H-L by removal of the C-11 hydroxyl group (Table 1) favors a free rotation at the C-8,14 bond as a prerequisite for optimal receptor binding. At present we prefer a model which is schematically represented by Figure 1 and which may be used for first synthetic trials.

The present data were achieved by use of a bioassay. However, for an in-depth structure—activity based drug design, a cell free test system is unavoidable. There seems to be only a minute amount of high-affinity receptor proteins in the insect. On the other hand, receptors must have a very general function due to the fact that azadirachtins interfered with each insect species tested so far. Histological autoradiography after injection of tritiated 22,23-dihydroazadirachtin A has shown that the compound is incorporated into the corpus cardiacum of locusts[23] in a highly specific way. This neurohemal organ secretes most of the organotropic neuropeptides in insects. This explains why we have not yet been successful in isolating any measurable amount of a high-affinity binding protein from insects.

Our future strategy will be twofold. We will try affinity chromatography to concentrate the binding proteins present in the insect tissue in trace amounts only. We will also prepare anti-azadirachtin antibodies for binding studies in competition with simple synthetic products. For both these approaches, we need an azadirachtin molecule with a reactive spacer. Promising results have been achieved in this direction.

ACKNOWLEDGEMENTS

We gratefully acknowledge the assistance in the *Epilachna* bioassays by Dr. Ch. Czoppelt and Mrs. E. Woker and in the preparation of some azadirachtinoids by Dr. C.S.S.R. Kumar. Samples of the azadirachtins H, I and K were generously provided by Professor T.R. Govindachari, Madras, and of azadirachtin L by Dr. H.-O Kalinowski, Giessen.

REFERENCES

1. SCHMUTTERER, H., ASCHER, K.R.S., REMBOLD, H. (eds.) 1981. Natural Pesticides from the Neem Tree. GTZ, Eschborn, 297 pp.
2. SCHMUTTERER, H., ASCHER, K.R.S. (eds.) 1983. Natural Pesticides from the Neem Tree (*Azadirachta indica* A. Juss) and Other Tropical Plants. GTZ, Eschborn, 587 pp.
3. SCHMUTTERER, H., ASCHER, K.R.S. (eds.) 1987. Natural Pesticides from the Neem Tree (*Azadirachta indica* A. Juss) and Other Tropical Plants. GTZ, Eschborn, 703 pp.
4. DREYER, D.L. 1968. Limonoid bitter principles. In: Progress in the Chemistry of Natural Products (L. Zechmeister, ed.). Springer Verlag, Wien New York, pp. 190-244.
5. TAYLOR, D.H.A. 1984. The Chemistry of the Limonoids from Meliaceae. In: Progress in the Chemistry of Natural Products (W. Herz, H. Grisebach, G.W. Kirby, eds.). Springer Verlag, Wien New York, pp. 1-102.
6. KUNDU, A.B., RAY, S., CHAKRABARTI, R., NAYAK, L., CHATTERJEE, A. 1985. Recent developments in the chemistry of

C_{26}-terpenoids (tetranortriterpenoids). J. Sci. Ind. Res. 44:256-278.

7. CONOLLY, J.D. (1983) Chemistry of the limonoids of the Meliaceae and Cneoraceae. In: Ann. Proc. Phytochem. Soc. Europe 22. Academic Press, London, pp. 175-213.
8. REMBOLD, H. 1989. Isomeric azadirachtins and their mode of action. In: Focus on Phytochemical Pesticides. (M. Jacobson, ed.). CRC Press, Boca Raton, pp. 47-67.
9. REMBOLD, H. 1989. The azadirachtins—their potential for insect control. In: Economic and Medicinal Plant Research. (H. Wagner, H. Hikino, N.R. Farnsworth, eds.). Academic Press, London, pp. 57-72.
10. REMBOLD, H. 1989. Azadirachtins. Their structure and mode of action. In: Insecticides of Plant Origin. (J.T. Arnason, B.J.R. Philogene, P. Morand, eds.). ACS Symposium Series 387:150-163.
11. REMBOLD, H., SHARMA, G.K., CZOPPELT, C., SCHMUTTERER, H. 1980. Evidence of growth disruption in insects without feeding inhibition by neem seed fractions. J. Plant Dis. Prot. 87:290-297.
12. MWANGI, R.W., REMBOLD, H. 1987. Growth-regulating activity of *Melia volkensii* extracts against the larvae of *Aedes aegypti*. In: Natural Pesticides from the Neem Tree (*Azadirachta indica* A. Juss) and other Tropical Plants. (H. Schmutterer, K.R.S. Ascher, eds.). GTZ, Eschborn, pp. 669-681.
13. MWANGI, R.W. 1982. Locust antifeedant activity in fruits of *Melia volkensii*. Entomologia exp. appl. 32:277-280.
14. ZEBITZ, C.P.W. 1984. Effect of some crude and azadirachtin-enriched neem (*Azadirachta indica*) seed kernel extracts on larvae of *Aedes aegypti*. Entomologia exp. appl. 35:11-16.
15. CHIU, S.F. 1985. Recent research findings on Meliaceae and other promising botanical insecticides in China. Z. PflKrankh. PflSchutz. 92:310-319.
16. SCHMUTTERER, H. 1989. Erste Untersuchungen zur insektiziden Wirkung methanolischer Blatt- und Rindenextrakte aus dem Philippinischen Niembaum *Azadirachta integrifolia* Merr. im Vergleich zu Extrakten aus *Azadirachta indica* (A. Juss) und *Melia azedarach* L. J. Appl. Ent. 108:483-489.
17. ERMEL, K., KALINOWSKI, H.-O., SCHMUTTERER, H. 1991. Isolierung und Charakterisierung von Marrangin, einer neuen, die Insektenmetamorphose störenden Substanz aus Samenkernen des Marrangobaumes *Azadirachta excelsa* (Jack). J. Appl. Ent. 112:512-519.

18. REMBOLD, H., FORSTER, H., CZOPPELT, Ch., RAO, P.J., SIEBER, K.-P. 1983. The azadirachtins, a group of insect growth regulators from the neem tree. In: Natural Pesticides from the Neem Tree (*Azadirachta indica* A. Juss) and other Tropical Plants. (H. Schmutterer, K.R.S. Ascher, eds.). GTZ, Eschborn, pp. 153-161.

19. GOVINDACHARI, T.R., SANDHYA, G., RAJ, S.P.G. 1992. Azadirachtins H and I: two new tetranortriterpenoids from *Azadirachta indica*. J. Nat. Prod. 55:596-601.

20. GOVINDACHARI, T.R., SANDHYA, G., RAJ, S.P.G. 1992. Structure of azadirachtin K, a new tetranortriterpenoid from *Azadirachta indica*. Indian J. Chem. 31B:295-298.

21. BILTON, J.N., JONES, P.S., LEY, S.V. 1988. Chemistry of insect antifeedants from *Azadirachta indica* (Part I): Conversion from the azadirachtin to the azadirachtinin skeletons.

22. RAJAB, M.S., BENTLEY, M.D. 1988. A new limonoid insect antifeedant from the fruit of *Melia volkensii*. J. Nat. Prod. 51:168-171.

23. SUBRAHMANYAM, B., REMBOLD, H. 1989. Effect of azadirachtin A on neuroendocrine activity in *Locusta migratoria*. Cell Tissue Res. 256:513-517.

Chapter Eight

RECENT ADVANCES IN THE ACETOGENINS OF ANNONACEAE

André Cavé,* Diego Cortes,*# Bruno Figadère,* Reynald Hocquemiller,* Olivier Laprévote,‡ Alain Laurens* and Michel Lebœuf*

* Laboratoire de Pharmacognosie, Unité associée au CNRS, Faculté de Pharmacie, 92296 Châtenay-Malabry Cedex, France

Laboratoire de Pharmacognosie, Faculté de Médecine-Pharmacie Université de Rouen, 76803 Saint Étienne du Rouvray, France

‡ Institut de Chimie des Substances Naturelles, CNRS 91198 Gif-sur-Yvette, France

Phytochemical Potential of Tropical Plants
Edited by K.R. Downum et al., Plenum Press, New York, 1993

INTRODUCTION

Annonaceae is a large family of tropical and subtropical trees, lianas and bushes. Economically, the family is of appreciable importance as a source of edible fruits, edible oils, and fragrant flowers for perfumery. Many members of this family are used in folk medicine for antiparasitic or antitumoral treatment of intestinal diseases. They contain isoquinoline alkaline alkaloids which have been studied extensively, as they possess some interesting biological properties such as dopaminergic receptors. Recently, a new class of biologically active neutral compounds, the acetogenins, has aroused considerable interest.

Acetogenins from Annonaceae are a series of C-35/C-37 natural products derived from fatty acids. Their general structure is characterized by a long alkyl chain bearing a terminal methyl-substituted unsaturated γ–lactone (sometimes rearranged to an acetonylated γ-lactone), one or two tetrahydrofuran rings, several oxygenated substituents (hydroxyl, acetoxyl, ketonic) and in some cases a double bond. To date, they have been isolated only from the Annonaceae. Annonaceous acetogenins exhibit a broad range of potent biological activities, such as antitumoral, cytotoxic, antiparasitic, antimalarial, antimicrobial, immunosuppressant, antifeedant and pesticidal properties. Because they possess significant bioactivity, they have attracted more and more attention, with the number of publications on the subject expanding rapidly. Uvaricin, the first acetogenin to be isolated, was described in 1982.[1] In 1990, J.K. Rupprecht, Y.H. Hui and J.L. McLaughlin published a review on acetogenins covering 31 compounds belonging to this series.[2] Since then, more than 40 new ones have been described.

According to the literature, only a few genera of Annonaceae contain acetogenins, but they are probably much more numerous, since we now have evidence that acetogenins are widely distributed in the Annonaceae family. Acetogenins were initially isolated from bark, but are also present in larger amounts in seeds and can be found in all parts of the plant.

CLASSIFICATION

Acetogenins of Annonaceae can be classified into three main groups according to the number and arrangement of the tetrahydrofuran ring(s): A, with one tetrahydrofuran, such as annonacin; B, with two adjacent tetrahydrofurans, such as uvaricin; C, with two non-adjacent tetrahydro-furans separated by a four-carbon chain, such as gigantecin.

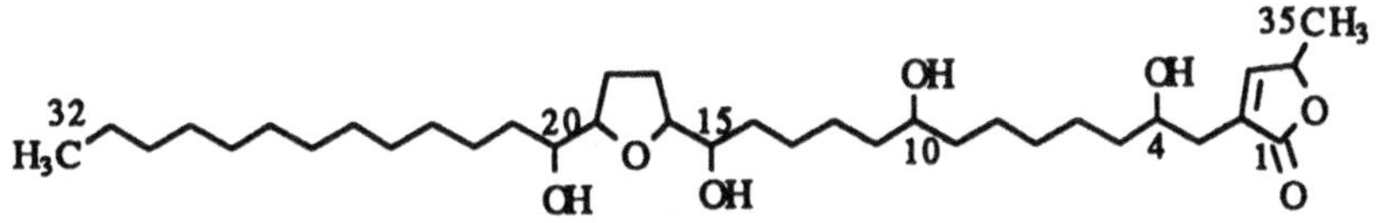

annonacin

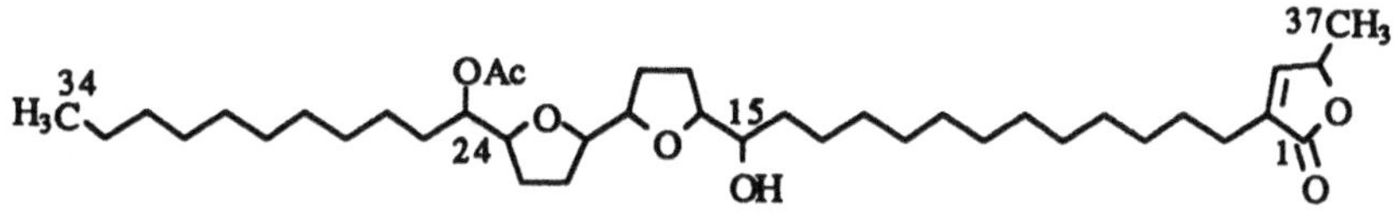

uvaricin

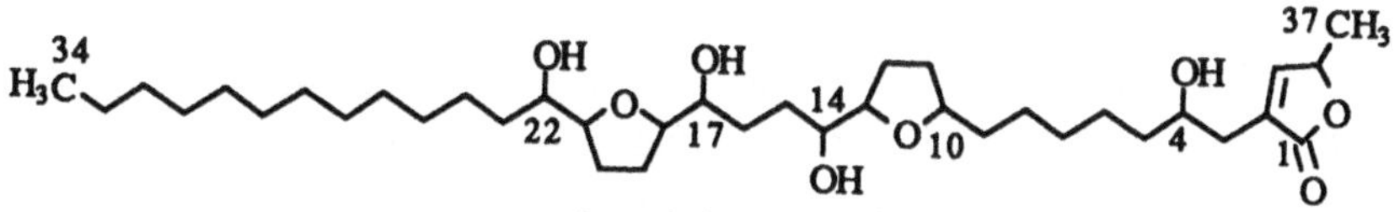

gigantecin

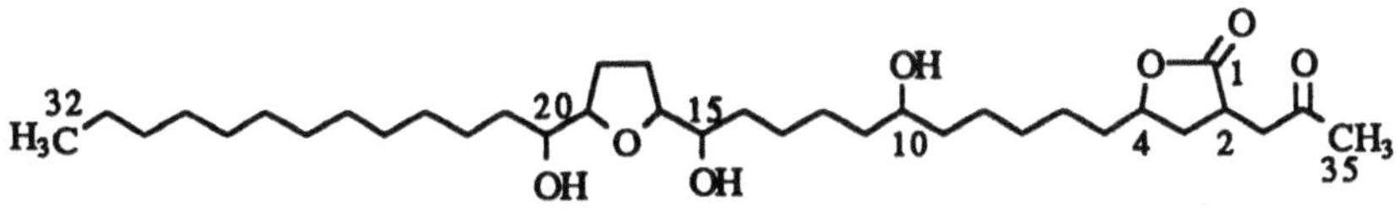

isoannonacin

For these three main classes, subtypes have been derived designated as A2, B2 and C2 and A3, B3 and C3. Subtypes A2, B2 and C2, or the "iso" series, are characterized by a saturated γ–lactone substituted by a acetonyl group, such as isoannonacin. Acetogenins belonging to these subtypes could be considered artefacts derived from 4-hydroxy-acetogenins by rearrangement in alkaline medium. We can present arguments for this assumption: first, "natural" isoacetogenins are a mixture of stereoisomers at C-2, which is unusual in nature; second, an isoacetogenin is easily obtained from the corresponding acetogenin after treatment of acetogenin by 2% KOH in *t*-BuOH.[3] Furthermore, it is not necessary to use a strong base to obtain isoacetogenin. A weak base, such as diethylamine, is sufficient to transform acetogenin to isoacetogenin at room temperature. By stirring a solution of acetogenin in dichloromethane containing 2% diethylamine, in the presence of silicagel, for 3 days at room temperature, transformation into isoacetogenin is complete. Simple elution from a silicagel column with such a solvent mixture is sufficient to transform acetogenins into isoacetogenins in about 60% yield.

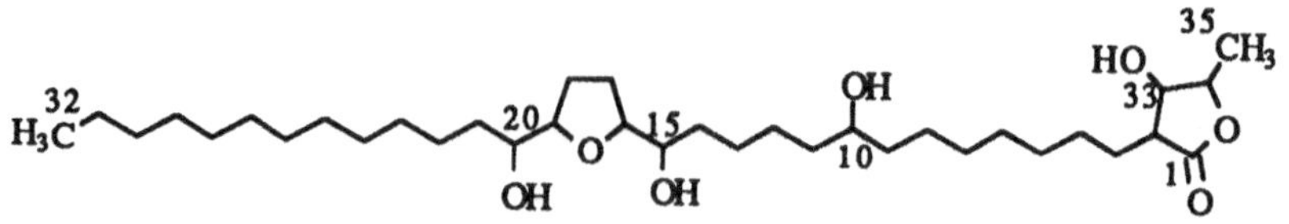

jetein

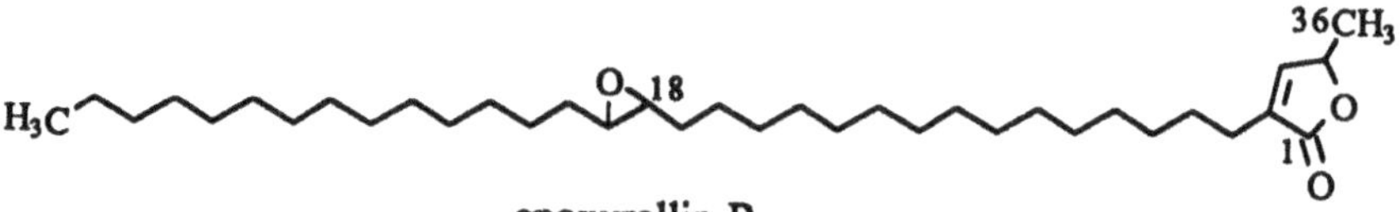

epoxyrollin-B

Subtypes A3, B3 and C3 are characterized by a saturated γ–lactone substituted by a hydroxyl β to the carbonyl. A monotetrahydrofuran (A3) example is jetein.[4] At present, there are only four known members of this subtype.[4]

Finally, a group of acetogenin-related compounds exists that possess no tetrahydrofuran ring, but which has a terminal unsaturated γ-lactone, such as epoxyrollin-B.[5] Until now, this group has consisted of only a few members, characterized by the presence of epoxides and, in some cases, double bonds along the chain. These compounds can be considered as biogenetic precursors of acetogenins.

EXTRACTION, ISOLATION AND DOSAGE METHODS

The extraction and isolation of acetogenins have become easier. Solvent extraction, solvent partition and chromatography are used, aided either by bioassay or by thin layer chromatography. Fractionation, based on the lethality to brine shrimp larvae according to J.L. McLaughlin,[6] is useful because of its rapidity, its low cost and its good correlation with the antitumor activity. Thin layer chromatography can also be used. Indeed, since most of the acetogenins are characterized by the presence of an unsaturated γ-lactone, they give a colored reaction with the Kedde s reagent[2] which can be used for staining the plates. The methods are complementary.

Purification of acetogenins is rather difficult, as they often exist as a complex mixture of components of similar polarities and they are very hard to crystallize. Analytical HPLC of acetogenins gives good results on reverse phase

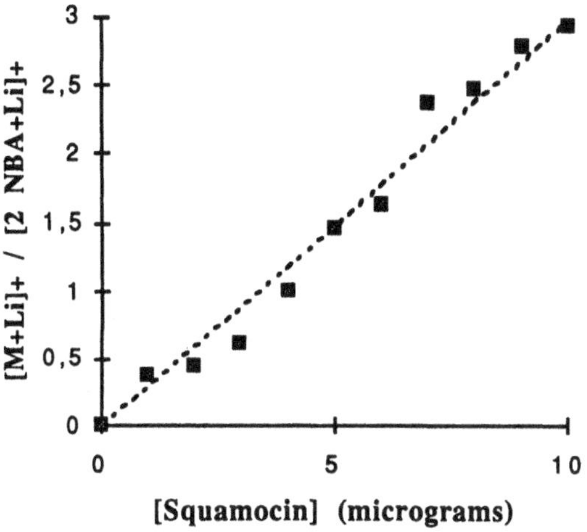

Fig. 1. $[M+Li]^+$ ion intensity vs. concentration of squamocin in the FAB matrix.

silica (μ-Bondapak C18 column, mobile phase: MeOH:H_2O or MeCN:H_2O, UV detection at 215 nm, flow rat: 1.5 ml/min). A mixture of MeCN:H_2O (85:15) permits the separation of the acetogenins of *Annona muricata*,[11] corossolone, corossolin, murisolin and solamin[7,8]: (rt=8.8 min, 7 min, 11 min and 27.3 min, respectively). We have also shown that it is possible to separate isomers. For instance, rolliniastatin-1 and rolliniastatin-2, which differ in stereochemistry at the tetrahydrofuran ring,[9,10] have been separated with retention times of 19.7 and 16.3 min, respectively (solvent system: MeCN:H_2O; 7:3).[11] We have developed a method of purification of acetogenins by high performance liquid chromatography and one for a preparative HPLC.

Dosage measurements of acetogenins include HPLC and mass spectrometry. This latter method makes use of the propensity of acetogenins to form complexes with lithium.[12] Quantitative and semi-quantitative measurements by Fast Atom Bombardment (FAB) mass spectrometry are often limited by the strong dependence between the desorption process itself (i.e. of the ion current) and the surface activity of the organic sample dissolved in the liquid matrix. When studying a mixture of compounds, the more surface-active products are detected at the expense of the others. For this reason, an impurity may be desorbed from the FAB matrix more easily than the major constituent. This is called the "suppression effect".[2]

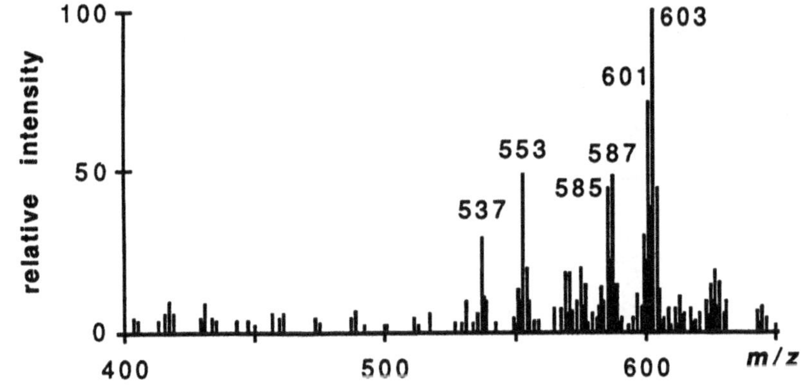

Fig. 2. FAB mass spectrum of a crude extract of *Annona muricata* seeds. FAB matrix: *m*-NBA + LiCl.

When lithium chloride is added to an FAB matrix of acetogenins, the relative intensity of the observed $[M+Li]^+$ ion peak is linearly dependent on the quantity of sample introduced. Figure 1 shows the results of an experiment in which squamocin was introduced into the matrix in increasing amounts. In this study, the dimeric lithiated form of the matrix component nitrobenzyl alcohol (*m*-NBA) was used as an internal standard.

When an equimolecular mixture of three acetogenins is dissolved in a *m*-NBA matrix to which LiCl is added, the three corresponding $[M+Li]^+$ species appear on the FAB spectrum with almost the same relative intensities. This result suggests that there is no suppression effect in the case of lithiated acetogenins. Consequently, the relative intensities of the $[M+Li]^+$ ion peaks of acetogenins in the admixture may reflect the relative composition in acetogenins of this mixture.

The high sensitivity obtained with lithium attachment/FAB conditions allows the direct analysis of a crude extract,[12] as shown in Figure 2 (Corossol extract). In this case, the analysis of each $[M+Li]^+$ ion peak by B/E (electric sector voltage/magnetic field) linked scanning is possible, giving some structural informations on each compound present in the mixture.

The ability of the acetogenins substituted at the C-4 position by a hydroxyl to lose a neutral fragment of 112 Da by unimolecular decomposition may also be used to increase the specificity of their detection in an extract. Indeed, the constant neutral loss scanning corresponding to the loss of 112 Da

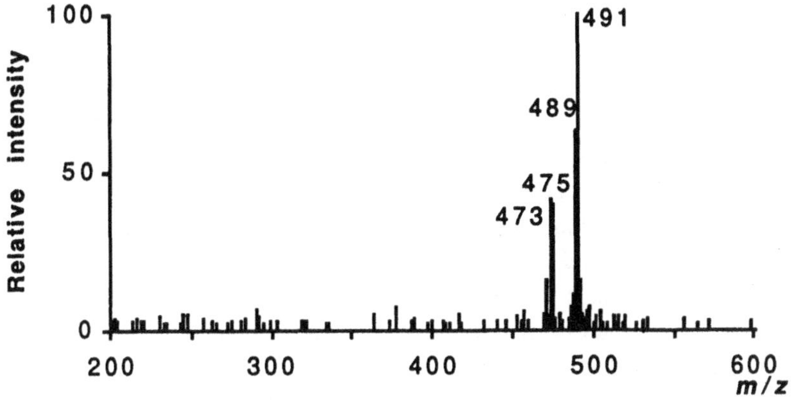

Fig. 3. Control neutral loss spectrum (112 Da) of a crude extract of *Annona muricata* seeds. FAB matrix: *m*-NBA + LiCl.

leads to the detection of only the fragment ions which are issued from this fragmentation pathway. Thus, the analysis of the crude extract of corossol by this linked scan mode gives a remarkably simple spectrum (Fig. 3). The comparison between the conventional FAB mass spectrum (Fig. 2) and the constant neutral loss spectrum of 112 daltons (Fig. 3) shows a spectacular reproductibility in the relative intensities of the ions. Lithium attachment on acetogenin under FAB ionization conditions may be used successfully either for the structure elucidation of a pure compound or for a semi-quantitative evaluation of acetogenin mixtures, particularly during the separation and purification steps.

STRUCTURAL ELUCIDATION

The elucidation of acetogenin structure is rather complex. Classical methods, such as UV, IR, proton NMR, ^{13}C NMR and mass spectrometry are essential in identifying structural subunits such as methyl γ-lactone and tetrahydrofuran rings, but the placement of the substituents along the carbon skeleton involves innovative mass strategies such as mass-tandem[5] or FAB-Li.[15]

The location of the tetrahydrofuran rings and oxygenated substituents in the alkyl chain requires preparation of the acetylated, silylated and TMS-d_9 derivatives and their analysis by conventional Electron Impact Mass

Spectrometry.[13] It is also desirable to determine the elemental composition of the major fragment ions obtained by electronic impact and chemical ionization mass spectrometry of derivatives through high-resolution measurements.

However, (i) the relatively low reproducibility of the fragment ion intensities from one compound to another, (ii) the eventual incomplete derivatization of the compounds, and (iii) the consecutive decompositions occurring in the EI ion source by way of loss of the oxygenated substituents make such an analytical method difficult and expertise is required.

In order to circumvent these problems, we developed a new mass spectrometric strategy for obtaining the structural information without any chemical derivatization and with a single sample-introduction into a conventional mass spectrometer. This is based on the strong affinity of acetogenins for alkali metal cations, in particular, lithium.[15,16]

The determination of the stereochemistry of the chiral centers is a complex problem. It is based on the NMR studies of Hoye and co-workers, but which allows deduction of the relative configuration only. Only two X-ray crystallographic analyses have been performed on acetogenin derivatives, namely rolliniastatin-1[9] and squamocin,[17] and neither established the absolute configuration. Chemical degradation was used to find the configuration of the methyl group of the lactone moiety of uvaricin.[18]

We present here the different stages of the determination of the structure of acetogenins. Some examples of recently-isolated acetogenins are given below, showing the ability of the different methods to solve structural problems in different parts of the molecules.

Molecular Weight

The molecular weight indicates the structural type of the acetogenin under investigation. A molecular weight in the 530-574 range is indicative of an epoxyacetogenin structure. The molecular weights of monotetrahydrofuran acetogenins usually fall between 578 and 600 and those of bis-THF compounds between 606 and 648. This observation holds for almost 70 natural acetogenins, with only 6 compounds outside the mass ranges described above.

Molecular weight determination requires the use of chemical ionization or FAB mass spectrometry. As stated previously, under FAB ionization conditions, acetogenins strongly interact with lithium. The FAB mass spectrum of an acetogenin, such as rolliniastatin-2, displays protonated molecules at *m/z* 623 accompanied by sodium adduct ions at *m/z* 645 confirming the molecular weight (622). While successive losses of water from the $[M+H]^+$ species are

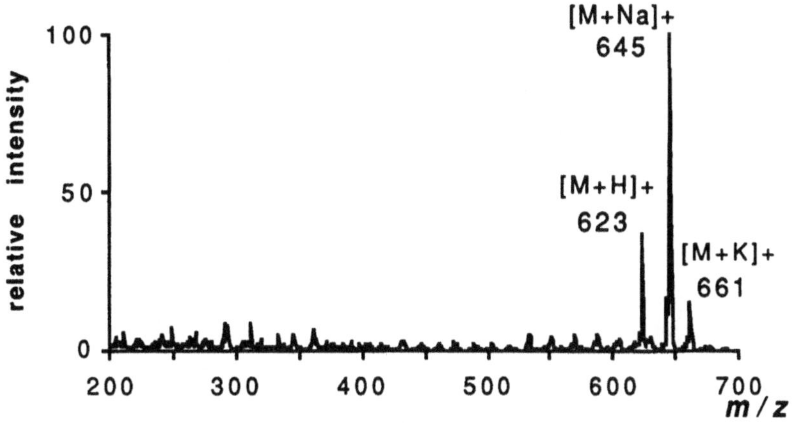

Fig. 4. FAB spectrum of rolliniastatin-2 (matrix: *m*-NBA).

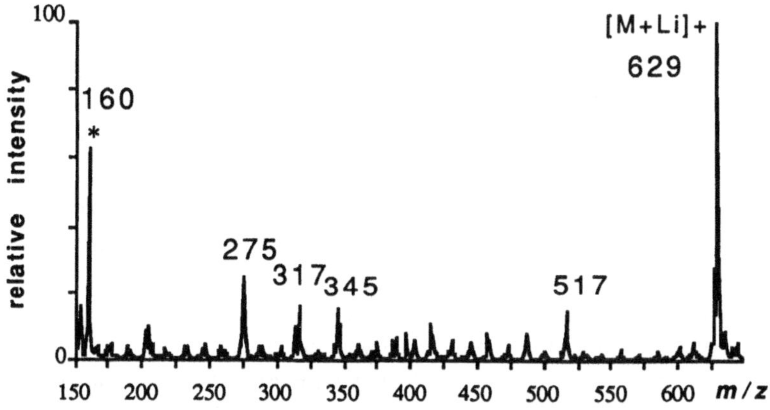

Fig. 5. FAB mass spectrum of rolliniastatin-2 (matrix: *m*-NBA + LiCl).

attributed to the presence of hydroxyl groups in the molecule, no skeletal fragmentation is observed in this spectrum (Fig. 4). In contrast, the mass spectrum obtained after addition of lithium chloride to the FAB matrix (*meta*-nitrobenzyl alcohol) is dominated by the $[M+Li]^+$ ion peak at *m/z* 629 (Fig. 5). Several fragment ion peaks are also observed in the spectrum. Thus simple examination of the FAB mass spectrum recorded in the presence of Li^+ cations yields unambiguously the molecular weight. Ten nanograms (around eighty picomoles) of underivatized compound are sufficient for such an analysis.[16]

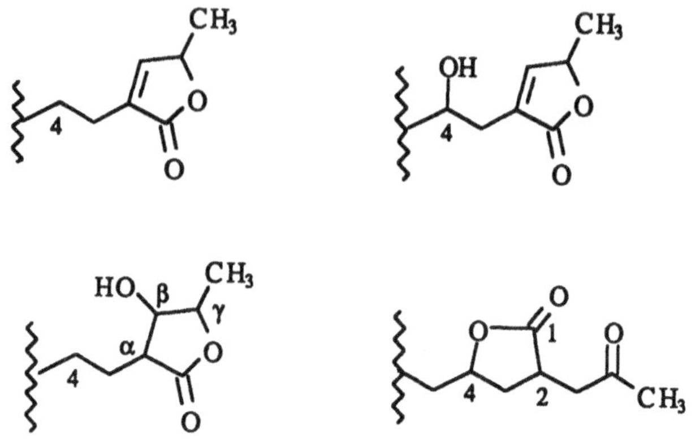

Fig. 6. Different types of lactone rings.

γ-Lactone Ring

The methyl-γ-lactone is characteristic of all the acetogenins of Annonaceae; four types are encountered (Fig. 6). The presence of this group is evidenced by IR and NMR spectroscopies. As mentioned above, a positive reaction with Kedde's reagent is indicative of an α,β-unsaturated γ-lactone. The IR spectrum affords helpful information allowing confirmation and characterization of a saturated γ-lactone. The presence of an unsaturated γ-lactone is indicated by a strong carbonyl absorption band at 1740-1750 cm^{-1} while that of a saturated γ-lactone appears at about 1770 cm^{-1}.

Unsaturated lactone ring. The ^{1}H NMR spectra show a characteristic pattern corresponding to the unsaturated γ-lactone ring (Fig. 7). The C-35 (or C-37) methyl group resonates as a doublet at 1.4 ppm with a coupling constant of about 7 Hz. The adjacent proton appears at about 5 ppm as a complex signal. The coupling constants are different according to the presence or absence of a substituent on the C-4 carbon. Finally, the presence of the olefinic proton is characterized by a signal at about 7 ppm with a coupling constant of about 1.7 Hz.

The presence of a hydroxyl group at C-4 induces some characteristic modifications as seen by comparing NMR data for corossolone and murisolin (Fig. 8). There is an effect on the chemical shifts of carbons C-2 and C-33 and evidently C-3 and C-4, but differences are more important on proton NMR spectra. The two protons of the C-3 methylene group are non-equivalent when

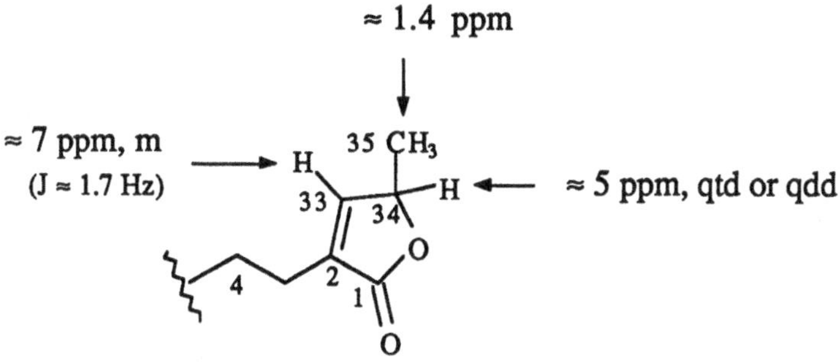

Fig. 7. ^{1}H NMR of unsaturated lactone ring.

the C-4 is substituted and resonate separately at about 2.4 and 2.5 ppm with different multiplicities. Non-equivalence of the C-3 protons can be explained by the existence of a hydrogen bond between the oxygen of the carbonyl and the hydrogen of the hydroxyl group.

Saturated lactone. There are two types of acetogenins with a saturated lactone, isoacetogenins and β-hydroxy-γ-methyl-γ-lactonic acetogenins. The saturated lactone of isoacetogenins is characterized by the presence of an acetonyl group at the C-2 position. This type of acetogenin can (and usually does) exist as a mixture of two diastereoisomers at the C-2 position. This is clearly evident on the NMR spectra of rollinone (Fig. 9) where all the signals of protons and carbons close to C-2 appear in pairs. As stated previously, this leads to the postulate that the isoacetogenins are probably artefacts which occur when 4-hydroxylated acetogenins are subjected to alkaline medium.

The second group of acetogenins with a saturated lactone are β-hydroxy-γ-methyl-γ-lactonic acetogenins. This recently-described group comprises, at present, 4 compounds: itrabin, jetein, laherradurin and otivarin all belonging to subtypes A3, B3 and C3.[4] These acetogenins are characterized by the presence of a hydroxyl β to the carbonyl on the saturated γ-lactone. Their lactone moieties present spectral characteristics that clearly distinguish them from the classical acetogenins (Fig. 10). In the ^{1}H NMR, an ABX system appears, corresponding to the lactone ring protons, H-α, H-β and H-γ. H-α (H-2) resonates at 2.55 ppm and couples with H-β (H-33) (Jα-β ≈ 5.5 Hz), which in turn exhibits a signal at 4.15 ppm, and shows a weak coupling with H-γ (H-34) (J β-γ < 1Hz); H-γ gives a signal at 4.50 ppm and couples with the *ipso* methyl (J = 7Hz). In the ^{13}C NMR, proton-carbon correlation experiments have proven the assignments.

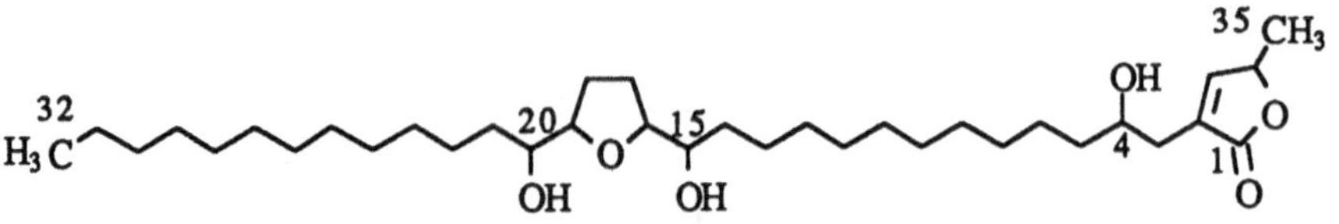

murisolin

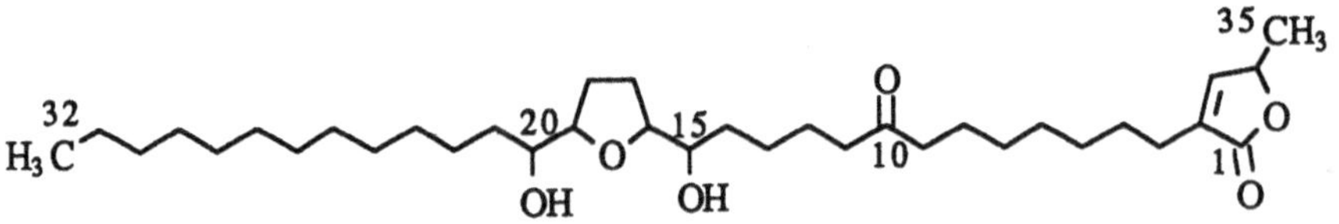

corossolone

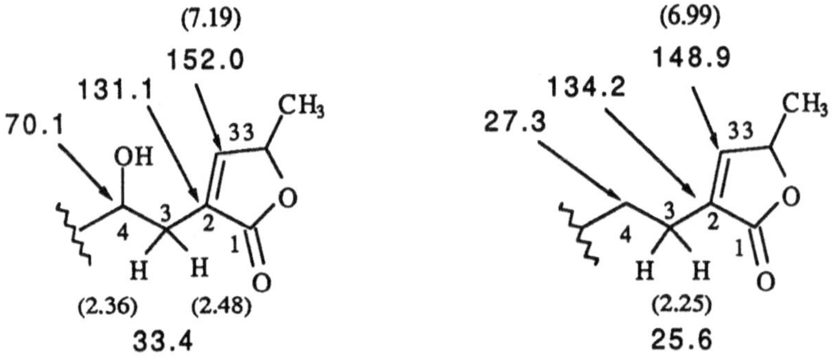

murisolin corossolone

Fig. 8. ^{1}H and ^{13}C NMR spectra of 4-hydroxy acetogenins.

Observation of a n.O.e. between the β and γ protons and between the α and β protons, and the values of the coupling constants, suggest a *cis* relationship between these two pairs of protons. The absolute configuration is depicted here as 34*S*, as deduced for desacetyl-uvaricin (Fig. 10).[2]

For the unsaturated 4-hydroxylated-g-lactone acetogenins, the electronic impact spectra show some characteristic ion peaks corresponding to the lactone moiety at *m/z* 141, due to a fragmentation a to the OH between C-4 and C-5. This fragment is accompanied by a fragment at *m/z* 123. The same fragmentations appear for the isoacetogenins.[2] If there is no hydroxyl at C-4, fragments of low intensity can be observed at *m/z* 111 and 129.[1,4]

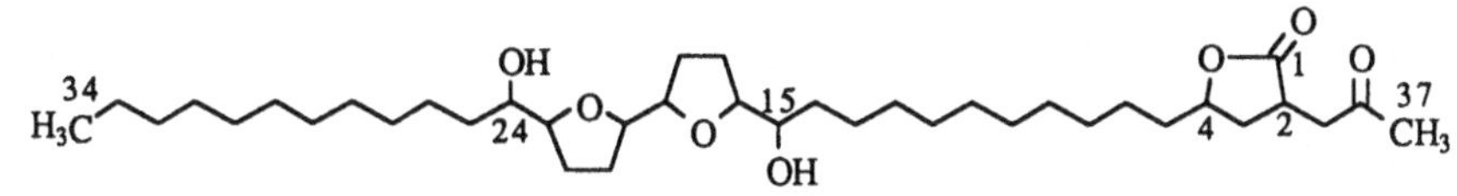

rollinone

^{1}H NMR

^{13}C NMR

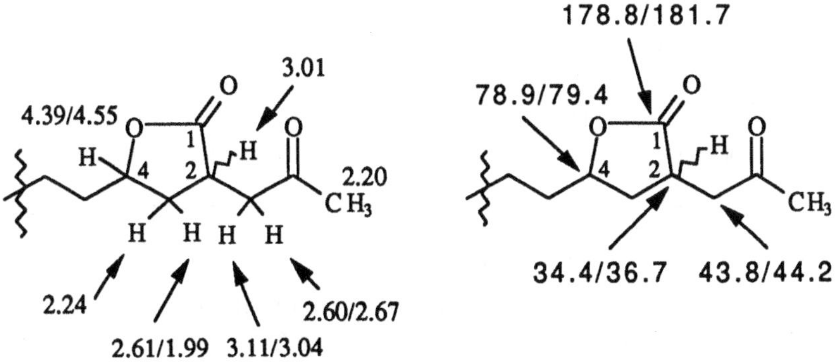

Fig. 9. ^{1}H and ^{13}C NMR of the saturated lactone of isoacetogenins.

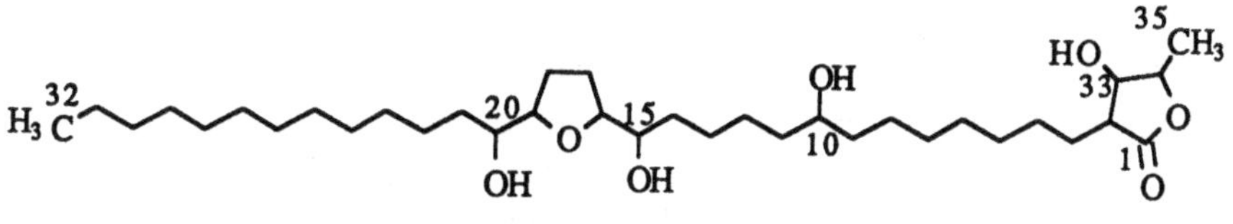

jetein

^{1}H NMR

^{13}C NMR

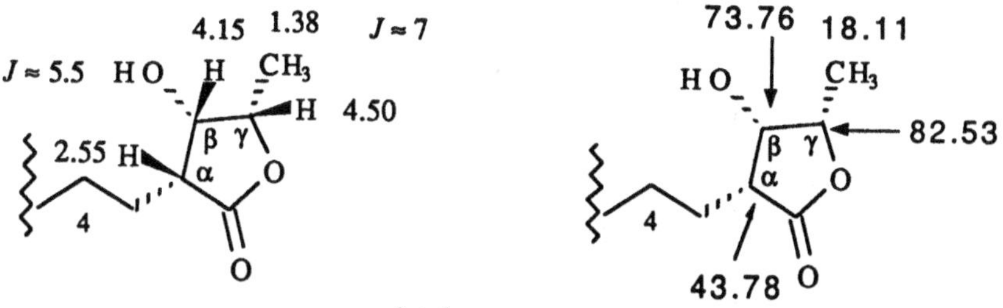

jetein

Fig. 10. ^{1}H and ^{13}C NMR of unsaturated lactone of subtype-3 acetogenins.

Table 1. Fingerprint of different lactonic types in FAB mass spectrometry (matrix: *m*-NBA + LiCl)

$[M+Li]^+$ ion structure: fragment ions (relative intensity arbitrary units)			
CHOH at position 4	$[M+Li-18]^+$ (1)	$[M+Li-44]^+$ (1)	$[M+Li-112]^+$ (1)
CH2 at position 4		$[M+Li-44]^+$ (1)	$[M+Li-112]^+$ (1)
iso-structure	$[M+Li-18]^+$(1)	$[M+Li-44]^+$ (5)	$[M+Li-58]^+$ (2)
OH at 33 (or 35)		$[M+Li-44]^+$ (1)	$[M+Li-116]^+$ (3)

The capability of acetogenins to complex lithium ions under FAB conditions may be used not only for molecular weight determination but also for obtaining structural information. Indeed, structurally significant fragment ions are obtained for the same sample in a second step. The linked scan of the electric sector voltage (E) and the magnetic field (B) of a conventional two-sector mass spectrometer such that the ratio B/E remains constant throughout the scan, allows the mass spectrometric study of a specific precursor ion present in the ion source. When the $[M+Li]^+$ ions are chosen as precursor ions, the fragment ions produced derive directly by unimolecular or collisionally-activated dissociation from the lithiated molecules. The B/E linked scan spectrum is free from ions issuing from consecutive decompositions; consequently, *only direct skeletal fragmentations* are observed and may be very quickly attributed to the structure of the investigated compound. Such a method yields characteristic fingerprints of each lactone type (Table 1).

Tetrahydrofuran Substitution Pattern

Skeletal type. The existence of one or two tetrahydrofuran rings within the classes of acetogenins is evidenced by the presence of ^{13}C NMR resonances due to tetrahydrofuran carbons between 79 and 83 ppm. Corresponding resonances on proton NMR at about 3.8 ppm are also proof of an α and α'-dihydroxylated tetrahydrofuran pattern.

The three types of acetogenins can usually be distinguished by examination of their ^{13}C NMR spectra (Fig. 11): -type A acetogenins show two carbon signals between 81 and 83 ppm, -type B acetogenins show four carbon

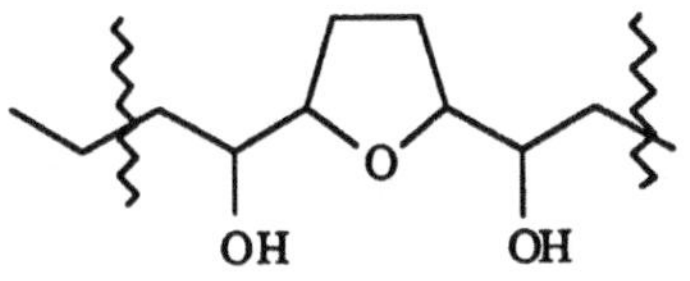

2 C between 81 and 83 ppm

type A

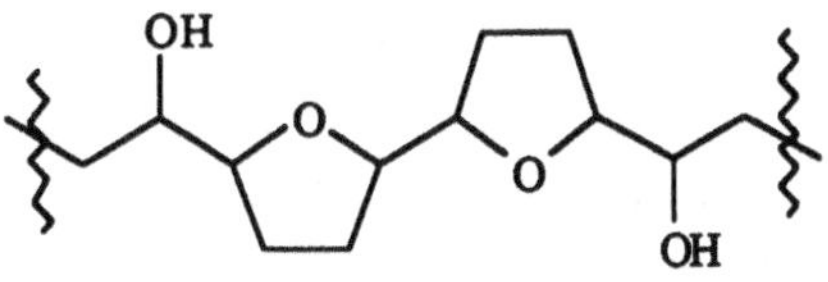

4 C between 81 and 83 ppm

type B

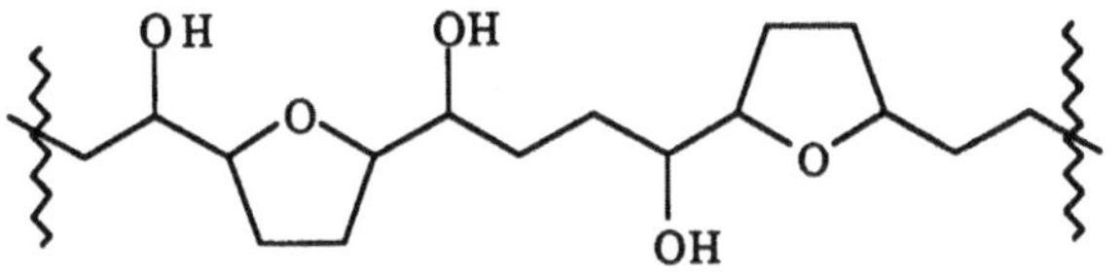

3 C between 81 and 83 ppm, 1C at 79 ppm

type C

Fig. 11. Different types of tetrahydrofuran pattern.

signals between 81 and 83 ppm, -type C acetogenins, show three carbon signals between 81 and 83 ppm and one at 79 ppm.Thus, according to the number of carbons which appear between 81 and 83 ppm, acetogenins can be assigned to groups A, B or C. The placement of the tetrahydrofuran ring(s) along the chain is deduced by mass spectrometry.

Stereochemistry. It is important to determine configuration. To date, absolute configuration has not been determined for any acetogenin. The relative configurations have been determined by X-ray analysis for rolliniastatin-1[9] and for squamocin,[17] and by diffraction studies on crystallized derivatives. Relative configurations have been proposed for the tetrahydrofuran moiety by Hoye *et al*[19] and by Born *et al.*[17]

We recently carried out a stereospecific synthesis of α,α'-dihydroxy monotetrahydrofuran compounds (Fig. 12), which produced the stereochemistries *threo-trans-threo* and *threo-trans-erythro*.[20] The points to stress are the differences in chemical shifts of carbons bearing hydroxyl groups and their geminal protons when *threo* and *erythro* configurations are compared (Fig. 12).

We shall consider some examples of monotetrahydrofuran and bistetrahydrofuran acetogenins.

Monotetrahydrofuran acetogenins. Usually the monotetrahydrofuran acetogenins have two hydroxyl groups α to the tetrahydrofuran. However, two monotetrahydrofuran acetogenins possessing only one hydroxyl α to the THF ring have been recently described, gigantetrocin and gigantriocin from *Goniothalamus giganteus*.[21] NMR spectroscopy, using homonuclear and heteronuclear correlation techniques, allows the characterization of the tetrahydrofuran pattern.

On the ^1H NMR spectrum (Fig. 13), the four protons adjacent to the oxygen atoms resonate as two multiplets signals at about 3.8 and 3.4 ppm, both

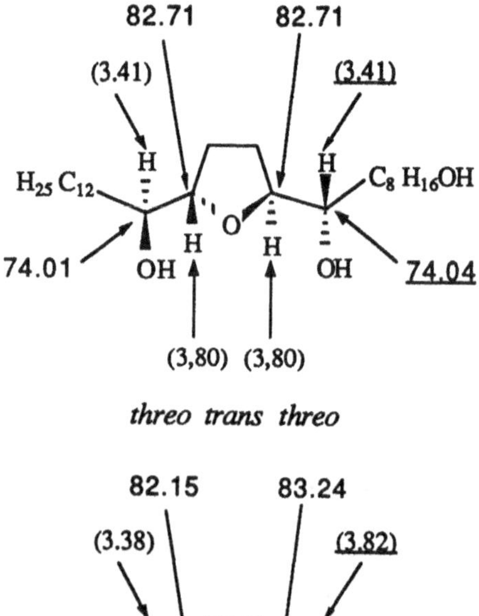

threo trans threo

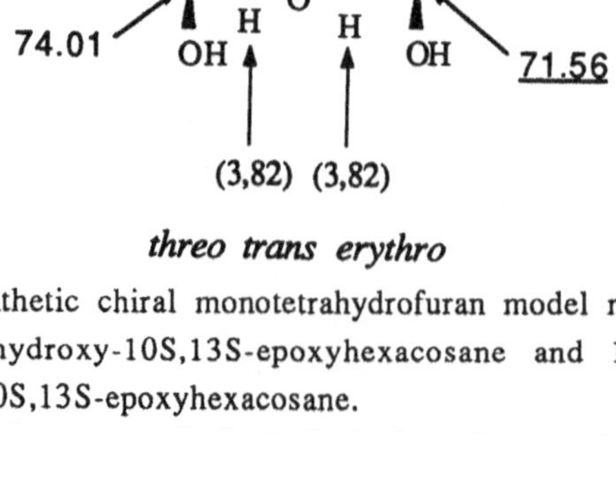

threo trans erythro

Fig. 12. Synthetic chiral monotetrahydrofuran model molecules: 1,9S,14S-trihydroxy-10S,13S-epoxyhexacosane and 1,9R,14S-trihydroxy-10S,13S-epoxyhexacosane.

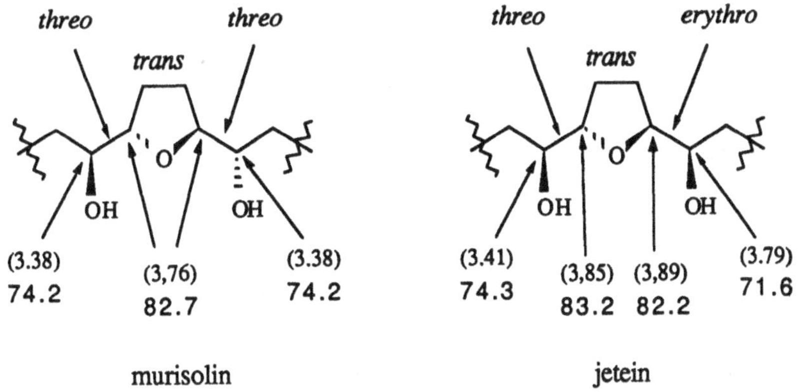

Fig. 13. Monotetrahydrofuran acetogenins.

correlated according to the COSY spectrum. The ^1H-^{13}C correlation spectrum allows correlation between the two multiplets and the two methine signals for tetrahydrofuran carbons at about 82.5 ppm and for two carbons bearing ahydroxyl group. They resonate at about 74 ppm (*threo-trans-threo*) or at about 71 and 74 ppm (*threo-trans-erythro*).

Bistetrahydrofuran acetogenins. Acetogenins with two tetrahydrofuran rings belong to type B, with the two adjacent tetrahydrofurans, or to type C, with the two tetrahydrofurans separated by four carbons, two of which adjacent to THF are hydroxylated. The complexity of NMR spectra signals is higher than for the monotetrahydrofuran acetogenins. The ^1H NMR of type B acetogenins shows two multiplets at about 3.8 and 3.4 ppm corresponding to six protons geminal to oxygen functions. As seen before, the ^{13}C NMR spectra are characterized by two resonance ranges, the first at about 81-83 ppm corresponding to furanic methines, the second at about 71-74 ppm corresponding to alcoholic methines. Homo- and heteronuclear correlation spectra, as well as the determination of the carbon multiplicity by ^{13}C NMR SPIN ECHO, are of prime importance for elucidation of the relative configuration of the bistetrahydrofuran pattern.

Whereas the work of Hoye *et al*[19] concentrated on the differences between ^1H NMR δ values for acetogenins and their acylated derivatives, we have noted that the differences were much more significant in the ^{13}C NMR spectra of these compounds, allowing classification of type-B bistetrahydrofuran acetogenins according to one of four different relative configurations: *threo-trans-*

threo-trans-threo as in asimicin,[19] *threo-trans-threo -trans-erythro* as in molvizarin,[22] *threo-cis-threo-cis-erythro* as in rolliniastatin-1,[19] and *threo-trans-erythro-trans-threo* as in trilobacin (Fig. 14).[23]

Placement of Tetrahydrofuran Ring(s)

We have recently developed a new method for determining the position of tetrahydrofuran (and accessory oxygenated groups) along the chain using classical CI mass spectrometry of dihydro derivatives obtained by catalytic hydrogenation of the double bond of the unsaturated γ-lactone. The fragments corresponding to the lactonic part increase by 2 amu while those corresponding to the other part of the molecule remain unaffected.[24] The comparison of spectra of parent acetogenins with those of dihydro derivatives is of great help in determining the position of the tetrahydrofuran ring(s) (Fig. 15).

Placement of tetrahydrofuran(s) may be also easily deduced from careful examination of the B/E linked mass spectra obtained from $(M+Li)^+$ species generated by FAB. We have already shown that this method simplifies the mass

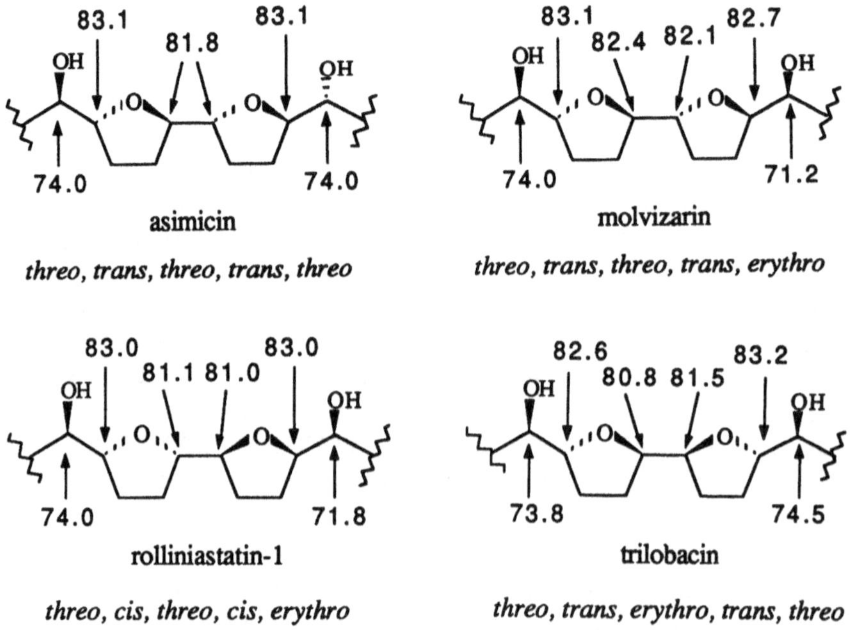

Fig. 14. Relative configuration of bistetrahydrofuran type B acetogenins.

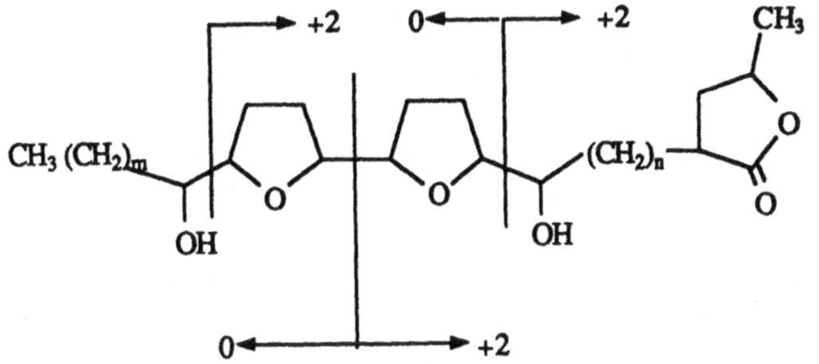

Fig. 15. CI ms comparative fragments of dihydroacetogenins.

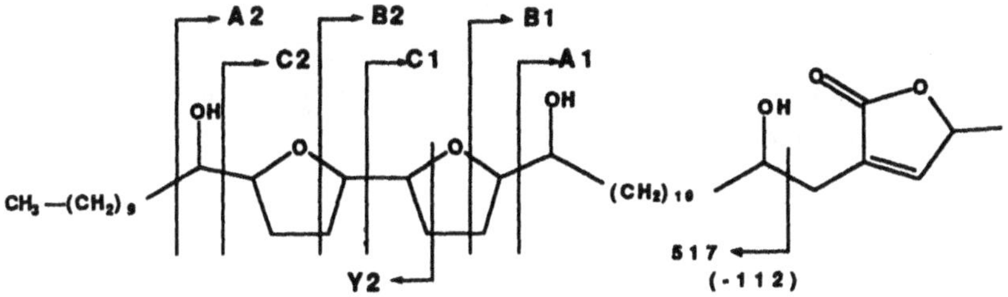

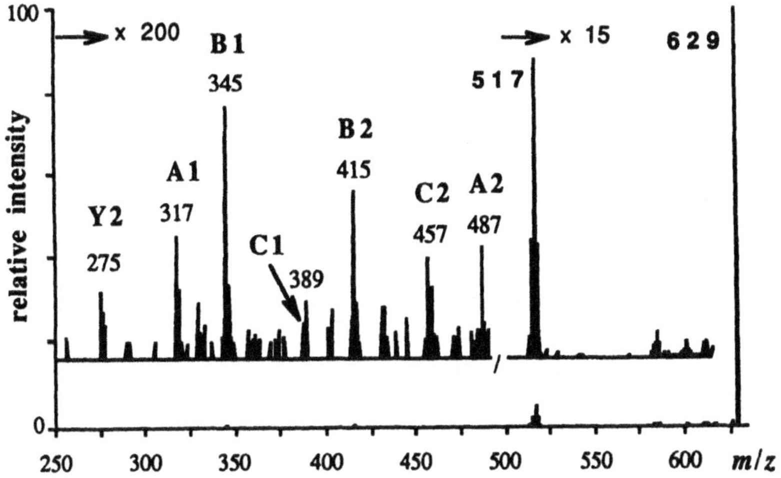

Fig. 16. B/E linked mass spectrum of $M+Li^+$ ions generated by FAB from rolliniastatin-2 (matrix: *m*-NBA + LiCl).

spectra (Figs. 4 and 5). Thus, the simple examination of the FAB mass spectrum recorded in presence of Li^+ cations yields unambiguously the molecular weight. Moreover, structurally significant fragment ions are obtained in a second step from the same sample, in the B/E linked scan spectrum free from ions issuing from consecutive decompositions, as stated previously. In the case of rolliniastatin-2, the fragmentation pattern corresponds to cleavage across the THF rings (B1, B2 and Y2 ions) and of their adjacent C-C bonds (A1, C1, C2 and A2 ions). All generated ions retain lithium. The presence of a hydroxyl at C-4 position results in a very strong loss of 112 Da (Fig. 16).

The reproductibility of the B/E spectra of the lithiated molecular species allows us to consider the observed fragmentations as real fingerprints of the acetogenins. Figure 17 shows the characteristic profiles obtained by B/E mass spectrometry on three lithiated acetogenins possessing two adjacent tetrahydrofuran rings (group B). Similar results were obtained with acetogenins belonging to the A and C groups, each structural type yielding a characteristic fragmentation pattern.

Epoxy-Acetogenins

As stated previously, a group of acetogenin related compounds have been described without any tetrahydrofuran ring but with epoxy group(s). The first reported were epoxyrollins A and B from *Rollinia ulei*.[5] Recently, we have isolated two additional new compounds from the seeds of *Annona muricata* diepomuricanin and epomuricenin-A.

Instead of a tetrahydrofuran moiety, diepomuricanin is characterized by the presence of two epoxy groups. The presence of an α,β-unsaturated γ-lactone and a long hydrocarbon chain with a terminal methyl group was deduced from the

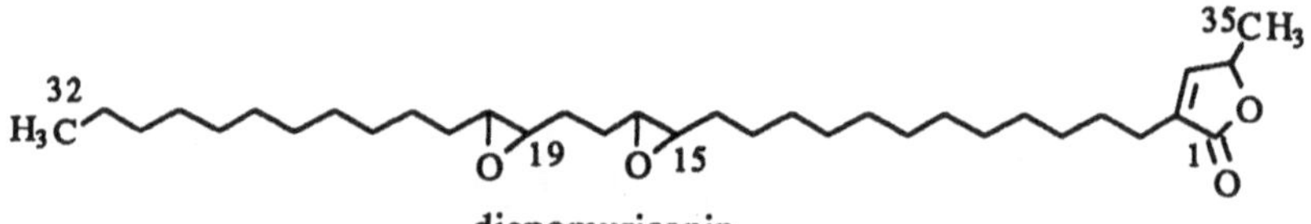

diepomuricanin

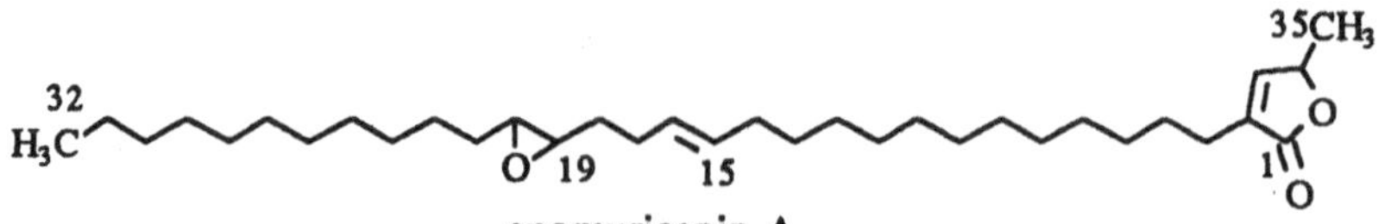

epomuricenin-A

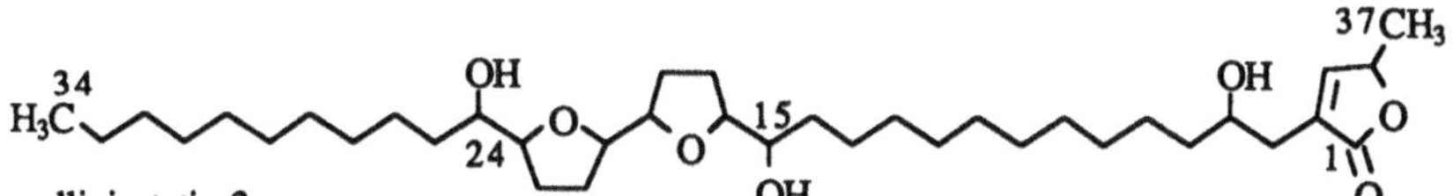

rolliniastatin-2

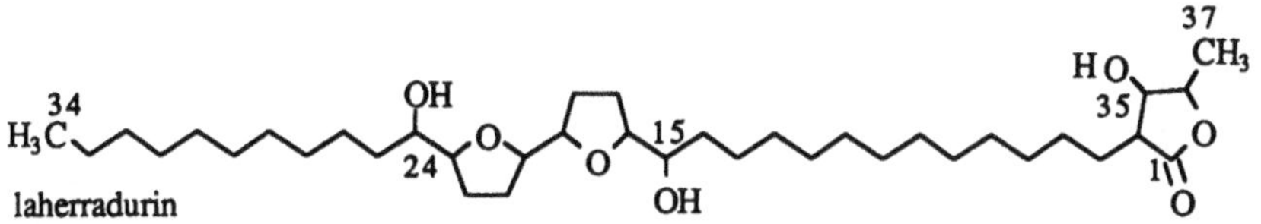

laherradurin

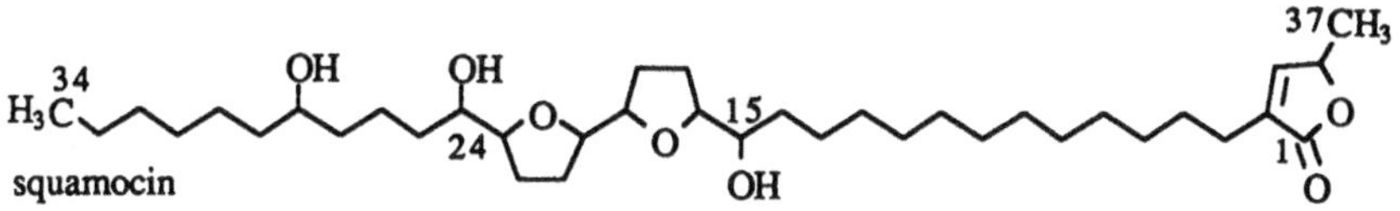

squamocin

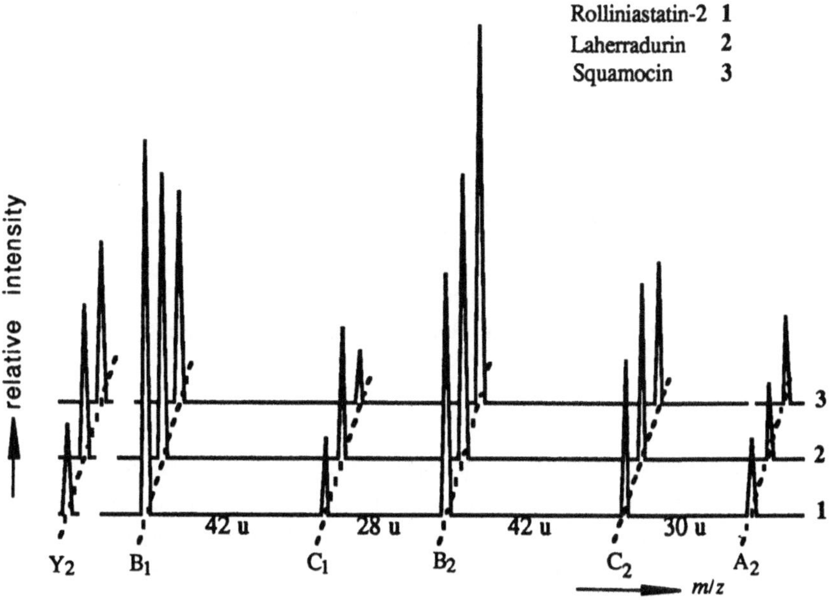

Fig. 17. Reconstructed *B*/*E* linked scan spectra of three lithiated acetogenins showing characteristic fingerprints of an adjacent *bis*-tetrahydrofuran system.

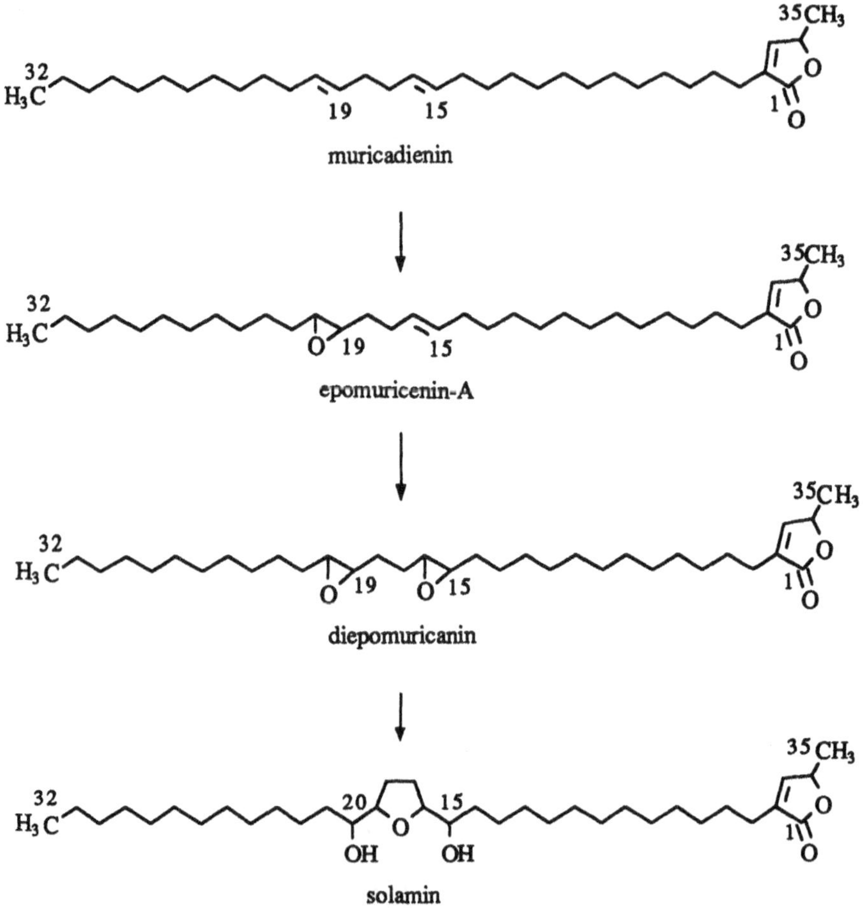

Fig. 18. Hypothetic biogenetic pathway to solamin.

IR, ^1H and ^{13}C NMR spectra by comparison with those of previously isolated acetogenins. The structural problem amounted to the determination of the two epoxy sites in the aliphatic chain, and this was solved by mass spectrometry.[15]

Fast-Atom-Bombardment, combined with mass spectrometry (using linked scan at constant B/E) of the lithium cationized acetogenin allowed determination of the position of the two epoxy rings. In this case we showed that analysis of metastable and collision-induced dissociations gave structurally useful information which could not be obtained by other spectroscopic methods such as NMR. We have also isolated from *Annona muricata* seeds epomuricenin-A which contains one epoxy group and one double bond. Determination of its structure has been performed according to the methods presented above.

The natural occurrence of such compounds is of particular interest for proposing a possible biogenetic pathway. The position of the double bond and of the epoxy group in epomuricenin-A, and the localization of the two epoxy groups in diepomuricanin, are such that they can lead chemically to solamin, existing in the same plant. Thus, the hypothesis that epomuricenin-A could be the precursor of diepomuricanin was supported by epoxidation of its double bond which led to diepomuricanin; when treated with 1N NaOH the latter furnished, as expected, solamin which was isolated as its diacetyl derivative.[25] Thus epomuricenin-A and diepomuricanin emerge as possible links in the biogenetic pathway of monotetrahydrofuran acetogenins from polyunsaturated fatty acid derivatives *via* epoxidation of olefinic functions followed by ring openings and ring closures. We are presently searching for lactonic dienic compounds such as the hypothetical muricadienin, which represents the missing link (Fig. 18).

STEREOSELECTIVE SYNTHESIS

Stereoselective synthesis of murisolin, a monotetrahydrofuran γ-lactone acetogenin, has been performed. We have developed a simple and practical route that could be easily generalized to the preparation of further mono-tetrahydrofuran γ-lactone acetogenins. This synthesis has been performed with a perfect control of the configuration of all asymmetric centers. The pathway used allows the molecule to be constructed with all possible absolute configurations of the stereogenic centers. This synthetic approach allows determination of absolute configuration of the stereogenic centers without ambiguity and provides substantial quantities of material for biological screening.

The retrosynthetic pathway (Fig. 19) used is based on the preparation of both chirons A and B which after coupling afforded a precursor of murisolin. The starting material used is the cheapest α-amino acid commercially available: L-glutamic acid. It leads to both the THF carbon skeleton and the lactone ring.[26]

Synthesis of Chiron A

The synthesis begins with L-glutamic acid which after deamination afforded (4*S*) 4-carboxy-4-butanolide (**1**) with a complete retention of the absolute configuration of the asymmetric center (Fig. 20). The optical purity of this product is 99%. Reduction of the carboxylic function was performed in high yield in three steps to obtain (4*R*) 4-methyl-4-butanolide (**4**) or γ-valerolactone. Alkylation at C-2 was effected by deprotonation followed by addition of allyl bromide. The *cis-trans* mixture (**5**) was directly oxidized to produce the mixture of *cis-trans* (**6**). This mixture was reacted with one equivalent of vinyl magnesium bromide to give the corresponding allylic alcohol, which after oxidation afforded the desired chiron A (**8**) as a *cis-trans* mixture. Whereas chiron A with the 4*S* configuration can be obtained starting from D-glutamic acid, it is cheaper to obtain it starting with L-glutamic acid after total inversion of the stereogenic center from (**1**) through the sequence described in Figure 21, and therefore allowing an access to both enantiomers of acetogenins.

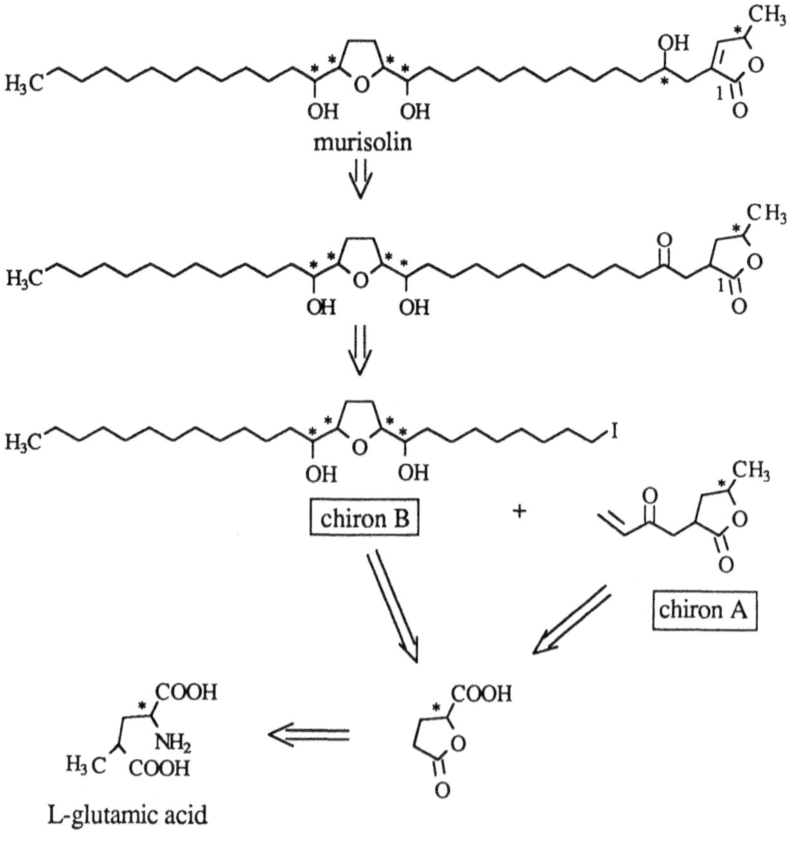

Fig. 19. Retrosynthetic scheme.

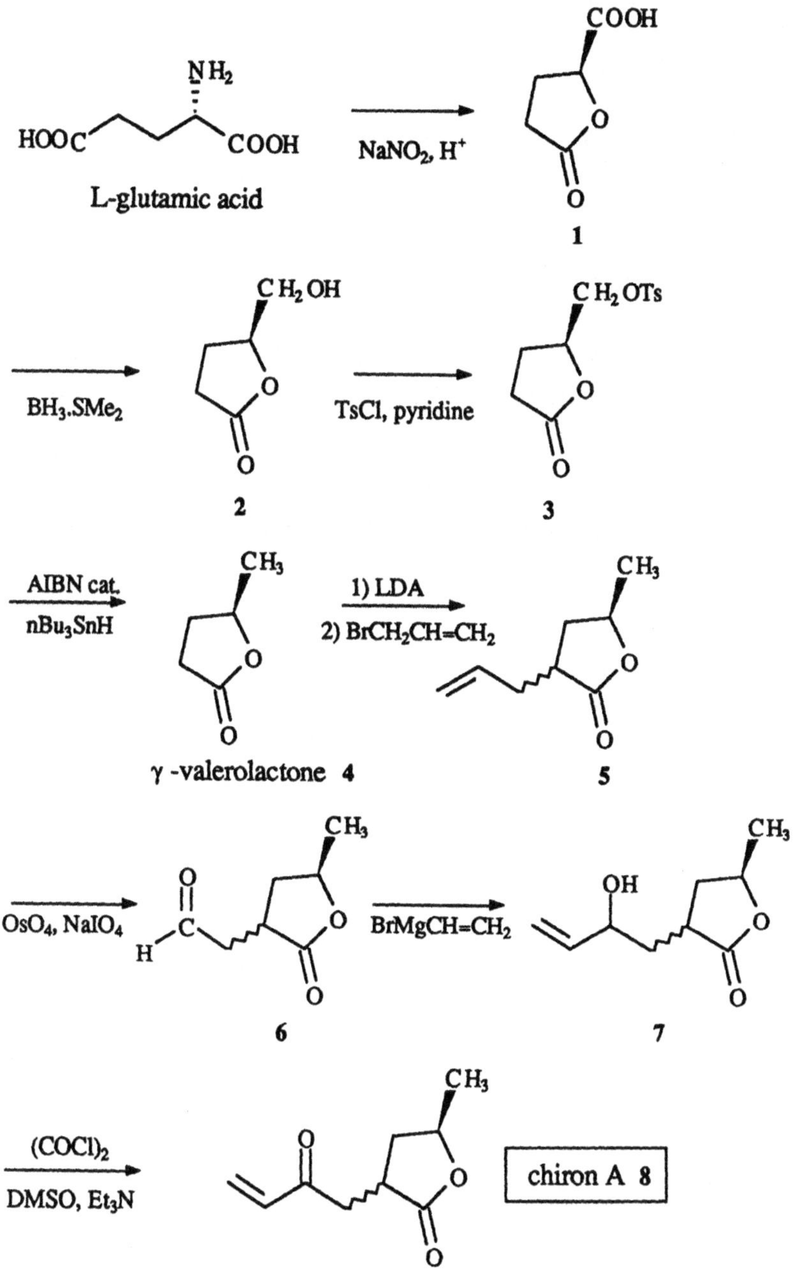

Fig. 20. Synthesis of chiron A: 2-(2'-oxo-3'-butenyl)-4-methyl-4-butanolide.

Synthesis of Chiron B

Preparation of the second chiron (Figs. 22 and 23) was also performed starting from L-glutamic acid *via* (4*S*) 4-carboxy-4-butanolide (**1**). Treatment of (**1**) with oxalyl chloride produced (**9**). Acylation of dodecyl manganese chloride with (**9**) afforded the desired ketone (**10**). The reduction of the carbonyl of the ketone was effected either with L-Selectride® to give the unique *threo* compound (4*S*, 5*S*),[27] or (+)-muricatacin, (**11**),[28] or by sodium borohydride to give as the main product the epimeric *erythro* alcohol (4*S*, 5*R*) (**12**). Both products have been obtained with an excellent optical purity.

After protection of the hydroxyl group of (**11**) as a silyl ether (**13**), the lactone was reduced to give the corresponding lactol which was directly acetylated with acetic anhydride. The corresponding acetal (**14**) was treated with trimethylcyanosilane to afford a mixture of the epimers (**15**) and (**16**) which were easily separated by flash chromatography. As proved by NMR study (n.O.e. experiment) compound (**15**) has the configuration 2*S* and (**16**) the configuration 2*R*. The overall yield for the last three steps was over 96%. Treatment of (**15**), with the requisite Grignard reagent, led to ketone (**17**), which was selectively reduced by L-selectride® to the corresponding alcohol (**18**) with 1'*S* configuration. Chiron B (**21**) was then obtained from (**18**) using a straightforward sequence of reactions, and thus with an overall yield of 10% and in 99% optical purity, in 13 steps from L-glutamic acid. The key feature of this

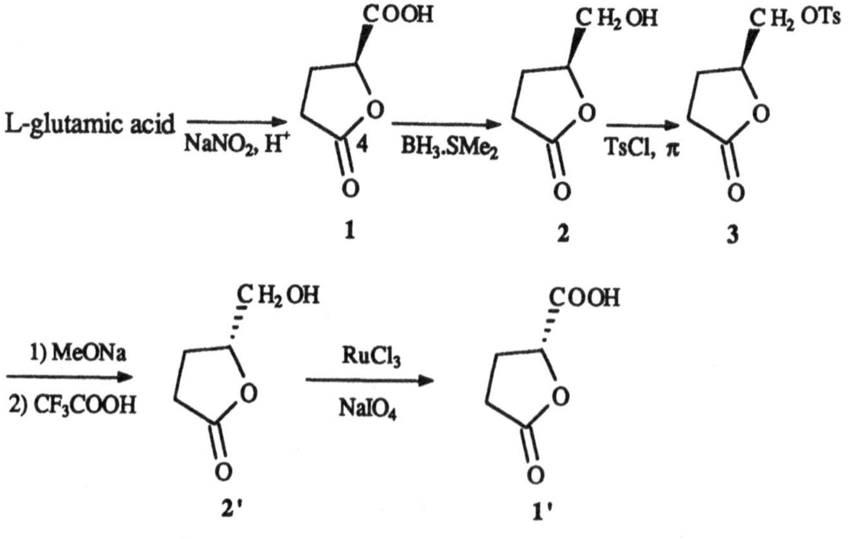

Fig. 21. Inversion of configuration at C-4 of chiron.

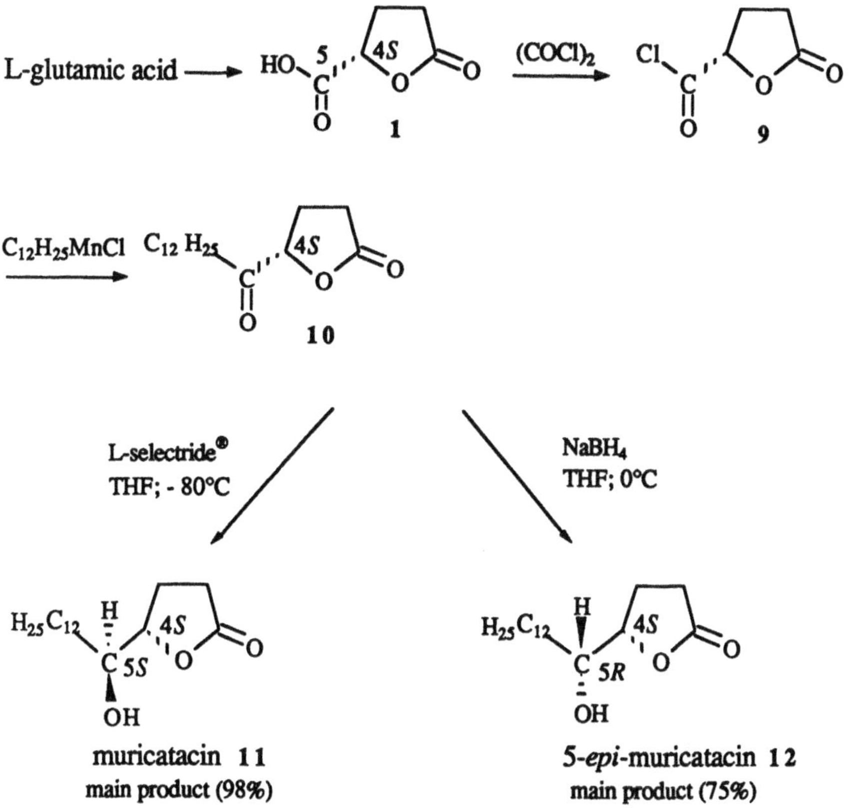

Fig. 22. Synthesis of chiron B: from glutamic acid to muricatacin.

method is the possibility of synthesizing all epimers of this chiron B from the same starting material. *Therefore access to all configurations across the THF ring is easily feasible.*

Coupling Reaction between Chirons A and B

The coupling reaction (Fig. 24) was performed through a single electron transfer (SET) type of reaction in good yield.[29] Then, after creation of the double bond in the lactone ring with stereoselective reduction of the carbonyl, murisolin was obtained in 17 steps from L-glutamic acid. At the present time, the last step has been performed successfully on a model molecule.

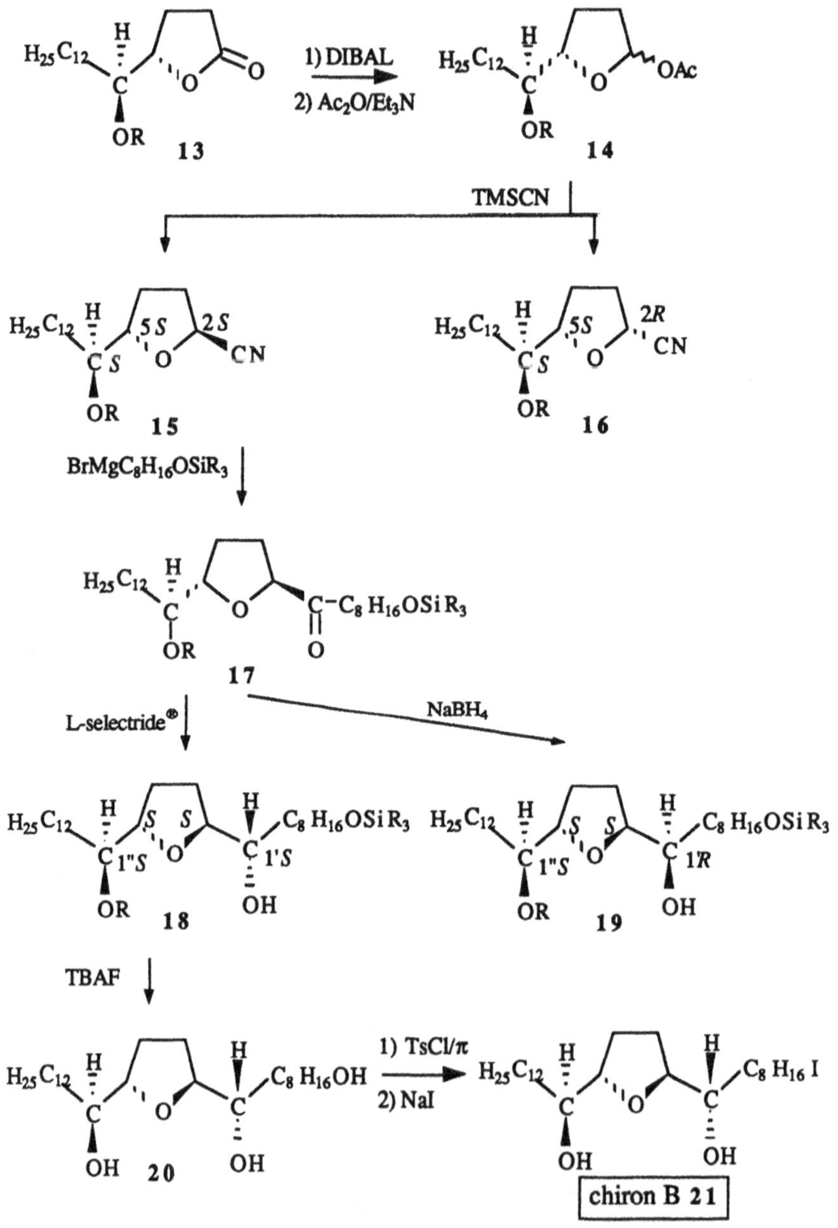

Fig. 23. Synthesis of chiron B: from muricatacin to chiron B: 9*S*,14*S*-dihydroxy-1-iodo-10*S*,13*S*-epoxyhexacosane.

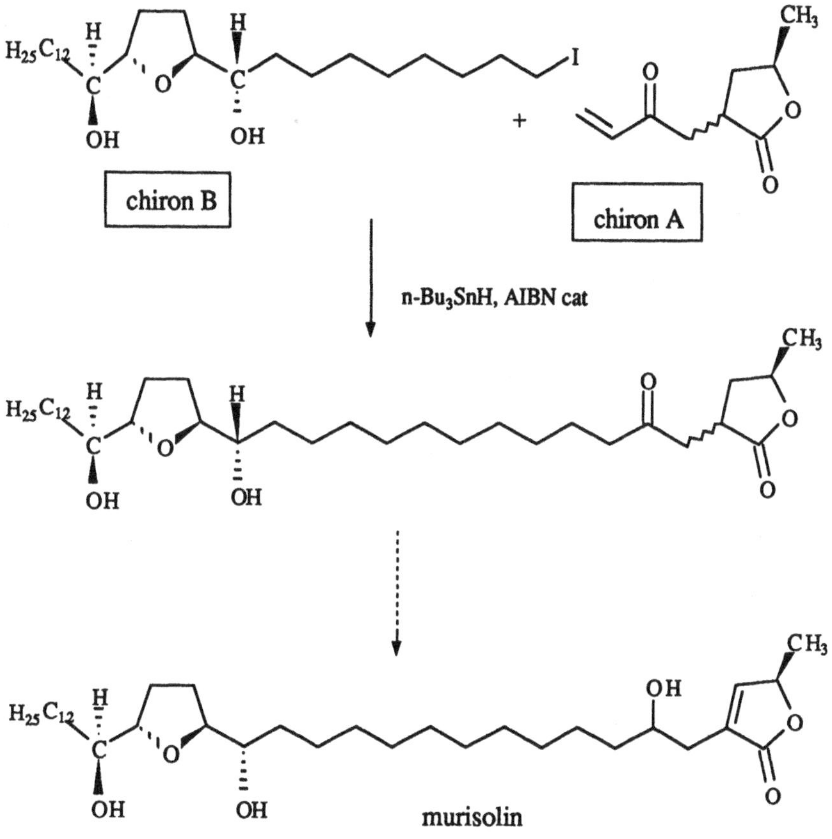

Fig. 24. Coupling of chiron A and chiron B.

BIOLOGICAL ACTIVITIES

The first isolated acetogenin, uvaricin, was proclaimed as a "new antitumor agent."[1] However, it presents a rather important *in vivo* cytotoxic activity against lymphocytic leukemia 3PS with a T/C of 160% at the dose of 1.4 mg/kg. Other acetogenins, such as rollinone from *Rollinia papilionella*, rolliniastatin-1 from *Rollinia mucosa* or asimicin isolated from *Asimina triloba* have shown the same type of activity. It is now evident that all the acetogenins possess, to a varying degree, cytotoxic activity. Against KB cells, this toxicity appears at a concentration between 10^{-1} and 10^{-5} µg/ml according to the nature of the functional groups and the stereostructure.

Table 2. *In vitro* cytotoxic activity of acetogenins from *Annona muricata* (ED_{50}; $\mu g. ml^{-1}$)

	K.B. cells	VERO cells
solamin	0.3	1
murisolin	0.1	0.1
corossolone	0.1	0.3
corossolin	0.003	0.03
annonacinone	0.01	0.1
annonacin	0.0001	0.01
vinblastine	0.01	> 3

Table 2 lists the monotetrahydrofuran acetogenins from *Annona muricata* which possess the same relative stereochemistry in the tetrahydrofuran pattern (*threo-trans-threo*). It should be stressed that activities are more pronounced for alcohol than for ketone derivatives, as shown by the couples annonacin/annonacinone and corossolin/corossolone. The presence of a hydroxyl on the chain seems to be important. Corossolin has an activity at a concentration of 3×10^{-3} whereas murisolin, with one hydroxyl group at C-4, has an activity at 10^{-1} and annonacin, with two hydroxyl groups at C-4 and C-10, has a stronger activity, 10^{-4}. Acetylation of the hydroxyl groups reduces the cytotoxic activity. It is also noteworthy that the lactonic nucleus has an effect on the activity since its opening strongly decreases activity, but the reduction of the double bond has very little influence.[30] For the bistetrahydrofuran acetogenins, the same structure/activity relationship is observed. It seems that for the same level of hydroxylation, the bistetrahydrofuran derivatives are more cytotoxic than their monotetrahydro-furan counterparts.

It is difficult to draw conclusions on the significance of the absolute configuration of the various different asymmetric carbons, since this has not so far been determined for any acetogenin. Relative stereochemistry, at least, is clearly important: for instance, the described activity[2] for asimicin and isomers, is from 10^{-3} to 10^{-5} while bullatacin has been claimed[3] an activity at 10^{-13} with the relative configuration *erythro-trans-threo-trans-threo*. In addition to these results of assays on KB cells, studies of acetogenins activity have been carried out on cell line panels of human tumors such as lung, colon, breast, skin and kidney. It looks as if a specificity for one or another cell line exists depending on the acetogenin structure. Thus, Cassady *et al*[31] have reported that isoannonacin and isoannonacinone belonging to the iso series are 10,000 times less active against leukemia cells and 1,000 times more active against colon tumor cells than the parent acetogenins, annonacin and annonacinone. Antitumor activity has been demonstrated on nude mice with transplanted tumor.[32]

Some acetogenins exhibit an antiparasitic activity. Some results have been published relating to activity against a 4-aminoquinoline-resistant strain of *Plasmodium falciparum*.[2] Uleicins from *Rollinia ulei*[33] possess significant *in vitro* activity against *Leishmania donovani* with a good therapeutic index. Similarly, acetogenins isolated from *Annona muricata* and from *Annona cherimolia* have a filaricidal activity against *Molinema dessetae*[34] (Table 3).

Among the monotetrahydrofuran series, the most potent acetogenin is annonacin; acetylation reduces its activity. The oxidation of the hydroxyl at C-10 into a carbonyl slightly increases the activity. The absence of a hydroxyl at C-4 appreciably reduces potency. Among the bistetrahydrofuran series, the activity of cherimolin is comparable to that of annonacinone.

Besides these activities, pesticidal activity has often been described particularly for annonin,[35] bullatacin,[3] asimicin,[2] squamocin,[36] sylvaticin,[38]

Table 3. Activity of acetogenins on infective larvae of *Molinema dessetae*

Compound	LD_{50} ($\mu g\ ml^{-1}$) Day 1	LD_{50} ($\mu g\ ml^{-1}$) Day 7
annonacin	0.66	0.08
annonacinone	0.52	0.28
corossolin	5.12	0.41
murisolin	1.5	0.30
cherimolin	0.67	0.04
otivarin	6.66	0.25
tetraacetylannonacin	10.2	1.20
diethylcarbamazine	> 500	330
ivermectine	1.3	0.27

and goniothalamicin.[37] It is notable that in some countries of South America, ground barks or seeds of some species of Annonaceae are spread on soils as pesticides.

Recently, we have demonstrated an interesting immunosuppressive activity on mixed lymphocit reaction (M.L.R.) of mouse system cells for acetogenins isolated from *Annona muricata*.[39] CI50 of annonacin is about 3 nM on this model. In comparison, ciclosporin used as a reference in this test is active at 10 nM.

REFERENCES

1. JOLAD, S.D., HOFFMANN, J.J., SCHRAM, K.H., COLE, J.R., TEMPESTA, M.S., KRIEK, G.R., BATES, R.B. 1982. Uvaricin, a new antitumor agent from *Uvaria acuminata*, Annonaceae. J. Org. Chem. 47:3151-3153.
2. RUPPRECHT, J.K., HUI, Y.-H., McLAUGHLIN, J.L. 1990. Annonaceous acetogenins: a review. J. Nat. Prod. 53:237-278.
3. HUI, Y.H., RUPPRECHT, J.K., LIU, Y.M., ANDERSON, J.E., SMITH, D.L., CHANG, C.J., McLAUGHLIN, J.L. 1989. Bullatacin and bullatacinone: two highly potent bioactive acetogenins from *Annona bullata*. J. Nat. Prod. 52:463-477.
4. CORTES, D., MYINT, S.L., LEBŒUF, M., CAVÉ, A. 1991. A new type of cytotoxic acetogenins: the tetrahydrofuranic β-hydroxy methyl γ-lactones. Tetrahedron Lett. 32:6133-6134.
5. LAPRÉVOTE, O., ROBLOT, F., HOCQUEMILLER, R., CAVÉ, A. 1990. Structural elucidation of two new acetogenins, epoxyrollins A and B, by tandem mass spectrometry. Tetrahedron Lett. 31:2283-2286.
6. MEYER, N.B., FERRIGNI, N.R., PUTNAM, J.E., JACOBSEN, L.B., NICHOLS, D.E., McLAUGHLIN, J.L. 1982. Brine shrimp: a convenient general bioassay for active plant constituents. Planta Med. 45:331-334.
7. MYINT, S.H., LAURENS, A., HOCQUEMILLER, R., CAVÉ, A., DAVOUST, D., CORTES, D. 1990. Murisolin: a new cytotoxic mono-tetrahydrofuran-γ-lactone from *Annona muricata*. Heterocycles 31:861-867.
8. CORTES, D., MYINT, S.H., LAURENS, A., HOCQUEMILLER, R., LEBŒUF, M., CAVÉ, A. 1991. Corossolone et corossoline, deux

nouvelles γ-lactones monotétrahydrofuraniques. Can. J. Chem. 69:8-11.

9. PETTIT, G.R., CRAGG, G.M., POLONSKY, J., HERALD, D.L., GOSWAMI, A., SMITH, C.R., MORETTI, C., SCHMIDT, J.M., WEISLEDER, D. 1987. Isolation and structure of rolliniastatin-1 from the south american tree *Rollinia mucosa*. Can. J. Chem. 65:1433-1435.
10. PETTIT, G.R., RIESEN, R., LEET, J.E., POLONSKY, J., SMITH, C.R., SCHMIDT, J.M., DUFRESNE, C., SCHAUFELBERGER, D., MORETTI, C. 1989. Isolation and structure of rolliniastatin-2: a new cell growth inhibitory acetogenin from *Rollinia mucosa*. Heterocycles 28:213-217.
11. GROMEK, D., LAURENS, A., HOCQUEMILLER, R., CORTES, D., CAVÉ, A. Unpublished results.
12. LAPRÉVOTE, O., GIRARD, C., DAS, B., LAURENS, A., CAVÉ, A. 1992. Désorption de complexes acétogénines-lithium par FAB: application à l'analyse semi-quantitative d'extraits végétaux. Communication to European Meeting of Mass Spectrometry, Metz, France, August 22-24. Analusis 20:540.
13. SMITH, D.L., LIU,Y.-M., WOOD, K.V. 1991. Modern Phytochemical Methods, Recent Advances in Phytochemistry 25. (N.H. Fischer, M.B. Isman, H.A. Stafford, eds.), Plenum Press, New York.
14. LAPRÉVOTE, O., ROBLOT, F., HOCQUEMILLER, R., CAVÉ, A., CHARLES, B., TABET, J.C. 1991. Structural elucidation of five stereoisomeric acetogenins, uleicins A-E, by tandem mass spectrometry. Phytochemistry 30:2721-2727.
15. LAPRÉVOTE, O., GIRARD, C., DAS, B.C., LAUGEL, T., ROBLOT, F., LEBŒUF, M., CAVÉ, A. 1992. Location of epoxy rings in a long chain acetogenin by fast-atom bombardment and linked scan (B/E) mass spectrometry of lithium cationized molecules. Rapid Commun. Mass Spectrom. 6:352-355.
16. LAPRÉVOTE, O., GIRARD, C., DAS, B., CORTES, D., CAVÉ, A. 1992. Formation of gas-phase lithium complexes from acetogenins and their analysis by fast atom bombardment mass spectrometry. Tetrahedron Lett. 33:5237-5240.
17. BORN, L., LIEB, F., LORENTZEN, J.P., MOESCHLER, H., NONFON, M., SÖLLNER, R., WENDISCH, D. 1990. The relative configuration of acetogenins isolated from *Annona squamosa*:annonin I (squamocin) and annonin VI. Planta Med. 56:312-316.

18. JOLAD, S.D., HOFFMANN, J.J., COLE, J.R., BARRY, C.E., BATES, R.B., LINZ, G.S., KONIG, W.A. 1985. Desacetyluvaricin from *Uvaria acuminata*, configuration of uvaricin at C-36. J. Nat. Prod. 48:644-645.
19. HOYE, T.R., ZHUANG, Z. 1988. Validation of the ^{1}H NMR chemical shift method for determination of stereochemistry in the *bis* (tetrahydrofuranyl) moiety of uvaricin-related acetogenins from Annonaceae, rolliniastatin-1 (and asimicin). J. Org. Chem. 53:5578-5580.
20. HARMANGE, J.C., FIGADÈRE, B., CAVÉ, A. 1992. Stereocontrolled synthesis of 2,5-linked monotetrahydrofuran units of acetogenins. Tetrahedron Lett. 33:5749-5752.
21. FANG, X.P., RUPPRECHT, J.K., ALKOFAHI, A., HUI, Y.H., LIU, Y.M., SMITH, D.L., WOOD, K.V., McLAUGHLIN, J.L. 1991. Gigantetrocin and gigantriocin: two novel bioactive annonaceous acetogenins from *Goniothalamus giganteus*. Heterocycles 32:11-17.
22. CORTES, D., MYINT, S.L., HOCQUEMILLER, R. 1991. Molvizarin and motrilin: two novel cytotoxic bis-tetrahydrofuranic γ-lactone acetogenins from *Annona cherimolia*. Tetrahedron 47:8195-8202.
23. ZHAO, G., HUI, Y., RUPPRECHT, J.K., McLAUGHLIN, J.L., WOOD, K.V. 1992. Additional bioactive compounds and trilobacin, a novel highly cytotoxic acetogenin, from the bark of *Asimina triloba*. J. Nat. Prod. 55:347-356.
24. CORTES, D., MYINT, S.L., HARMANGE, J.C., SAHPAZ, S., FIGADÈRE, B. 1992. Catalytic hydrogenation of acetogenins. Tetrahedron Lett. 33:5225-5226.
25. LAPRÉVOTE, O., LAUGEL, T., ROBLOT, F., LEBŒUF, M., CAVÉ, A. 1993. Epomuricenin, a new epoxyacetogenin isolated from *Annona muricata*. Phytochemistry (in press).
26. HARMANGE, J.C., FIGADÈRE, B., HOCQUEMILLER, R. 1991. Enantiospecific preparation of the lactone fragment of murisolin. Tetrahedron Asymmetry 2:347-350.
27. FIGADÈRE, B., HARMANGE, J.C., LAURENS, A., CAVÉ, A. 1991. Stereospecific synthesis of muricatacin: a biologically active acetogenin derivative. Tetrahedron Lett. 32:7539-7542.
28. RIESER, M.J., KOZLOWSKI, J.F., WOOD, K.V., McLAUGHLIN, J.L. 1991. Muricatacin: a simple biologically active acetogenin

derivative from the seeds of *Annona muricata seeds* (Annonaceae). Tetrahedron Lett. 32:1137-1140.

29 FIGADÈRE, B., HARMANGE, J.C., HAI, L.X., CAVÉ, A. 1992. Synthesis of 2,33-dihydro-4-oxo-murisolin: conjugate addition of primary alkyl iodides to α,β-unsaturated ketones. Tetrahedron Lett. 33:5189-5192.

30. MYINT, S.H., CORTES, D., LAURENS, A., HOCQUEMILLER, R., LEBŒUF, M., CAVÉ, A., COTTE, J., QUÉRO, A.M. 1991. Solamin, a monotetrahydrofuranic γ-lactone acetogenin. Cytotoxic activity of the acetogenins from *Annona muricata* seeds. Phytochemistry 30:3335-3338.

31. CASSADY, J.M., BAIRD, W.M., CHANG, C.J. 1990. Natural products as a source of potential cancer chemotherapeutic and chemopreventive agents. J. Nat. Prod. 53:23-41.

32. JOSSANG, A., DUBAELE, B., CAVÉ, A., BARTOLI, M.H., BERIEL, H. 1991. Annomontacine: une nouvelle acétogénine γ-lactone-monotétrahydrofuranique cytotoxique de l'*Annona montana*. J. Nat. Prod. 30:967-971.

33. CAVÉ, A., HOCQUEMILLER, R., LAPRÉVOTE, O. 1989. Utilisation d'acétogénines en thérapeutique en tant que substances antiparasitaires. F Patent 1048 N° 88 09 674.

34. BORIES, C., LOISEAU, P., CORTES, D., MYINT, S.H., HOCQUEMILLER R., GAYRAL, P., CAVÉ, A., LAURENS, A. 1991. Antiparasitic activity of *Annona muricata* and *Annona cherimolia* seeds. Planta Med. 57:434-439.

35. MOESCHLER, H.F., PFLUGER, W., WENDISCH, D. 1987. US Patent N° 4 689 232 issued August 25.

36. KAWAZU, K., ALCANTARA, J.P., KOBAYASHI, A. 1989. Isolation and structure of neoannonin, a novel insecticidal compound from the seeds of *Annona squamosa*. Agric. Biol. Chem. 53:2719-2722.

37. ALKOFAHI, A., RUPPRECHT, J.K., SMITH, D.L., CHANG, C.J., McLAUGHLIN, J.L. 1988. Goniothalamicin and annonacin: bioactive acetogenins from *Goniothalamus giganteus* (Annonaceae). Experientia 44:83-85.

38. MIKOLAJCZAK, K.J., MADRIGAL, R.V., RUPPRECHT, J.K., HUI, Y.H., LIU, Y.M., SMITH, D.L., McLAUGHLIN, J.L. 1990. Sylvaticin:a new cytotoxic and insecticidal acetogenin from *Rollinia sylvatica* (Annonaceae). Experientia 46:324-327.

39. LAURENS, A., DUTARTRE, P., CORTES, D., HOCQUEMILLER, R., CAVÉ, A. 1992. Immunomodulating activity of annonacin isolated from *Annona muricata* seeds. Communication to 18th International IUPAC Symposium on the Chemistry of Natural Products, Strasbourg, France, 30 August-4 September.

Chapter Nine

PHYTOCHEMICAL DIVERSITY IN THE ORDER RUTALES

Peter G. Waterman

Phytochemistry Research Laboratories
Department of Pharmaceutical Sciences
University of Strathclyde, Glasgow G1 1XW, Scotland, U.K.

INTRODUCTION

The order Rutales, which is taken here to include the families Rutaceae, Meliaceae, Simaroubaceae, Cneoraceae and Ptaeroxylaceae, is, for its size (Table 1), among the richest and most diverse sources of secondary metabolites in the Angiospermae. In this paper I intend to review the range of metabolites that are known to be produced within the order, highlighting those for which distribution

Phytochemical Potential of Tropical Plants
Edited by K.R. Downum et al., Plenum Press, New York, 1993

is either wholly or largely restricted to the order. This review is not meant to be an exhaustive, extensive listing of the occurrence of some classes of compounds as provided by Waterman and Grundon[1] up to 1983 and recently updated by Hegnauer.[2]

The classes of compounds discussed can be broadly divided into the following: (i) alkaloids, (ii) coumarins, (iii) flavonoid aglycones and glycosides, (vi) chromones and acetophenones and (v) highly oxidised tetracyclic triterpenes (limonoids and quassinoids). The chromones and acetophenones are derived from the acetate pathway, the limonoids and quassinoids from mevalonate and the flavonoids from a combination of shikimate, phenylpropanoid, and acetate pathways. The coumarins and the vast majority of the alkaloids originate primarily through the shikimate route (anthranilic acid, tryptophan, tyrosine, phenylalanine and cinnamic acid).

Table 1. Major taxonomic groupings within the Rutales

Family	Approximate number of:	
	Genera	Species
Rutaceae[a]	150	1600
Flindersioideae	2	25
Spathelioideae	3	20
Dictyolomatoideae	1	2
Meliaceae	50	1400
Ptaeroxylaceae	2	8
Simaroubaceae[b]	20	150
Cneoraceae	2	3

[a]The figures are for the three large sub-families of the family (according to Engler), the Rutoideae, Toddalioideae and Aurantioideae. It is now widely believed that the Rutoideae and Toddalioideae are not taxonomically viable. The three small subfamilies are listed separately.
[b]The Irvingiaceae and Kirkiaceae are not included.

THE ROLE OF THE HEMITERPENE SUBSTITUENT (USUALLY, 3-METHYLBUT-2-ENYL)

The 5-carbon hemiterpene unit plays an important role in the proliferation of secondary metabolites, particularly in the Rutaceae. It is found as both C- and O-bonded substituents on alkaloid, coumarin, flavonoid, chromone and acetophenone skeletons. A C-bonded hemiterpene generally occurs at an electron-rich site either *ortho* or *para* to an oxygen function on an aromatic nucleus, where it is directed through a process of nucleophilic addition of an electrophile produced from 3,3-dimethylallyl-pyrophosphate. The positioning of prenyl substitution appears to be enzyme controlled.[3]

Structural diversity from prenylated intermediates may arise in several ways. The 5C, or more rarely 10C (geranyl), substituent can undergo a large number of oxidation, reduction, dehydration and esterification reactions which commence with the epoxidation of the double bond. Figure 1a illustrates the type of reaction sequence that can occur. The facile nature of the epoxide means that great care has to be taken in isolation and purification processes. For example, generation of the commonly isolated 2,3-dihydroxy-3-methylbutyl moiety can occur through the injudicious use of acid or simply by separations performed on acidic silica gel columns. A further variant that is sometimes seen is the 1,1-dimethylallyl unit which arises through a Claisen re-arrangement of a corresponding 3-methylbut-2-enyloxy precursor (Fig. 1b).

A widespread phenomenon in the Rutaceae is the cyclization of the 3-methylbut-2-enyl unit to give either 2-(1-hydroxyisopropyl)dihydrofuran or 2,2-dimethylpyran systems. The former then commonly loses the 3-carbon isopropanol substituent to give a furan ring (Fig. 2). Furocoumarins and furoquinoline alkaloids are very characteristic metabolites of the Rutaceae. Pyranocoumarins and pyranoquinoline alkaloids are frequent but less common than their furano counterparts; the pyran system, however, is far more commonly found in acridone and carbazole alkaloids, acetophenones and chromones.

A number of dimeric alkaloids and coumarins have been isolated in which dimerization occurs through Diels-Alder (cyclo 4+2) condensation reactions of prenyldienyl side chains with suitable unsaturated systems. The most diverse group of such compounds are alkaloids; indole, quinoline, tryptamine and hordenine bases all being involved.[4] Such dimerizations also occur in the coumarins, such as phebalarin (**1**), and in chromones. The recently isolated bosistoaflavanone dimer (**2**) represents a different type of dimerization in which an isobutyl group is lost leaving a simple methylene bridge.[5] The 'parent' of **2** would be a dimer such as acrimerin-A (**3**).[2]

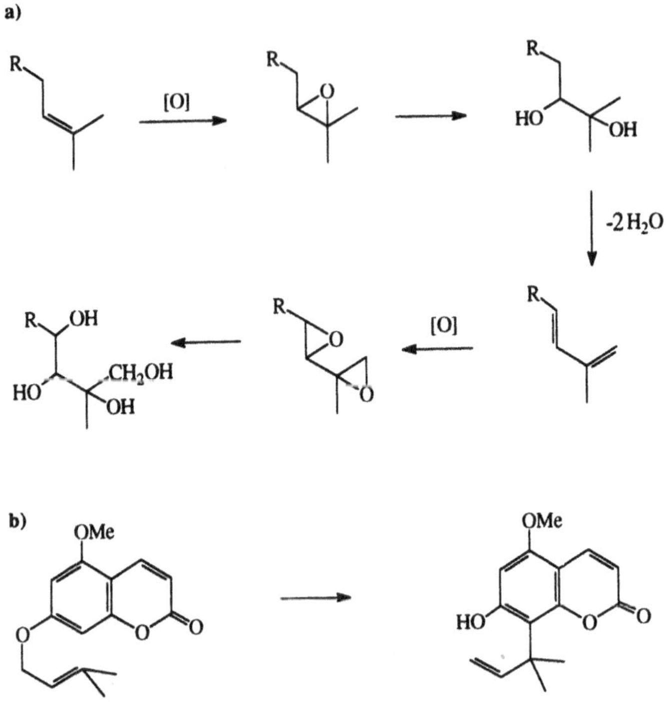

Fig. 1. (a) Oxidation/reduction changes in the hemiterpene 3-methylbut-2-enyl; (b) Claisen rearrangement of a 3-methylbut-2-enyloxy group to a 1,1-dimethylallyl group.

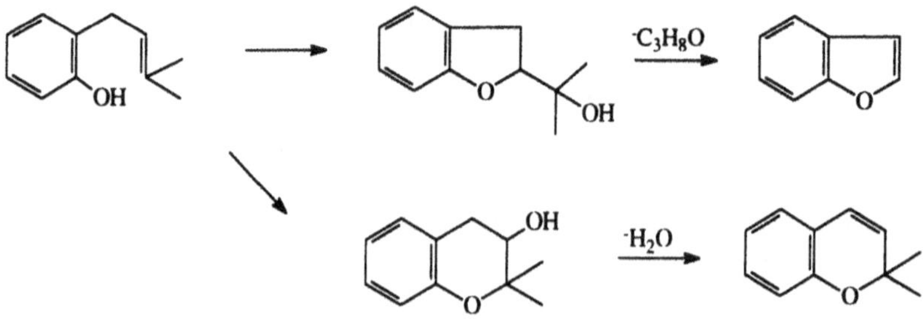

Fig. 2. Formation of pyran and furan ring systems from a hemiterpene precursor.

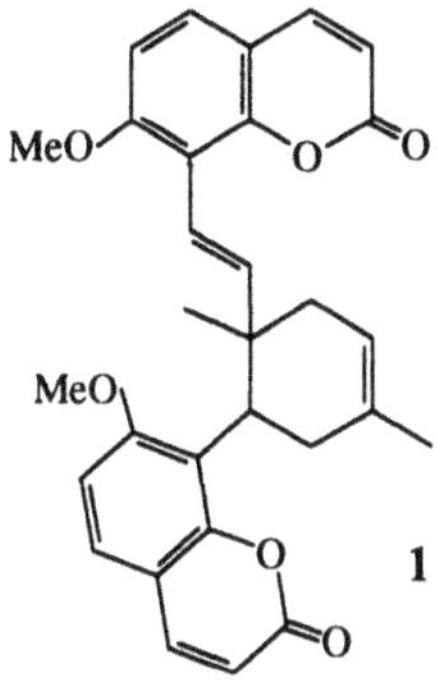

1

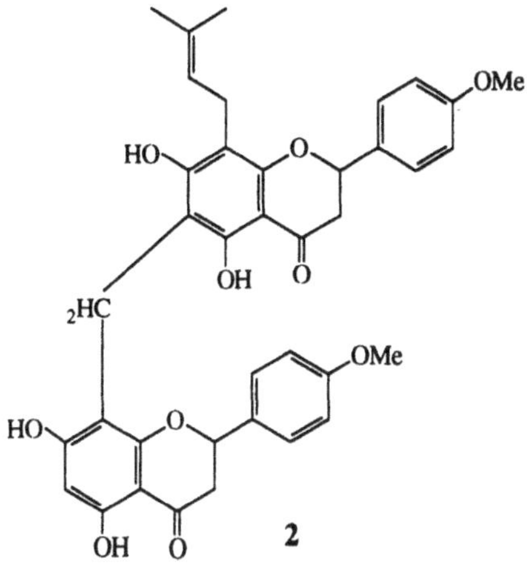

2

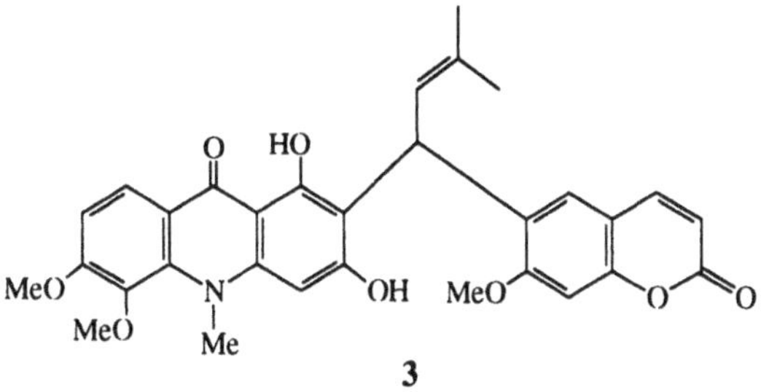

3

ALKALOIDS

The range of predominant structural types produced within the Rutaceae is very extensive. The largest group, and one that is, for the most part, confined to the Rutales, are alkaloids based on anthranilic acid as the nitrogen source. Less common but still abundant in several genera are the tyrosine-derived isoquinoline alkaloids, which are more widely distributed among the families of the Annonales and Berberidales.[6] Tyrosine is also the precursor for a number of amides. Tryptophan-derived alkaloids are present in the form of canthinones and carbazoles, which are found in both Rutaceae and Simaroubaceae. Of uncertain biogenetic origin are the small group of 2-prenylindole derivatives (carbazoles) and corresponding 3-prenylindoles. Lastly, there are a few alkaloids (the imidazoles) that originate from the amino acid histidine.

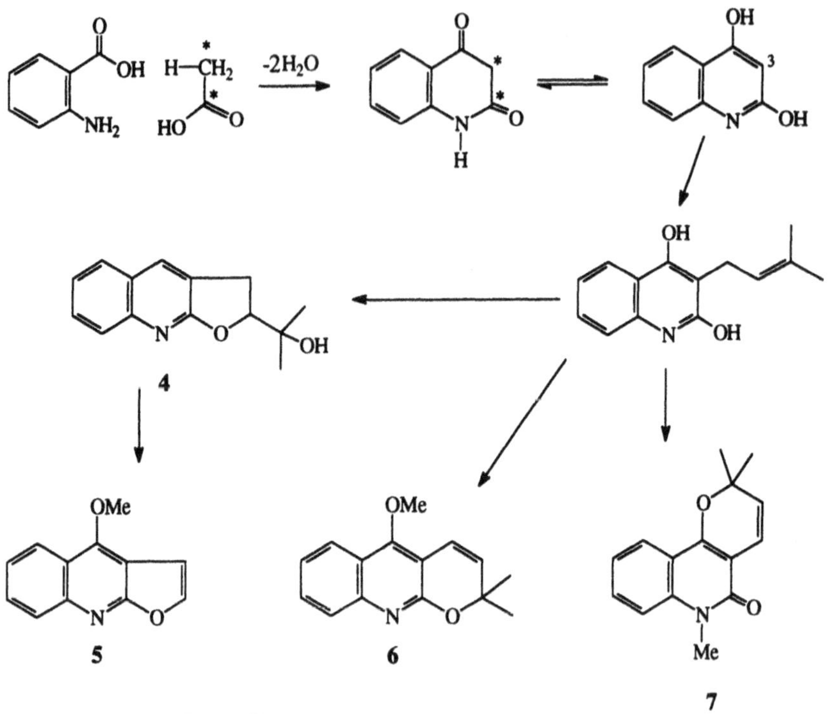

Fig. 3. Formation of furo- and pyranoquinolines.

The Anthranilic Acid Pathway

The ability to use anthranilic acid directly in the biosynthesis of alkaloids is a phenomenon that is, among Angiosperms, largely restricted to the Rutales. The general process appears to involve a second component linking to the carboxylic acid group of anthranilic acid, the simplest being an acetate, which leads to a 2,4-dioxygenated quinoline (Fig. 3). Prenylation occurs readily at the highly activated C-3 position of this quinoline and can lead to dihydrofuroquinolines (**4**), furoquinolines (**5**), linear pyranoquinolines (**6**) and angular pyranoquinolines (**7**). Angular furoquinolines have rarely been detected. The distribution of simple 2-quinolones, 4-quinolones and the more complex furo- and pyranoquinolines have been covered in reviews by Mester[7] and Grundon[8] and methods for their isolation and identification by Gray.[9] It would seem that very few genera in the Rutaceae do not produce members of this class. The most widespread are compounds oxygenated at C-2, C-4 and substituted in the aromatic by methoxyl, methylenedioxy or, more rarely, hydroxyl, C-prenyl or *O*-prenyl groups. Aberrant patterns of prenylation can lead to alkaloids such as lemobiline (**8**) while partial reduction in the aromatic ring produces alkaloids such as perfamine (**9**).

Alkaloids with a 4-quinolone nucleus and a C-2 substituent are also produced by replacing the acetate unit in the biosynthetic processes by another component. These include a long-chain unit, such as is found in acutine (**10**) and evocarpine (**11**) or a mixed alkyl-aryl substituent, as in rutaverine (**12**).

Lastly, within the simple anthranilate alkaloids are the acridones, in which the combination of anthranilic acid with a polyketide made up of three acetate units leads to the tricyclic nucleus which is typically oxygenated at C-1 and C-3 (Fig. 4). The acridones are fairly widely distributed in the Rutaceae and undergo all of the types of substitution that are common in the family. To date they have been isolated only fron the Rutaceae. Acronycine (**13**), originally isolated from *Baurella simplicifolia* (= *Acronychia baueri*), is the best known of the acridones because of its wide spectrum cytotoxicity.[10] Recent work,

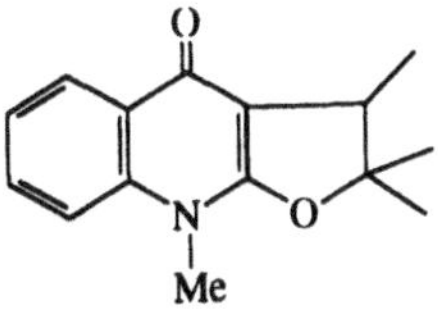

8

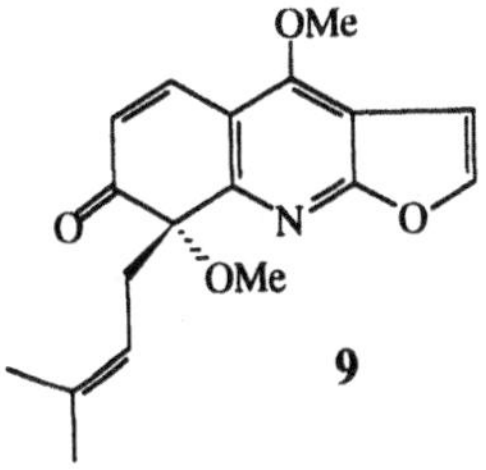

9

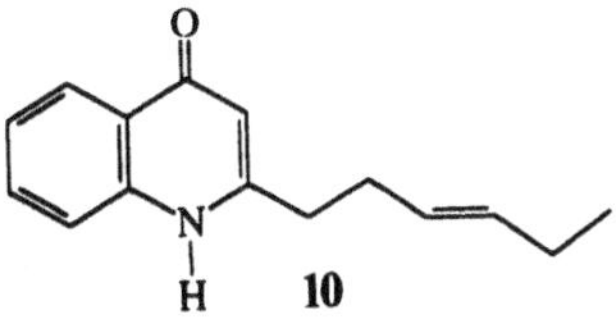

10

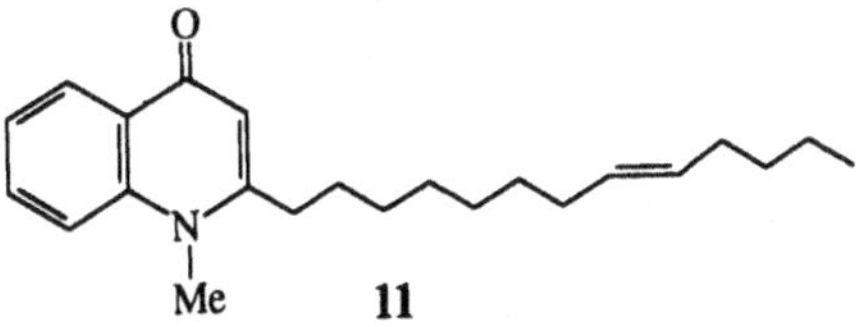

11

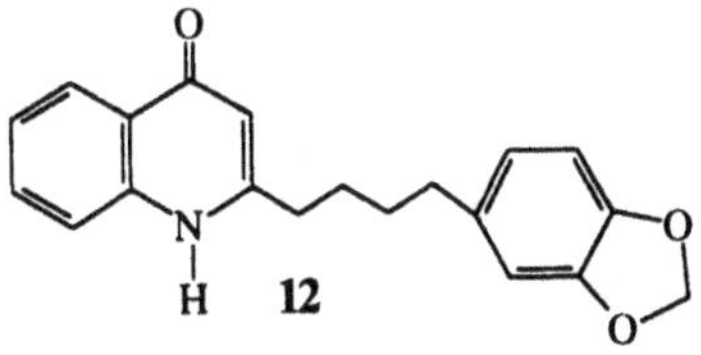

12

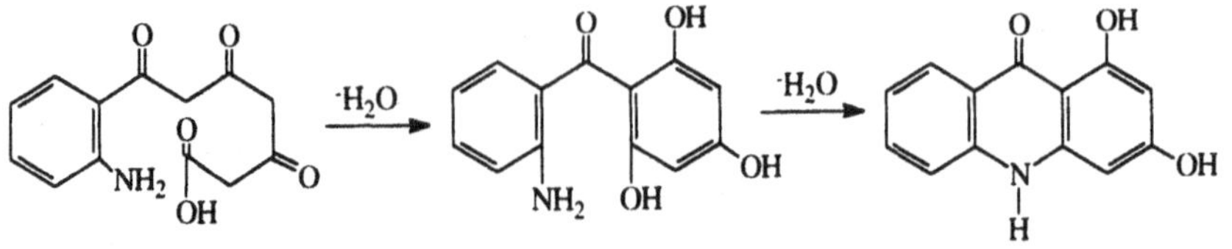

Fig. 4. Formation of the acridone nucleus.

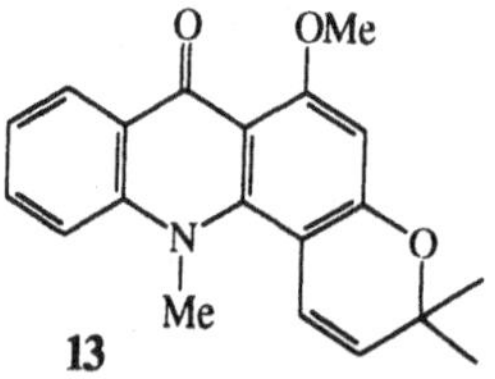

13

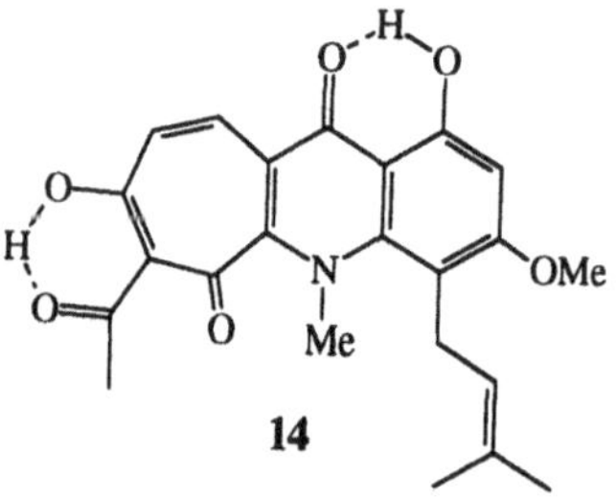

14

primarily by Wu and his co-workers, has shown that the roots of *Citrus* species are particularly rich sources of acridones, including citropone-B (**14**)[11] in which ring expansion has occurred.

With the exception of acronycine, few of this group of alkaloids have shown significant biological activity. The available information up to 1988 on quinolines, including furoquinolines, has been reviewed by Grundon.[8] Simple furoquinolines are reported to be mutagenic,[12] and some acridones exhibit anti-spasmolytic activity.[13] The phototoxicity and resulting mutagenicity of furoquinolines is less pronounced than that of the corresponding furocoumarins.

Tyrosine-Derived Alkaloids and Amides

In contrast to the products of the anthranilate pathway, tyrosine-derived alkaloids are not common in the Rutales. None of the alkaloid types produced is unique to the Rutales, each being more characteristic of plant families such as the Annonaceae, Papaveraceae and Berberidaceae. To date their presence is recorded in five genera of the Rutaceae (*Zanthoxylum*, *Toddalia*, *Phellodendron*, *Tetradidium* and *Fagaropsis*) where they appear to be ubiquitous and their presence has considerable systematic significance.[14] The most commonly encountered are benzo[c]phenanthridines such as chelerythrine, (**15**) which occur

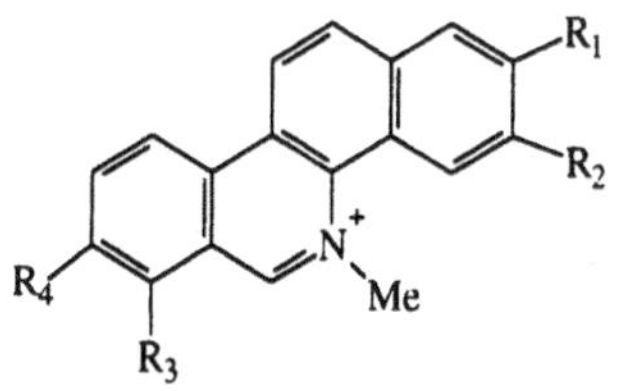

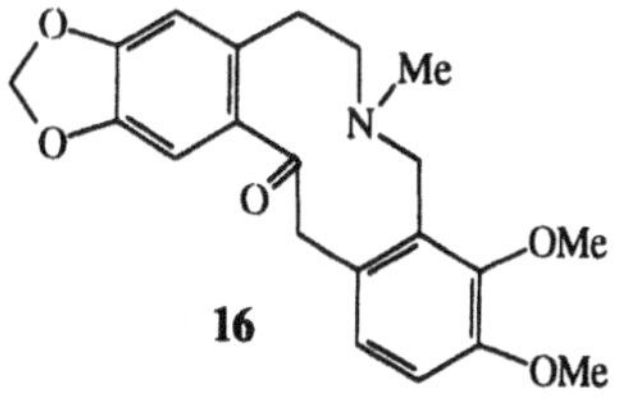

16

15 $R_1R_2 = O\text{-}CH_2\text{-}O$; $R_3 = R_4 = OMe$

17 R_1R_2; $R_3R_4 = O\text{-}CH_2\text{-}O$

18 $R_1 = OH$; $R_2 = R_3 = R_4 = OMe$

in high concentrations in the bark, particularly the root bark, of *Zanthoxylum* species (often as the dihydro derivative). Other types encountered include protopines (α-allocryptopine, **16**), aporphines and berberines.[7] There is one report of a benzo[c]phenanthridine and a pyranoquinoline, from *Xylocarpus granatum* of the Meliaceae.[15] Some Meliaceae and Rutaceae are difficult to distinguish in the field and it would clearly be valuable to re-examine and confirm the presence of these alkaloids in *Xylocarpus*.

Many of the benzo[c]phenanthridine alkaloids have pronounced antibiotic activity; sanguinarine (**17**) is incorporated in some toothpastes for its specific activity against dental plaque.[16] Several benzo[c]phenanthridines have also attracted interest because of their anti-cancer activity, notably fagaronine (**18**).[17]

Amides made up of the combination of phenylethylamine and C_6C_3 or C_6C_1 units have been found sporadically in the Rutaceae. A typical example is (**19**) from *Zanthoxylum armatum*.[7] A most surprising recent addition to the alkaloids of the Rutales are the erythrina-type compounds, such as dysoxyline (**20**) and dsyazecine (**21**), which have been reported from species of *Dysoxylum* (Meliaceae).[18] The authenticity of the plant material involved in the studies of *Dysoxylum* has been questioned, but presuming these alkaloids do occur, it is interesting to consider their biosynthetic relationship to the amides such as (**19**).

One additional type of alkaloid found in the Rutaceae which appears to originate from tyrosine are the oxazoles, typified by halfordinol (**22**). Experimental evidence is lacking but oxazoles appear to represent a condensation between nicotinic acid and tyramine.

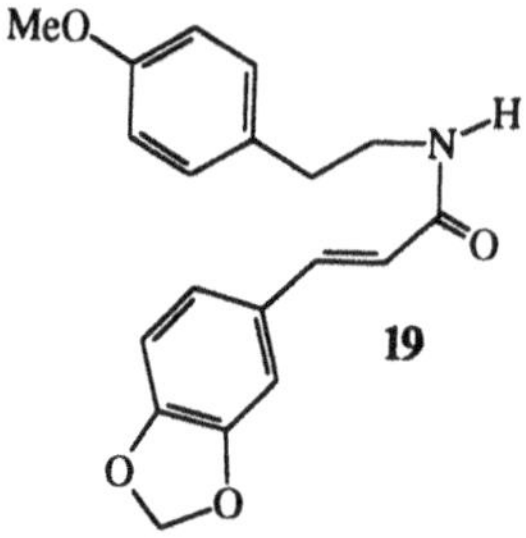

19

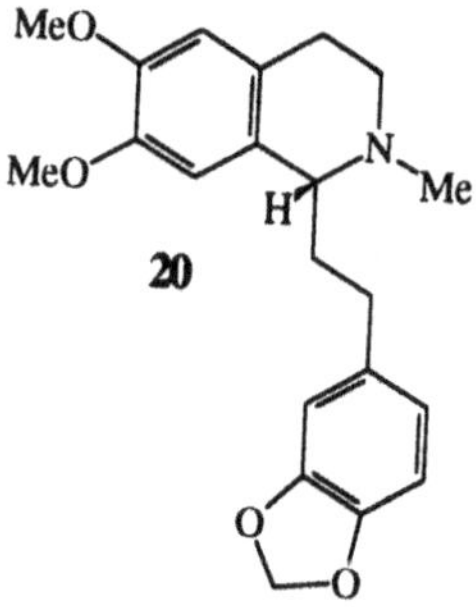

20

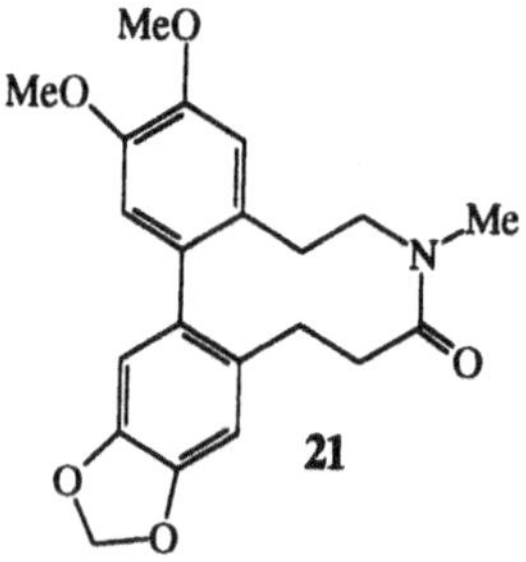

21

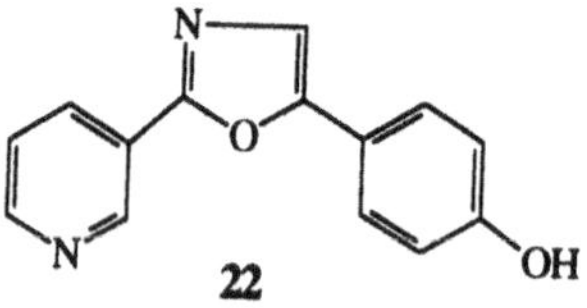

22

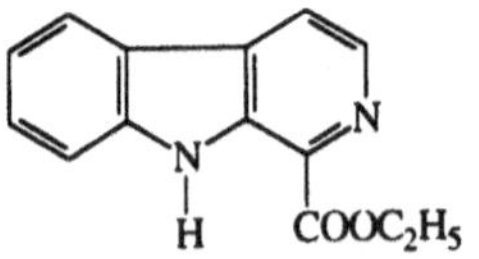

23

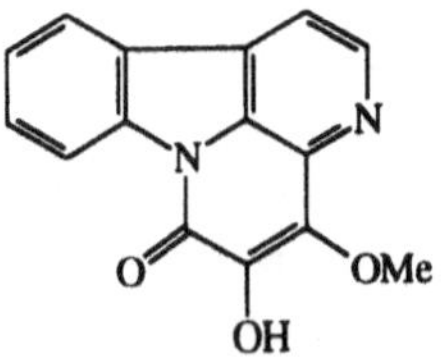

24

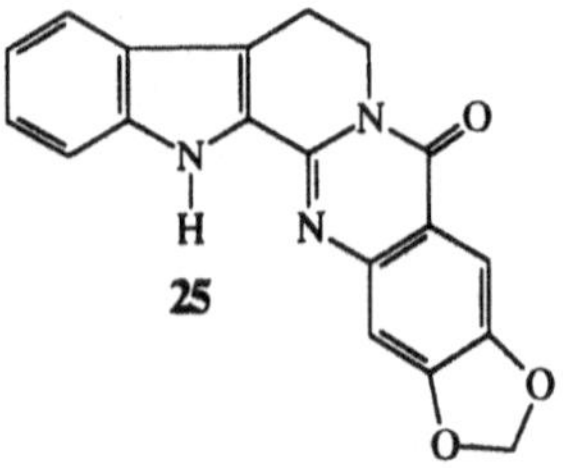

25

Tryptophan and Indole-Derived Alkaloids

Alkaloids containing a tryptamine nucleus occur as widely in the Simaroubaceae as in the Rutaceae. The current impression is that simple carboline alkaloids such as kumujian-A (**23**), and canthinones such as nigakinone (**24**), are actually more common in the Simaroubaceae than the Rutaceae. In contrast the β-indoloquinazolines like euxylophoricine-C (**25**), which represent an interesting combination of anthranilic acid and tryptamine, have been isolated only from Rutaceae. Structural diversity among canthinones is far greater in the Simaroubaceae than in the Rutaceae. However, where canthinones are found in the Rutaceae (in *Zanthoxylum* and *Pentaceras*) yields seem to be high, whereas in many Simaroubaceae the quantities produced are small. Canthinones have been shown to have some antibiotic and cytotoxic activity.[19]

Another striking group of alkaloids which are known only from the Rutales are the 3-methylcarbazoles of which heptazoline (**26**) is a typical example. The origin of these alkaloids remains unresolved but the most plausible route would seem to involve cyclisation of a 2-prenylindole. Carbazoles occur commonly in *Clausena, Glycosmis, Micromelum* and *Murraya* sect. Bergera[20] (Rutaceae) and are otherwise known only from *Ekebergia senegalensis* (Meliaceae).[21] Other alkaloids found in *Murraya paniculata* (*Murraya* sect. Murraya) possess a 3-prenylindole skeleton, a typical example

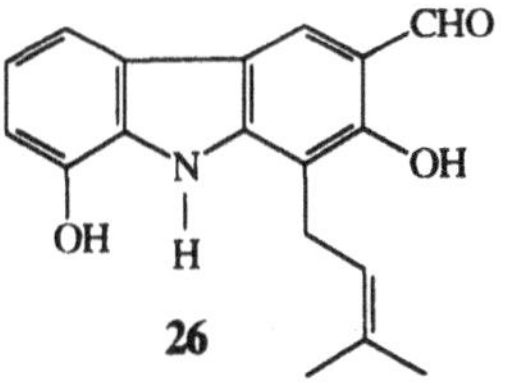

26

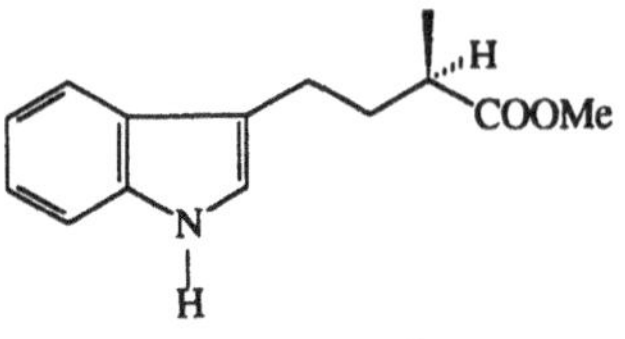

27

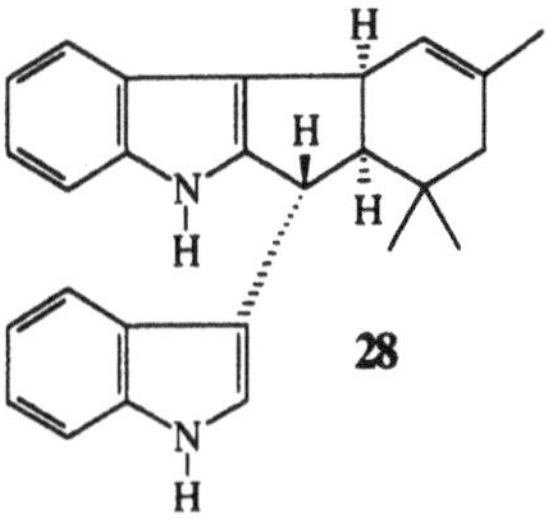

28

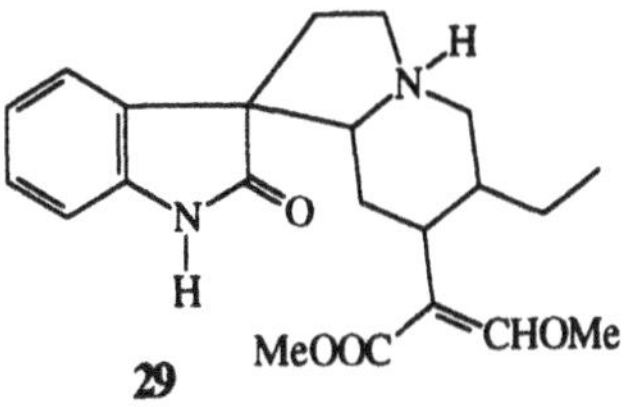

29

being paniculidine-A (**27**).[22] The dimer yuehchukene (**28**) is a Diels-Alder cyclo-addition product of 3-prenylindole and has received considerable attention because of its pronounced anti-implantation activity.[23] Yuehchukene is also known to occur in the genera *Merrilia* and *Micromelum* and the distribution of 2-prenylindole (carbazole) and 3-prenylindole (yuehchukene) derivatives appears to be an excellent taxonomic character in the Rutaceae sub-tribe Clauseneae.[20] Kinoshita *et al.*[22] have proposed a common biogenetic origin for the carbazoles and 3-prenylindoles in the 3-prenylation of a co-enzyme bound Schiff's base of tryptophan (Fig. 5).

One of the most surprising reports for alkaloids in the Order has been of the isolation of the oxindoles rhyncophylline (**29**) and isorhyncophylline, together with canthinone and carbazole alkaloids, from the roots of *Hannoa*

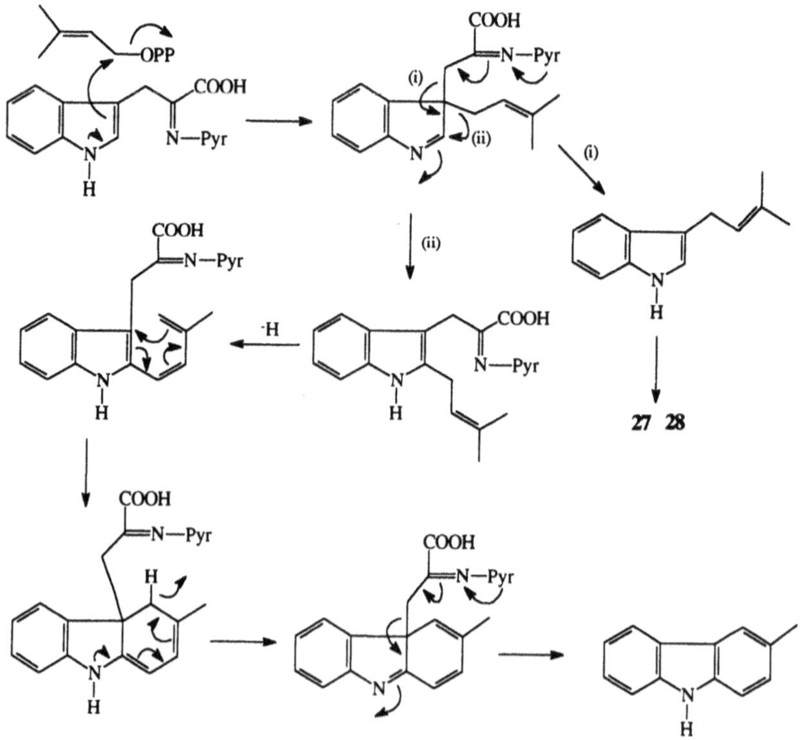

Fig. 5. Possible routes to the 3-prenylindole and carbazole alkaloids (after Kinoshita *et al.*[22]).

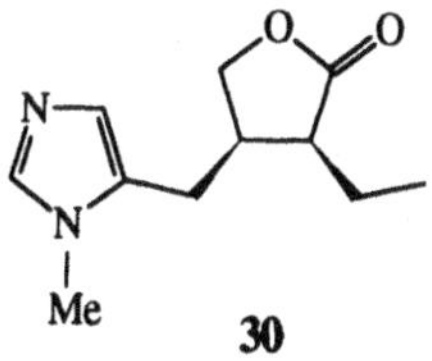

30

klaineana (Simaroubaceae).[24] Of three samples of roots investigated oxindoles were found in only one but that was also found to contain a canthinone. The presence of oxindoles in the Rutales is surprising and this report needs to be substantiated.

Imidazoles

The imidazole alkaloids have long been known[7] from *Pilocarpus* and *Casimiroa* and, in the Rutaceae, remain confined to those two genera. The best known of this small class is pilocarpine (**30**) which is used in western medicine as an opthalmic cholinergic drug. Pilocarpine is presumed to arise from the amino acid histidine but, surprisingly, this has yet to be confirmed.

COUMARINS

The 7-hydroxycoumarin nucleus is formed through the cyclisation of a (*Z*)-2,4-diglucosyloxycinnamic acid. All coumarins found in the Rutales possess 7-oxygenation. The most common further developments are oxygenation at C-6 or prenylation at C-6 or C-8, prenylation seemingly being once again typical of only the Rutaceae within the Order. Coumarins oxygenated at C-6 and C-7 with unusual further prenylation, such as obliquin (**31**), occur in both Ptaeroxylaceae and Cneoraceae.[25] There are some interesting differences between C-6 and C-8 prenylated coumarins. The former are widely distributed in the Rutaceae and commonly give rise to linear furo- and pyranocoumarins. The latter, in contrast, seem to be concentrated in a relatively few taxonomic groupings, notably *Murraya* and its allies,[20] and *Phebalium*,[26] and rarely give rise to the more widely occurring angular furocoumarins.[25]

Another interesting group of coumarins are those based on a 5,7-oxygenation pattern. Dipetalolactone (**32**) and allied 6,8-prenylated derivatives have been found extensively in *Zanthoxylum*[25] and in genera of the Australian

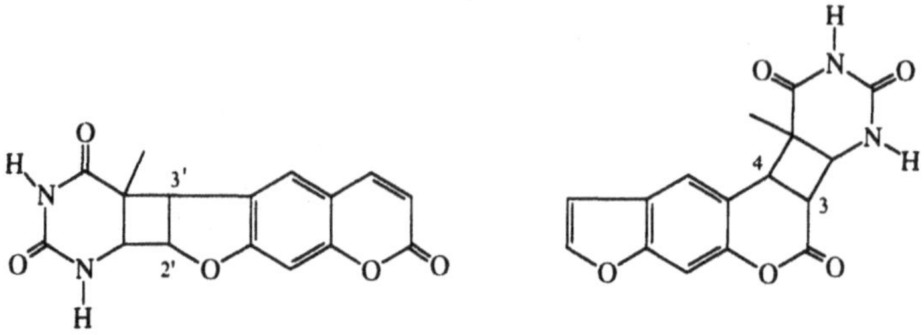

Fig. 6. Photoadduct compounds from interaction of thymine with furocoumarins.

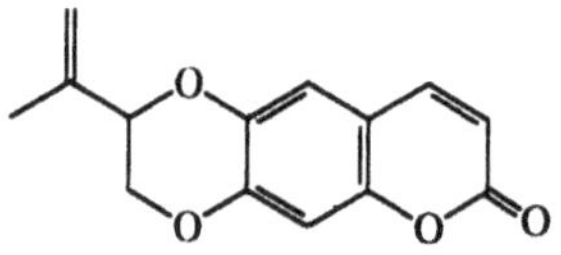

31

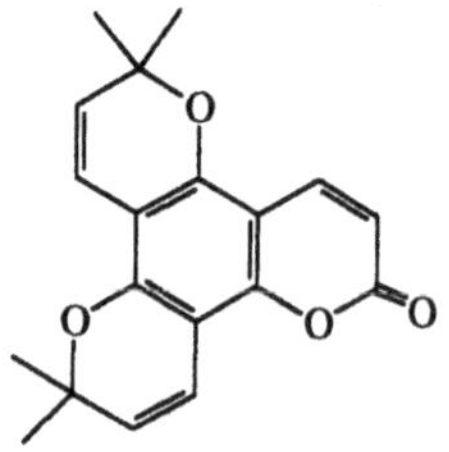

32

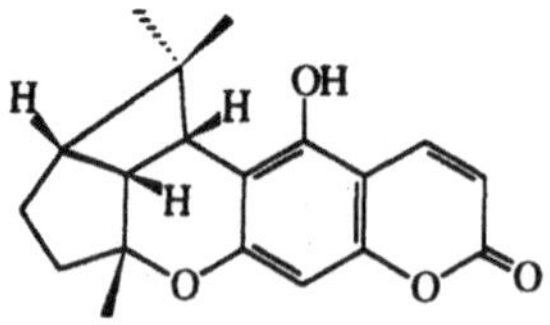

33

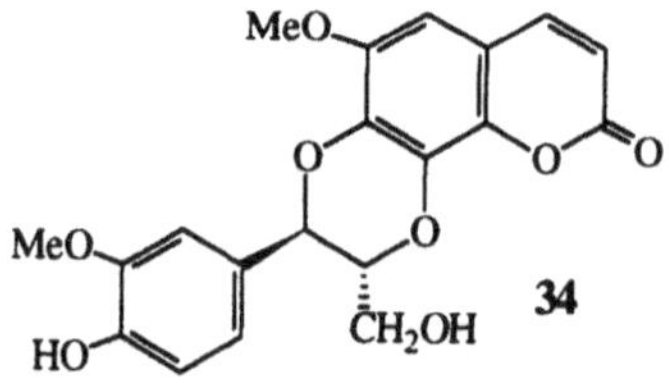

34

Boronieae (*Eriostemon*,[27] *Philotheca*,[27] *Geleznowia*[28] and *Asterolasia* [unpublished]). In another species of *Eriostemon*, *E. brucei*, a series of 6-geranyl-5,7-oxygenated coumarins occurs where multiple cyclizations of the 10C side-chain lead to complex structures such as eriobrucinol (**33**).[29,30]

Coumarins exhibit considerable biological activity.[31] In addition to antibiosis they are able to act as feeding deterrents to invertebrates and vertebrates and have allelopathic properties. Furocoumarins are well known for their ability to cause photodermatitis and furocoumarin-rich species of *Phebalium* in Australia, *Melicope* (formerly *Pelea*) in Hawaii, and *Ruta* in Europe are all recorded as acting as 'blister bushes'. These skin reactions can be caused by simply brushing against foliage that is rich in furocoumarins situated on or near the surface of the leaf (in oil glands for example). This phototoxicity is thought to be caused by the UV-mediated cross-linkage between C-3/C-4 or C-2'/C-3' of a furocoumarin with the DNA component thymine (Fig. 6). The coumarino-lignan cleomiscosin-A (**34**) which has been obtained from simaroubaceous species showed good *in vitro* activity against the P-388 lymphocytic leukemia system.[32]

FLAVONOIDS

In comparison to studies on other metabolites the flavonoids of the Rutales have been rather neglected. The most widespread appear to be rhamnoglucosides of flavanones and flavones with hesperidin (**35**) being particularly common. Hesperidin possesses a rutinosyl (glucose-6-rhamnosyl) disaccharide whereas its congener neohesperidin has the neohesperidosyl (glucose-2-rhamnosyl) disaccharide but is otherwise identical. Whereas hesperidin has no appreciable taste, neohesperidin is one of a group of intensely bitter flavanones, the occurrence of which creates a considerable problem to the producers of orange juice.[33] Furthermore, whereas neohesperidin is bitter, the corresponding flavone (neodiosmin) is tasteless, and the corresponding dihydrochalcone has a sweet taste. The distribution of these glycosides in *Citrus* hybrids has been reviewed.[34]

Some genera within the Rutaceae (*Citrus* and its allies, *Murraya, Merrillia, Micromelum, Melicope*) produce poly-oxygenated flavanones, flavones and flavonols.[35] In most cases these compounds are highly methylated, the most extreme example being exoticin (3,5,6,7,8,3',4',5'-octamethoxyflavone) from *Murraya exotica*. In one genus, *Melicope*, methylenedioxy substituents have been found in several flavonoids (e.g., melinervin, **36**). The occurrence of methylenedioxy substituents in flavonoids is a rare event in nature.[35] The flavone eupatorin (**37**), which is found in appreciable amounts in the fruits of *Merrillia caloxylon*, is reported to be cytotoxic.[36]

Less highly oxygenated C-prenylated flavanols such as phellamurin (**38**) have long been known from the Asiatic genus *Phellodendron*, and from *Phebalium*.[35] More recently C-6 and C-8 prenylated flavanones and chalcones, some of which are reminiscent of flavonoids of the *Millettieae* (Leguminosae), have begun to be found in *Citrus* species.[37] In some cases the prenyl group has cyclised, as in citrunobin (**39**). The latest additions to this group of prenylated flavanones have been found in the Australian species *Bosistoa brassii*[5] the most interesting compound being the dimer (**2**).

In a recent study of another Australian taxon, *Brombya platynema*, an unusual group of piperonyl naphthalene derivatives was obtained, of which brombyin-I (**40**) is a typical example.[38] These compounds are clearly not flavonoids, but brombyin-V (**41**) suggests a route of formation from a cinnamic acid-derived piperonyl unit and a polyketide chain; this makes them 'biosynthetic allies' of flavonoids. Formation of the naphthalene ring system found in brombyin-I, -II, -III, -IV and -VI is thought to proceed by means of a Diels-Alder 4+2 cycloaddition (Fig. 7).[38] These compounds appear to share a common origin with the mostachans, such as moskachan-D (**42**), which were isolated from *Ruta angustifolia*.[39]

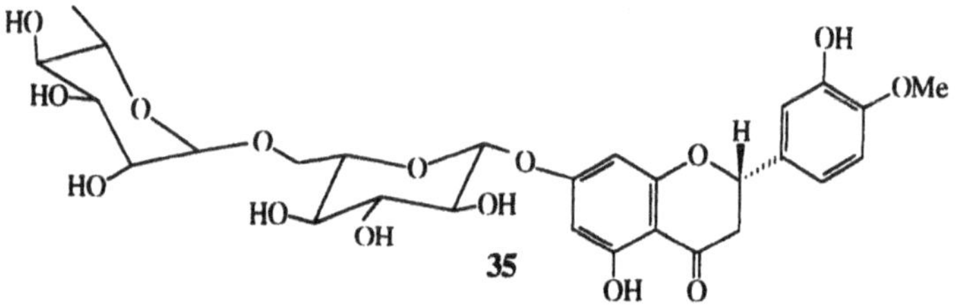

35

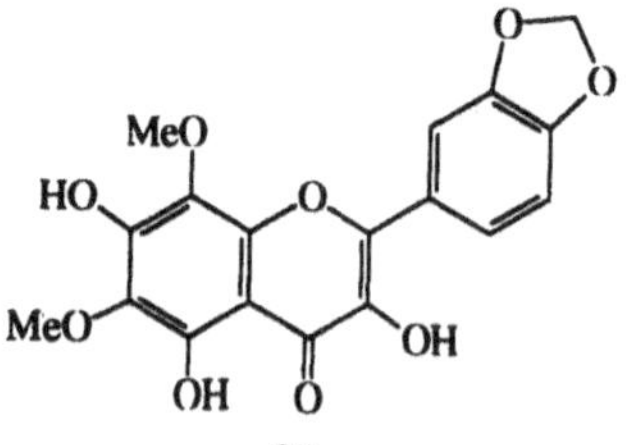

36

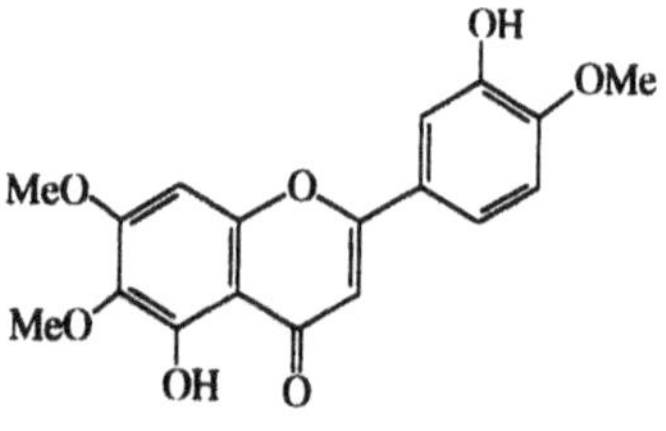

37

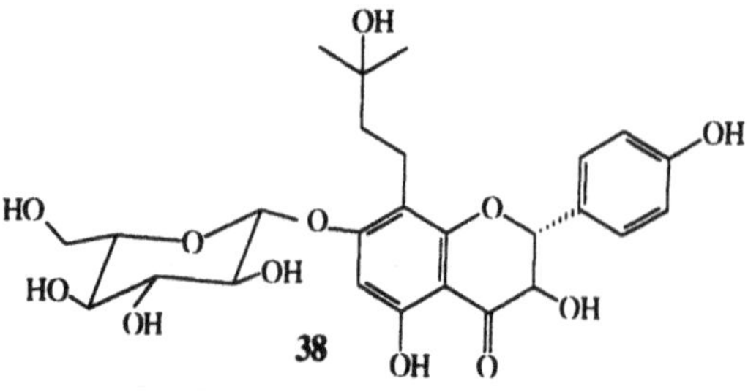

38

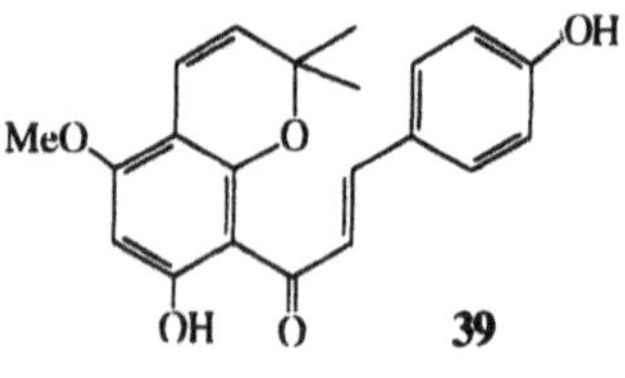

39

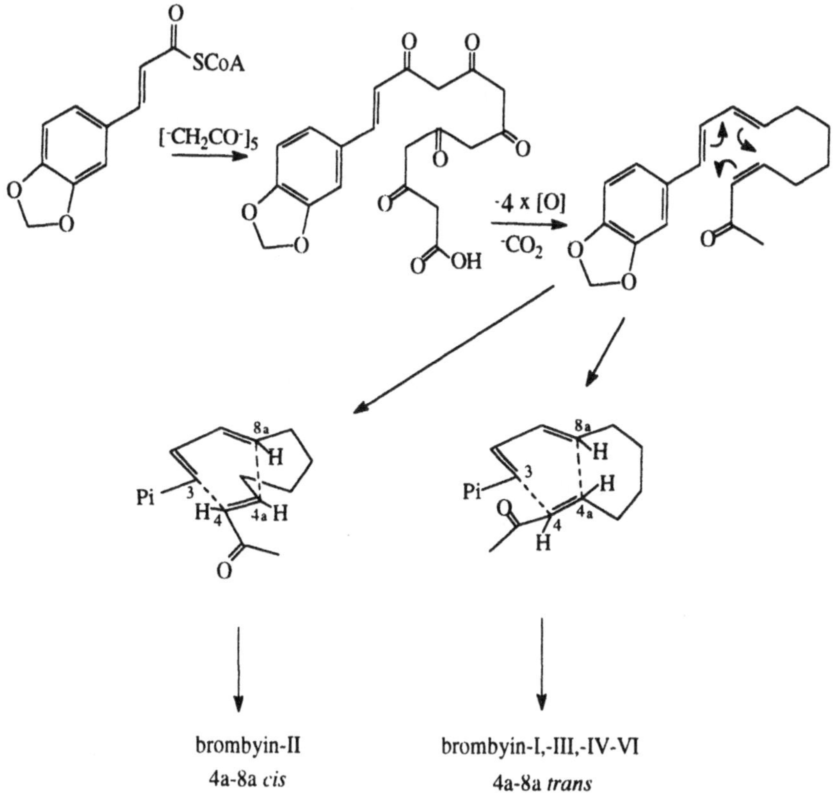

Fig. 7. Possible route for the formation of the brombyins.

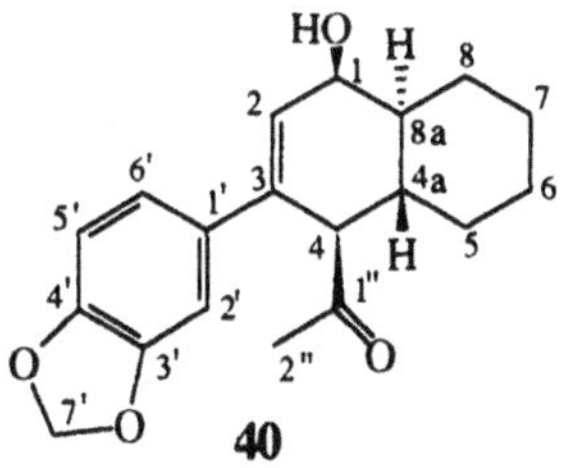

40

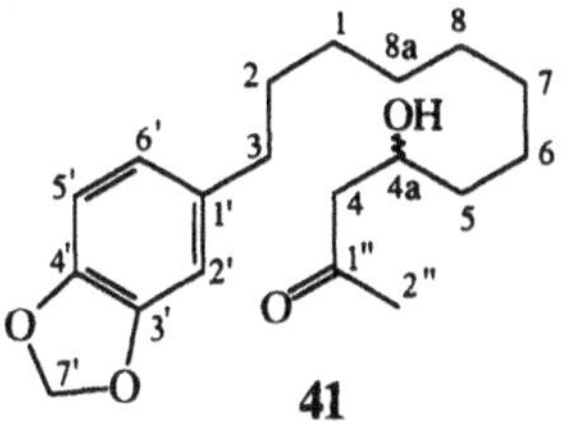

41

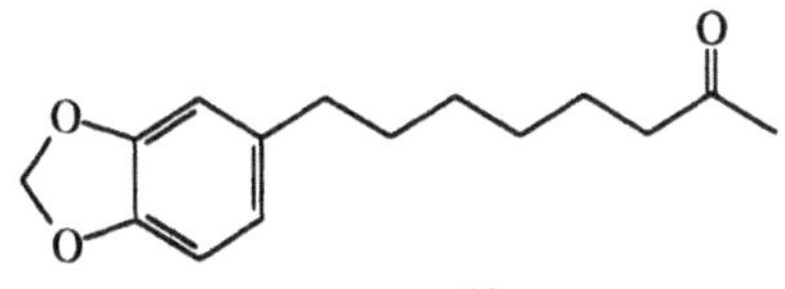

42

ACETOPHENONES AND CHROMONES

The formation of acetophenones and chromones occurs through the cyclization of polyketides made up of four and five acetate/malonate units, respectively (Fig. 8). Both groups are characterised by prenylation of the aromatic nucleus and subsequent modification of the hemiterpenoid group. This leads to series of structures parallelling those seen among the coumarins. For example, the dipyrano- system of dipetalolactone (**32**) is mirrored on octandrenalone (**43**) and spatheliabischromene (**44**).

The distribution of acetophenones has recently been reviewed.[40] They appear to occur predominately within the Rutaceae, the prenylated compounds being located almost entirely in the South-east Asian and Australian genera *Acradenia*, *Acronychia*, *Melicope* (incl. *Euodia*) and *Zieria*. The most significant exception to this generalization is the compound hydroperoxide-Ha (**45**) which was obtained by Liu *et al.*[41] from *Harrisonia abyssinica*, a species currently assigned to the Simaroubaceae but with chemistry that strongly suggests it is misplaced (see below for further evidence supporting this contention). Hydroperoxide-Ha has insect antifeedant properties.

Distribution of chromones is in striking contrast to that of acetophenones. Gray[25] reviewed their occurrence until 1983, when they appeared to be largely concentrated in the two small families Cneoraceae and Ptaeroxylaceae, in *Harrisonia* and in *Spathelia* which belongs to the small sub-

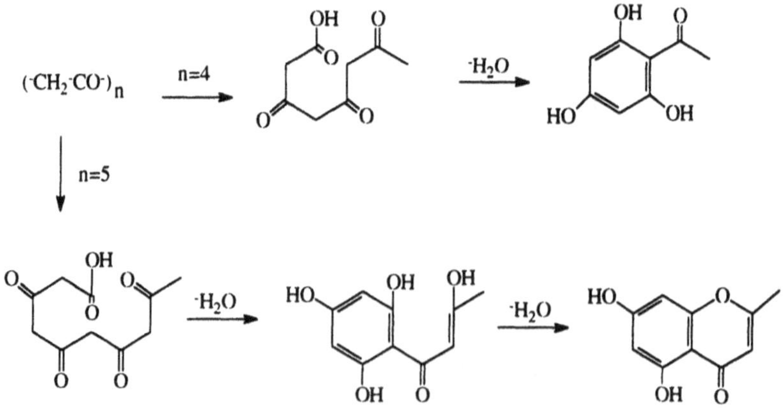

Fig. 8. Biosynthetic route to acetophenones and chromones from acetate.[40]

family Spathelioideae of the Rutaceae. Structural diversity in the chromones isolated from these taxa is relatively small; karenin (**46**) shows an atypical cyclisation of a hemiterpenoid substituent. The non-prenylated chromone derivative rohitukine (**47**) from *Amoora rohituka* (Meliaceae) is unique is possessing a piperidine substituent at C-8.[25] More recently, typical chromones have been reported from the sole species of the Rutaceae sub-family Dictyolomatoideae, *Dictyoloma incanescens*,[42] and from the typical rutaceous species *Skimmia laureola*.[43]

LIMONOIDS AND QUASSINOIDS

The limonoids and allied quassinoids are, together with the anthranilic acid derived alkaloids, the most characteristic metabolites of the Rutales. They have attracted considerable interest because of their fascinating structural diversity and their wide range of biological activity. The latter point is well borne out by the number of references made to them by other authors in this volume.

Limonoids and quassinoids share a common origin in the tetracyclic triterpenes tirucallane (**48**, 20-*S*, H-20a) and, more rarely, euphane (**48**, 20-*R*).[44] Modification takes the form of oxidation, often with ring fission and

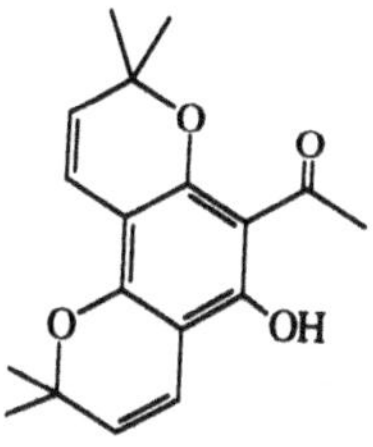

43

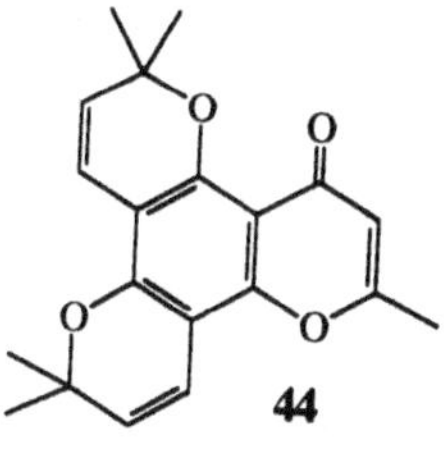

44

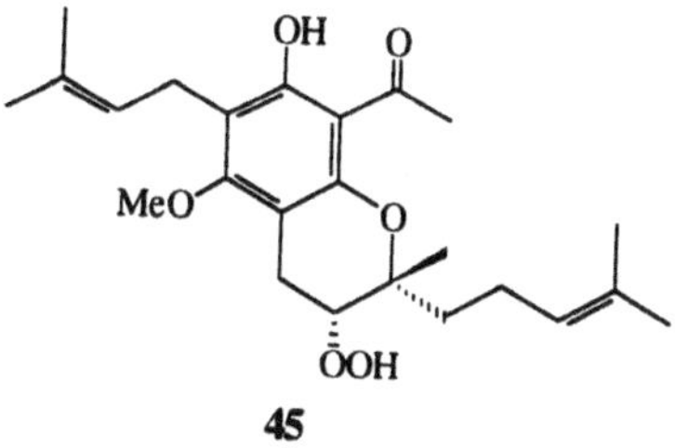

45

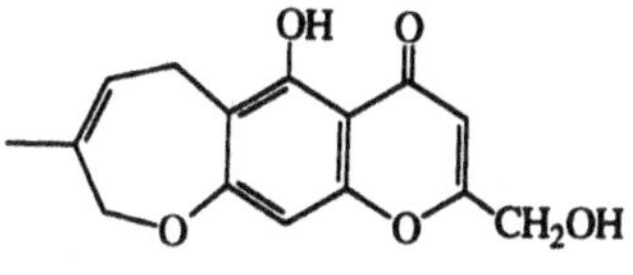

46

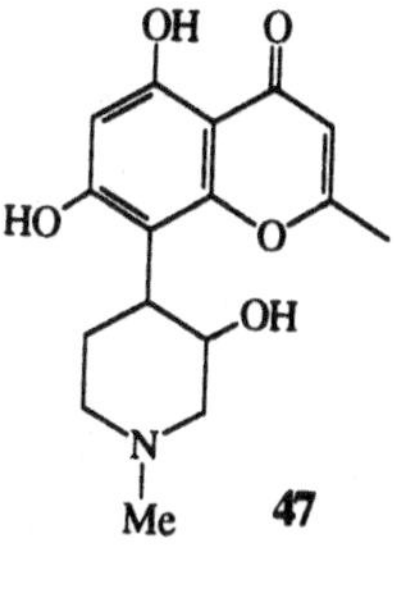

47

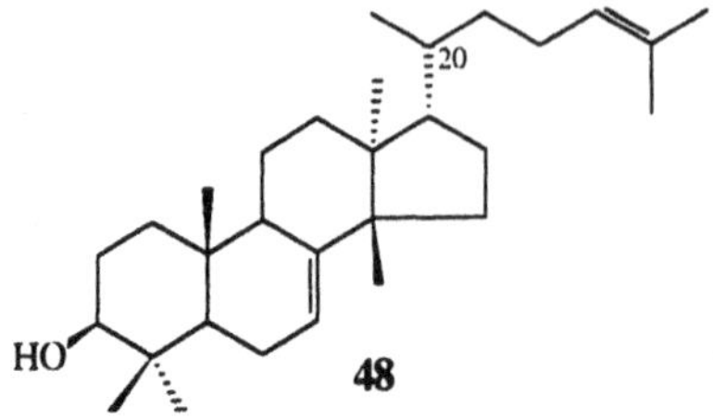

48

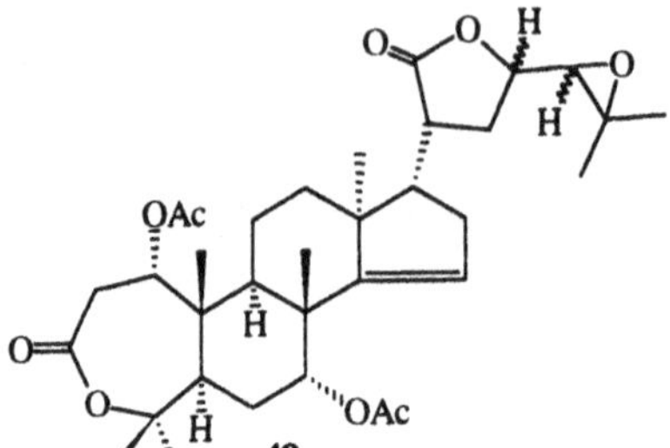

49

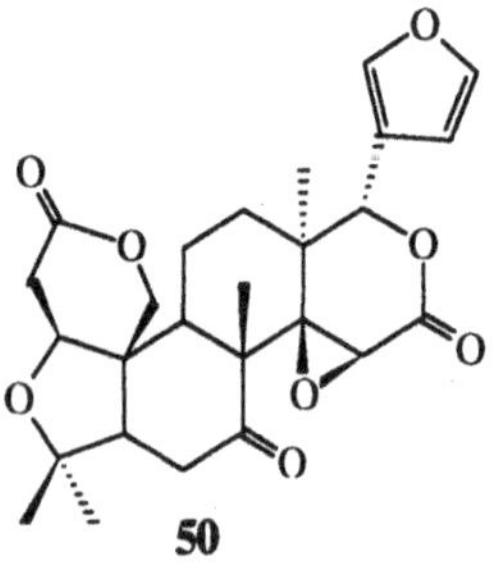

50

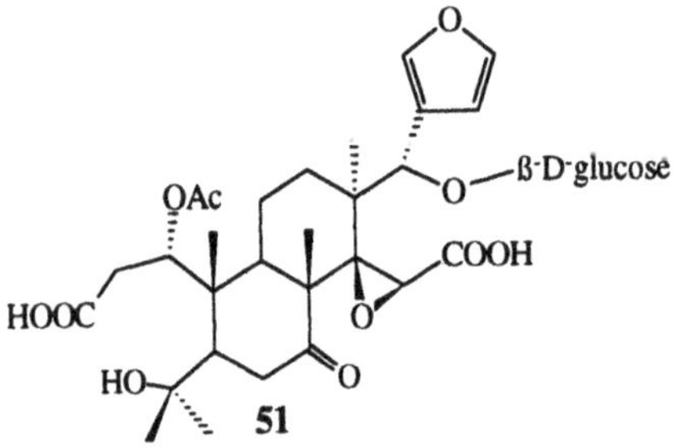

51

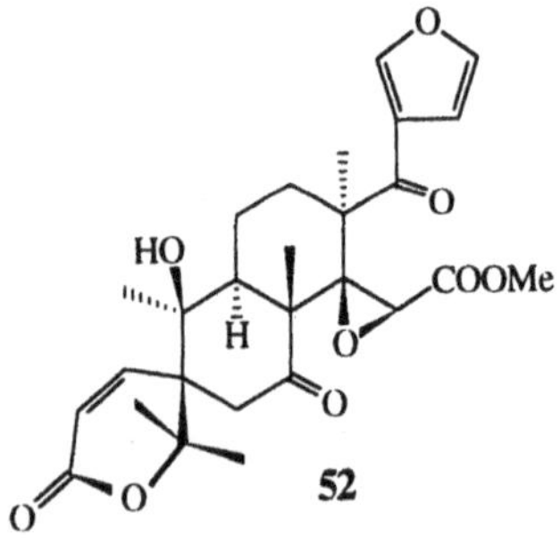

52

recyclization to give heterocyclic systems, and can involve the C-17 side chain and any of the four rings of the parent triterpene.

Changes in the side chain most commonly proceed through oxidation of the C-21 methyl and lead to a 21,23 or more rarely, 21,24 oxide. The protolimonoid phebaloparvilactone (**49**), recently reported from *Phebalium squamulosum* ssp. *parvifolium* (Rutaceae)[45] illustrates partial modification of the C-17 side-chain. After 21,23 cyclization there is subsequent loss of the four carbon unit C-24 to C-27 leading to the furan ring system that is found in most limonoids. Phebaloparvilactone (**49**) illustrates two further common modifications that are widely exploited in limonoid biosynthesis. The first of these is the so-called 'apo-rearrangement' brought about by oxidation of the 7/8 double bond and, through a shift of Me-30, leads to either a 14/15 double bond or, more rarely, the conversion of the C-18 methyl to a cyclopropane (Fig. 9). The second is the oxidative fission of ring-A between C-3 and C-4 with a subsequent reclosure to give the 7-membered A-ring. A similar oxidation at C-16 followed by fission and reclosure is also commonly found in ring-D in both Rutaceae and Meliaceae (Fig. 9).

The wide variety of structural types found among the limonoids is generally the product of further oxidative ring opening that may involve any of the rings of the original triterpene. In the Rutaceae there is commonly extensive modification of ring-A which, in limonin (**50**), involves oxidation of the 19-methyl and two cyclizations that require linking C-3 to C-19 and C-1 to C-4 through ether bridges (Fig. 10). Limonin is one of the major bitter principles of *Citrus* and like the flavonoids can have a major influence on the palatability of orange juice.[33] In recent years there have been increasing numbers of reports of limonin and related limonoids occurring as acids, with either or both ring-A and ring-D in the acyclic state. There is now evidence that glucosides, such as nomilinic acid 17-*O*-β-D-glucoside (**51**), are the most abundant limonoids in some *Citrus* tissues.[46] These compounds may be important in limonoid biosynthesis or transport and are reported to have only a slightly bitter taste.[46]

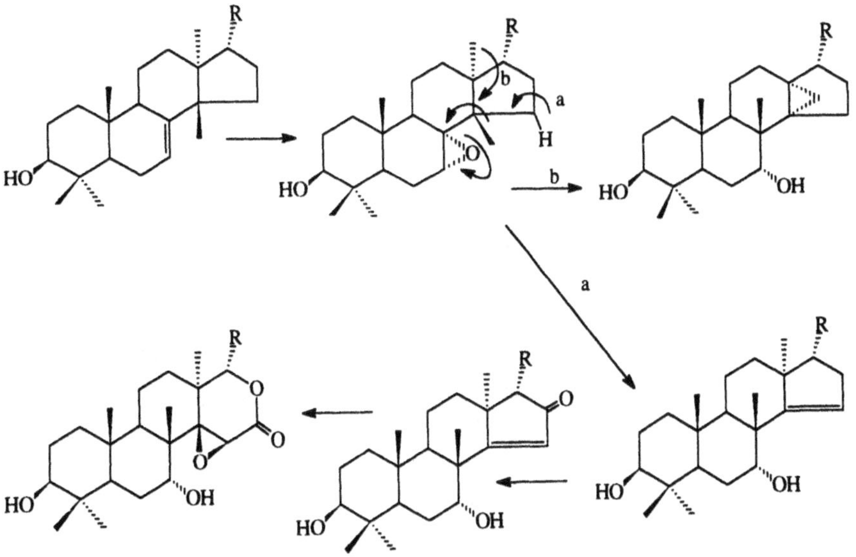

Fig. 9. 'Apo-rearrangement' and ring-D expansion in limonoids.

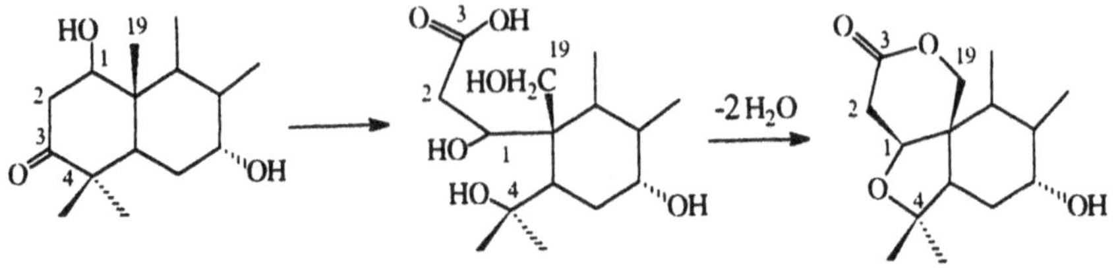

Fig. 10. Fission of ring-A and formation of limonin.

The limonin and nomilin types of limonoid are typical of the Rutaceae (including the Spathelioideae) and identical or very similar compounds such as pedonin (**52**) are known to occur in *Harrisonia* species,[47,48] which once more separates that genus, at least chemically, from the rest of the Simaroubaceae.

In the Meliaceae the degree of complexity in limonoid structures becomes greater with fission of rings B and C occurring widely as well as the A and D ring modifications found in the Rutaceae. Two examples that illustrate the extent to which such modifications can occur are rohitukin (**53**) and azadirachtin (**54**). Recently limonoids in which ring-B fission has occurred have also been recorded in the Rutaceae, in *Hortia*[2] and *Flindersia* (i.e., flindercarpin, **55**).[49] Finally, the small family Cneoraceae exhibits the most extensively modified group of limonoids in which a further carbon (C-16) has been lost and the D-ring has become a furan. Tricoccin-S_{14} (**56**) is a typical example.

The range of biological activity encountered in the limonoids has recently been reviewed.[50] Anti-feedant properties against insects have been widely studied and many limonoids exhibit activity. Of these, azadirachtin is the most well known and has been commercially exploited to some extent. C-*seco* limonoids appear to be the most active in currently used bioassays. A number of limonoids show appreciable *in vitro* activity in cytotoxicity assays but it has proved impossible to demonstrate useful *in vivo* activity. Limonoids also exhibit antifungal, antibacterial and antiviral effects and toxicity against a range of pathogenic protozoal organisms.

Reports of the occurrence of limonoids outside the Rutales *sensu stricto* have appeared twice in the literature. Cedrelone (**57**), first isolated from *Cedrela toona* (Meliaceae), has been reported from *Balsamodendron pubescens* of the Burseraceae.[51] This needs to be substantiated, but if true would only be a minor surprise as the Burseraceae are often linked taxonomically with the Rutales. The isolation of jangomolide (**58**) and coumarins from *Flacourtia jangomas*

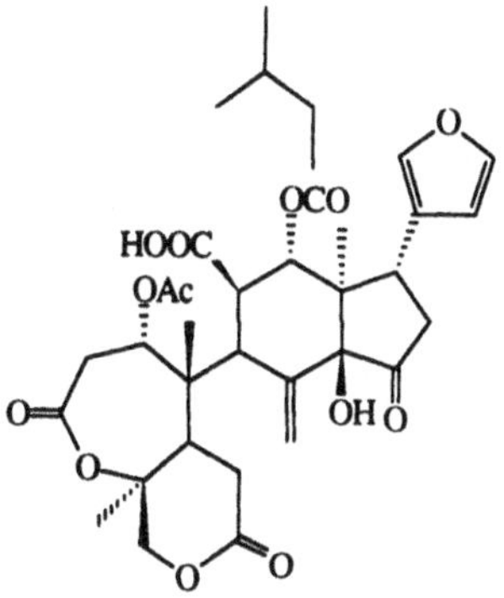

53

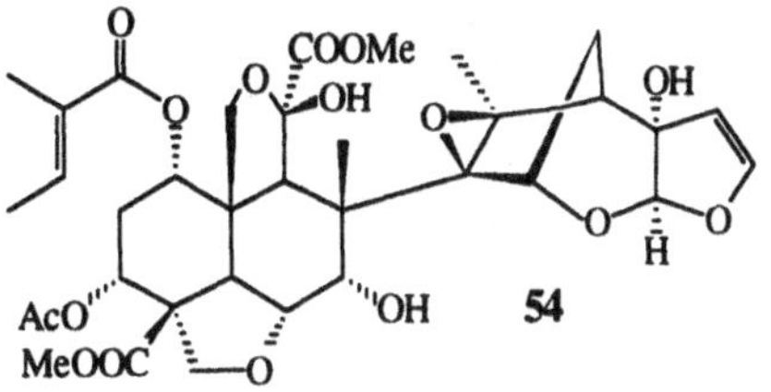

54

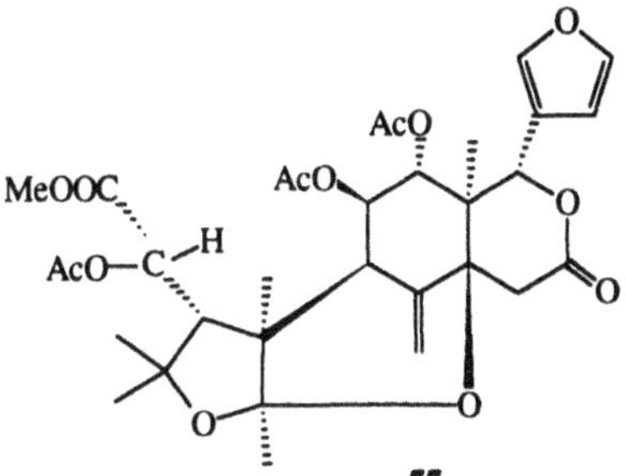

55

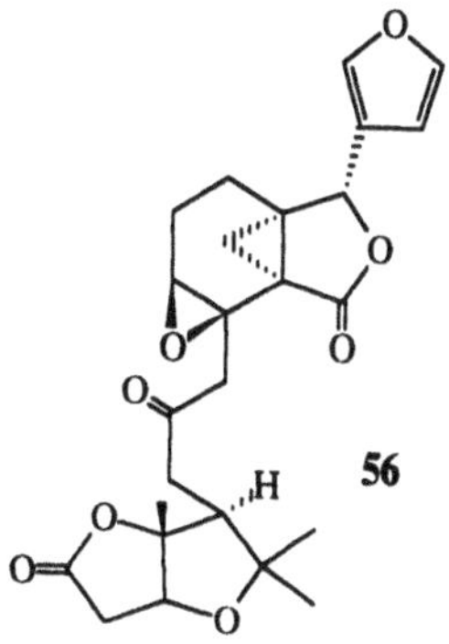

56

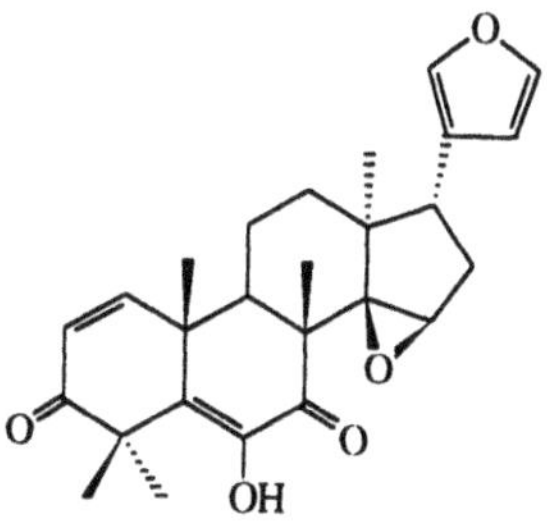

57

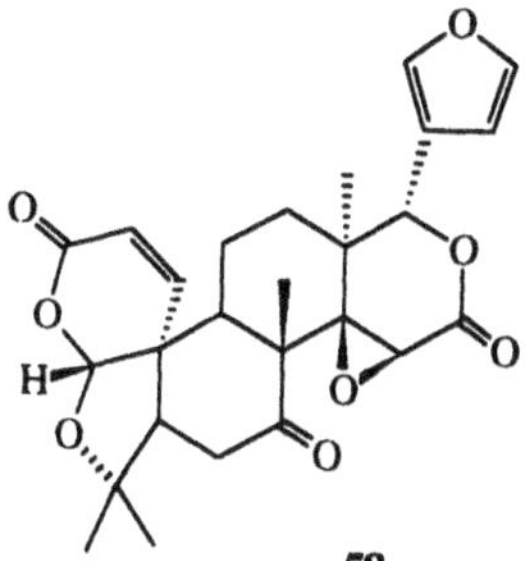

58

(Flacourtiaceae)[52] is rather more perplexing as this family has no obvious affinity to the Rutales. The co-occurrence of both of these typically rutalean groups of metabolites in *F. jangomas* represents an extraordinary coincidence that also requires further investigation to confirm the identity of the plant material involved and the correct placement of *F. jangomas* in the Flacourtiaceae.

Quassinoids are restricted to the Simaroubaceae *sensu stricto* where they replace other limonoids completely, except in the aberrant genus *Harrisonia* (which seems to lack quassinoids). The tetracyclic quassinoid skeleton is derived from a C_{26} protolimonoid in which ring-D has been oxidized to give the ring-opened form of the lactone. This then undergoes cyclization linking C-16 to the

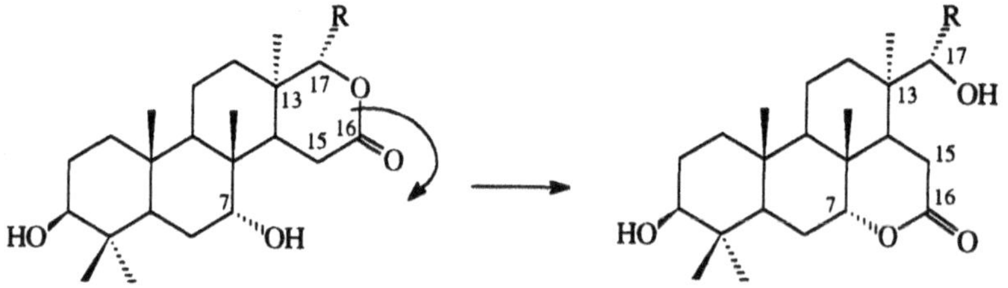

Fig. 11. Formation of the quassinoid nucleus.

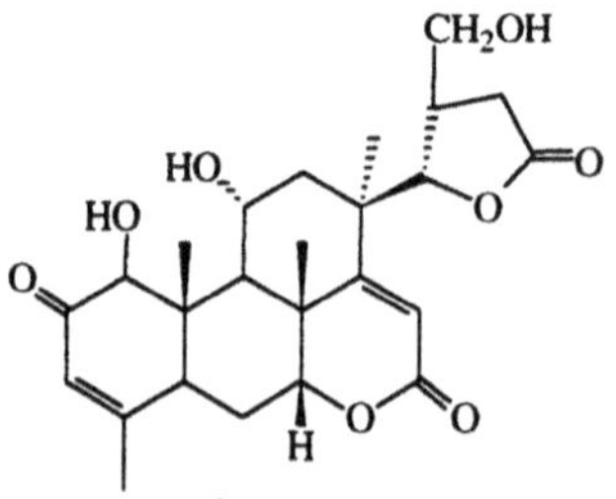

59

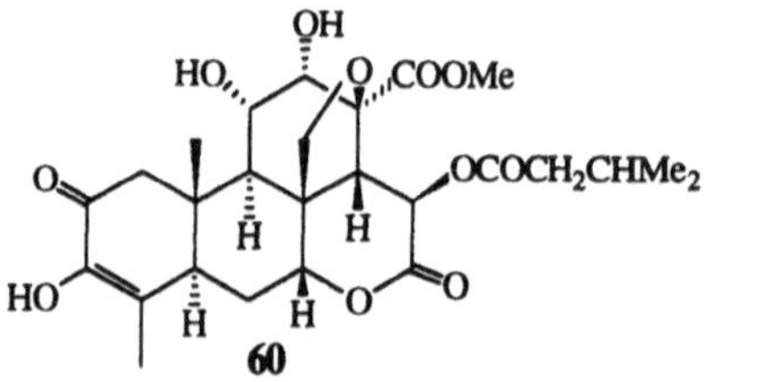

60

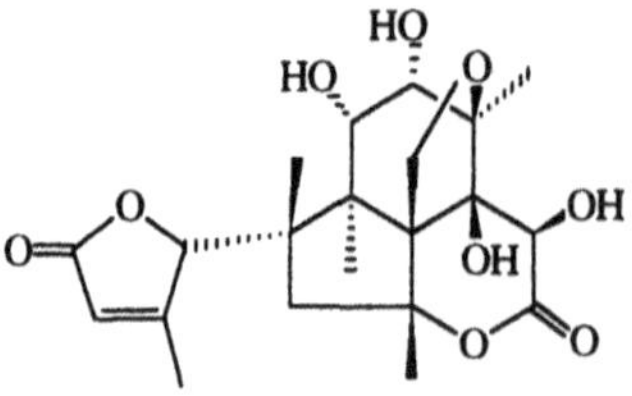

61

oxygen on C-7 (Fig. 11). The furan system together with C-17 is generally lost and there can be further modification, notably in ring-A, so that the bulk of quassinoids end up with C_{20}, C_{19} or even C_{18} skeletons.[53,54] Three typical quassinoids, in order of increasing modification, are soulameolid (**59**), which retains part of the original C-17 side-chain, bruceine-A (**60**), where the C-17 side-chain has been lost, and yadanziolid-D (**61**), where extensive A/B ring modifications have occurred. Quassinoids exhibit biological activity that is comparable to the limonoids but in terms of cytotoxic and, in particular, antimalarial and amoebicidal activity they are attracting greater interest.[55]

MISCELLANEOUS

Many species, notably in the Rutaceae, are rich in volatile oils. Mono- and sesquiterpenes and phenylpropenes are well represented in many of these products. In comparison, reports of diterpenes remain rare, although whether this is because of a lack of production or because they have not been specifically sought is not clear. Dimeric pimarane diterpenes are reported from *Dysoxylum* (Meliaceae) and labdane and clerodane derivatives from several species of both Rutaceae and Meliaceae.[2] The toxic picrotoxane group of compounds such as picrodendrin-F (**62**),[2,56] isolated from *Picrodendron* species (probably Simaroubaceae), are probably of diterpene origin.

Lignans are quite commonly found in the Rutales and cover a range of different structural types, but with nothing to date of any great novelty. Sulphur-containing compounds occur sporadically in canthinone and imidazole alkaloids (Rutaceae),[7] monoterpenes (Rutaceae),[1,2] polysulphides (Meliaceae),[7] and very recently,[57] in antifungal cinnamides (e.g. **63**).

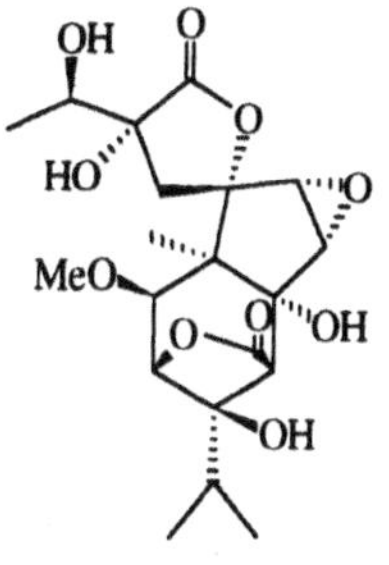

62

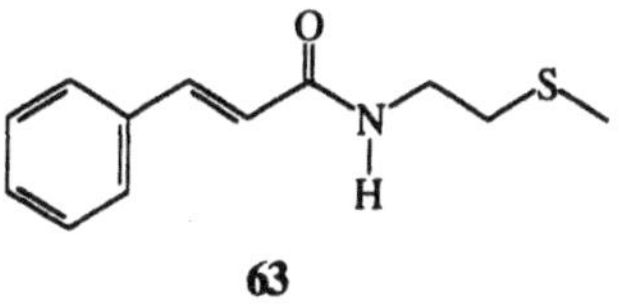

63

GENERAL COMMENTS

The Rutales *sensu stricto* are, numerically, not a particularly large taxonomic group yet they represent a prolific source of secondary metabolites of which most of the major classes are either wholly or largely confined to the Order or which, like the coumarins, reach a high degree of diversity within it.

There are three factors contributing to this diversity. First, the order appears to be in a state of flux for metabolites originating wholly or partly from the shikimic acid pathway. The employment of phenylalanine and tyrosine in the formation of alkaloids seems to be in the process of being supplanted by the use of phenylalanine in coumarin production in the Rutaceae and to have been discarded almost entirely in the Meliaceae and Simaroubaceae. However, nitrogenous secondary metabolites have continued to be produced in the Rutaceae and Simaroubaceae through the exploitation of anthranilic acid and, to a more limited extent, tryptophan (for which anthranilic acid is a precursor). The great alkaloid diversity of the Rutales comes through the 'experimentation' with a wide range of different substituting groups to convert these two amino acids into alkaloids.

The second factor is the evolution of limonoids and quassinoids, products of the acetate pathway. Whereas the individual steps involved in converting a tetracyclic triterpene into a limonoid, such as oxidation of methyl groups, oxidative ring fission or lactone formation, are all widespread and common in nature it is only in this order that they are all brought together and used in unison to produce this range of highly oxygenated compounds.

Third is the enormous additional capacity for diversity that arises from the extensive employment of the hemiterpene group. As noted previously, however, care must be used in interpreting what is real and what is an artifact.

Because of its biochemical diversity the order Rutales has attracted much attention. It would, however, be wrong to assume that this means that there is little left to be done. Many phytochemical studies have been very specifically directed at a single metabolic class. It is a safe assumption that even the most extensively worked species have not been investigated exhaustively. The order still has a great deal to offer.

REFERENCES

1. WATERMAN, P.G., GRUNDON, M.F. (eds.). 1983. Chemistry and Chemical Taxonomy of the Rutales. Academic Press. London.

2. HEGNAUER, R. 1991. Chemotaxonomie der Pflanzen, Vol. 9. Springer Verlag, Basel.
3. GRAY, A.I., WATERMAN, P.G. 1978. Coumarins in the Rutaceae. Phytochemistry 17:845-864.
4. WATERMAN, P.G. 1986. The dimeric alkaloids of the Rutaceae derived by Diels-Alder addition. In: Alkaloids: Chemical and Biological Perspectives, Vol. 4, (S.W. Pelletier, ed.), John Wiley and Sons, New York, pp. 331-387.
5. PARSONS, I.C., GRAY, A.I., WATERMAN, P.G., HARTLEY, T.G. 1992. A new triterpene and four new flavonoids from the leaves of *Bosistoa brassii*. J. Nat. Prod. In press.
6. GOMES, C.M.R., GOTTLIEB, O.R. 1980. Alkaloid evolution and angiosperm systematics. Biochem. Syst. Ecol. 8:81-87
7. MESTER, I. 1983. Structural diversity and distribution of alkaloids in the Rutaceae. In: Chemistry and Chemical Taxonomy of the Rutales (P.G. Waterman, M.F. Grundon, eds.), Academic Press, London, pp. 31-96.
8. GRUNDON, M.F. 1988. Quinoline alkaloids related to anthranilic acid. In: The Alkaloids, Vol. 30, (A. Brossi, ed.), Academic Press, New York, pp. 341-439.
9. GRAY, A.I. 1992. Quinoline alkaloids. In: Methods in Plant Biochemistry, Vol. 8, Nitrogen and Sulphur Containing Compounds (P.G. Waterman, ed.), Academic Press, London. In press.
10. GERZON, K., SVOBODA, G.H. 1983. Acridone alkaloids: Experimental antitumor activity of acronycine. In: The Alkaloids (A. Brossi, ed.), Vol. 21, Academic Press, New York, pp. 1-28.
11. McPHAIL, A.T., JU-ICHI, M., FUJITANI, Y., ONOUE, M., WU, T-S., Furukawa, H. 1985. Isolation and structure of citropone-A and -B from *Citrus* plants, first examples of naturally occurring homo-acridone alkaloids containing a tropone ring system. Tetrahedron Lett. 26:3271-3272.
12. PAULINI, H., WAIBEL, R., SCHIMMER, O. 1989. Mutagenicity and structure-mutagenicity relationships of furoquinolines, naturally occurring alkaloids of the Rutaceae. Mutation Res. 227:179-186.
13. YANG, M-S., WU, T-S., WANG, C-H. 1987. Citracridone-I: A new antispasmodic from root barks of *Citrus depressa*. Planta Medica. 143-147.
14. NG, K-M., BUT, P.P-H., GRAY, A.I., HARTLEY, T.G., KONG, Y-C., WATERMAN, P.G. 1987. The biochemical systematics of

Tetradium, *Euodia* and *Melicope* and their significance in the Rutaceae. Biochem. Syst. Ecol. 15:587-593.

15. CHOU, F.Y., HOSTETTMANN, K., KUBO, I., NAKANISHI, K., TANIGUCHI, M. 1977. Isolation of an insect antifeedant N-methylfindersine and several benzo[c]phenanthridine alkaloids from east African plants; a comment on chelerythrine. Heterocycles 7:967-977.
16. SOUTHARD, G.L. 1985. Plaque disclosing agent. U.S. Patent No. U.S. 4,517,172.
17. ARISAWA, M., PEZZUTO, J.M., BEVELLE, C., CORDELL, G.A. 1984. Potential anticancer agents XXXI. N-demethylation of fagaronine. J. Nat. Prod. 47:453-458.
18. ALADESANMI, A., KELLEY, C.J., LEARY, J.D. 1983. The constituents of *Dysoxylum lenticellare*. I. Phenylethylisoquinoline, homoerythrine and dibenzazecine alkaloids. J. Nat. Prod. 46:127-131.
19. OHMOTO, T., KOIKE, K. 1989. Canthin-6-one alkaloids. In: The Alkaloids (A. Brossi. ed.), Vol. 36, Academic Press, New York, pp. 135-170.
20. KONG, Y-C., BUT, P.P-H., NG, K-H., LI, Q., CHENG, K-F, WATERMAN, P.G. 1988. *Micromelum*: A key genus in the chemosystematics of the Clausenae. Biochem. Syst. Ecol. 16:485-489.
21. LONTSI, D., AYAFOR, J.F., SONDENGAM, B.L., CONNOLLY, J.D., RYCROFT, D.S. 1985. The use of two-dimensional long-range δ_H/δ_C correlation in conjunction with the one-dimensional proton-coupled ^{13}C NMR spectrum in the structural elucidation of eckeberginine, a new carbazole alkaloid from *Ekebergia senegalensis* (Meliaceae). Tetrahedron Letters 26:4249-4252.
22. KINOSHITA, T., TATARA, S., HO, F-C., SANKAWA, U. 1989. 3-Prenylindoles from *Murraya paniculata* and their biogenetic significance. Phytochemistry 28:147-151.
23. KONG, Y-C., NG, K-H., WAT, K-H., WONG, A., LAU, I-F., CHENG, K-F., BUT, P.P-H., CHANG, H-T. 1985. Yuehchukene—a novel anti-implantation indole alkaloid from *Murraya paniculata*. Planta Medica 304-307.
24. LUMONADIO, L., VANHAELEN, M. 1984. Indole alkaloids from *Hannoa klaineana* roots. Phytochemistry 23:453-455.
25. GRAY, A.I. 1983. Coumarins and chromones of the Rutales. In: Chemistry and Chemical Taxonomy of the Rutales (P.G. Waterman, M.F. Grundon, eds.), Academic Press, London, pp. 97-146.

26. QUADER, M.A., EL TURBI, J.A., ARMSTRONG, J.A., GRAY, A.I., WATERMAN, P.G. 1992. Coumarins and their taxonomic value in the genus *Phebalium*. Phytochemistry 31:3083-3089
27. RASHID, M.A., ARMSTRONG, J.A., GRAY, A.I., WATERMAN, P.G. 1991. Pyranocoumarins as chemotaxonomic markers in *Eriostemon coccineus* and *Philotheca citrina*. Phytochemistry 30:4033-4035.
28. RASHID, M.A., ARMSTRONG, J.A., GRAY, A.I., WATERMAN, P.G. 1991. 5,7-Dioxygenated coumarins from the aerial parts of *Geleznowia verrucosa* (Rutaceae). Biochem. Syst. Ecol. 19:698.
29. RASHID, M.A.,ARMSTRONG, J.A., GRAY, A.I., WATERMAN, P.G. 1992. Protobruceols: New 6-*C*-monoterpenyl-5,7-oxygenated coumarins from *Eriostemon brucei*. Nat. Prod. Lett. 79-84.
30. RASHID, M.A.,ARMSTRONG, J.A., GRAY, A.I., WATERMAN, P.G. 1992. Further new tetra- and pentacyclic 6-*C*-monoterpenyl-5,7-dioxycoumarins from *Eriostemon brucei* and *E. brucei* ssp. *cinereus*. Phytochemistry 31:3583-3588.
31. BERENBAUM M.R. 1991. Coumarins. In: Herbivores Their Interaction with Secondary Plant Metabolites (G.A. Rosenthal, M.R. Berenbaum, eds.), 2nd Edn., Vol. 1, Academic Press, New York, pp. 221-249.
32. ARISAWA, M., HANDA, S.S., McPHERSON, D.D., LANKIN: D.C., CORDELL, G.A., FONG, H.H.S., FARNSWORTH, N.R. 1984. Cleomiscosin A from *Simaba multiflora*, *Soulamea soulameoides* and *Matayba arborescens*. J. Nat. Prod. 47:300-307.
33. MAIER, V.P. 1983. Chemistry and significance of selected *Citrus* limonoids and flavonoids. In: Chemistry and Chemical Taxonomy of the Rutales (P.G. Waterman, M.F. Grundon, eds.), Academic Press, London, pp. 319-342.
34. KAMIYA, S., ESAKI, S., KONISHI, F. 1979. Flavonoids in *Citrus* hybrids. Agric. Biol. Chem. 43:1529-1536.
35. HARBORNE, J.B. 1983. The flavonoids of the Rutales. In: Chemistry and Chemical Taxonomy of the Rutales (P.G. Waterman, M.F. Grundon, eds.), Academic Press, London, pp. 147-173.
36. LEWIS, J.R. 1983. Biological activity of some rutaceous compounds. In: Chemistry and Chemical Taxonomy of the Rutales (P.G. Waterman, M.F. Grundon, Eds.), Academic Press, London, pp. 301-318.
37. WU, T-S. 1989. Flavonoids from root bark of *Citrus sinensis* and *C. nobilis*. Phytochemistry 28:3558-3560.

38. PARSONS, I.C., GRAY, A.I., HARTLEY, T.G., WATERMAN, P.G. 1992. Four further brombyins from the stem bark of *Brombya platynema* and comments in the biosynthesis of the brombyins. Phytochemistry In press.
39. BORGES del CASTILLO, J., SECUNDINO, M., RODRIGUEZ LUIS, F. 1986. Four aromatic derivatives from *Ruta angustifolia*. Phytochemistry 25:2209-2210.
40. QUADER, A., ARMSTRONG, J.A., GRAY, A.I., HARTLEY, T.G., WATERMAN, P.G. 1991. Chemosystematics of *Acradenia* and general significance of acetophenones in the Rutaceae. Biochem. Syst. Ecol. 19:171-176.
41. LIU, H-W., KUSUMI, T., NAKANISHI, T. 1981. A hydroperoxy-chroman with insect antifeedant properties from an African shrub. Characterization of fully substituted aromatic structures. Chem Commun. 1271-1272.
42. CAMPOS, A.M., KHAC, D.D., FETIZON, M. 1987. Chromones from *Dictyoloma incanescens*. Phytochemistry. 26:2819-2823.
43. RAZDAN, T.K., QADRI, B., HARKAR, S., WAIGHT, E.S. 1987. Chromones and coumarins from *Skimmia laureola*. Phytochemistry 26:2063-2069.
44. CONNOLLY, J.D. 1983. Chemistry of the limonoids of the Meliaceae and Cneoraceae. In: Chemistry and Chemical Taxonomy of the Rutales (P.G. Waterman, M.F. Grundon, Eds.), Academic Press, London, pp. 175-213.
45. QUADER, M.A., GRAY, A.I., WATERMAN, P.G., LAVAUD, C., MASSIOT, G., SADLER, I.H. 1991. Phebaloparvilactone: A new protolimonoid from the leaves of *Phebalium squamulosum* ssp. *parvifolium*. Tetrahedron. 47:3611-3620.
46. BENNETT, R.D., HASEGAWA, S., HERMAN, Z. 1989. Glucosides of acidic limonoids in *Citrus*. Phytochemistry. 28:2777-2781.
47. HASSANALI, A., BENTLEY, M.D., SLAWIN: A.M.Z., WILLIAMS, D.J., SHEPHARD, R.N., CHAPYA, A.W. 1987. Pedon. In: a spiro tetranortriterpenoid insect antifeedant from *Harrisonia abyssinica*. Phytochemistry 26:573-575.
48. BYRNE, L.T., TRI, M.V., PHUONG, N.M., SARGENT, M.V., SKELTON, B.W., WHITE, A.H. 1991. Perforatin: A novel tetranor-triterpenoid from *Harrisonia perforata*. Aust. J. Chem. 44:165-169.
49. BREEN G.W., PICKER, K., RASTIN: C.L., SKELTON, B.W., TAYLOR, W.C., WHITE, A.H. 1984. Chemical constituents of

Flindersia XXIII. Determination of the structure of flindercarpin-2 as methyl-(*S*)-(acetoxy){[3'*S*-(3α,3aα,4α,5β,6α,6aβ,7β,10aβ,11aβ)]-5',6'-di-acetoxy-7'-(furan-3"-yl)-2',2',3a',6a',11a'-penta-methyl-12'-methylene-9'-oxo-2',3',3a',4',5',6',6a',-9',10',11a'-decahydro-7'*H*-4',10a'-methano-10a'*H*-furo[2,3-*b*]pyrano[3,4-*g*]-oxocin-3'-yl}acetate. Aust. J. Chem. 37:1461-1471.

50. CHAMPAGNE, D.E., KOUL, O., ISMAN, M.B., SCUDDER, G.G.E., TOWERS, G.H.N. 1992. Biological activity of limonoids from the Rutales. Phytochemistry 31:377-394.
51. JOSHI, B.S., HEGDE, V.R. 1979. Extractives of *Balsamodendron pubescens*: Isolation and a new synthesis of siderin. Proc. Indian Acad. Sci. 88A:185-190.
52. AHMAD, J., WIZARAT, K., SHAMSUDDIN: K.M., ZAMAN, A., CONNOLLY, J.D. 1984. Jangomolide, a novel limonoid from *Flacourtia jangomas*. Phytochemistry 23:1269-1270.
53. POLONSKY, J. 1983. Chemistry and biological activity of the quassinoids. In: Chemistry and Chemical Taxonomy of the Rutales (P.G. Waterman, M.F. Grundon, Rds.), Academic Press, London, pp. 247-266.
54. SIMAO, S.M., BARREIROS, E.L., da SILVA, M.F.G.F., GOTTLIEB, O.R. 1991. Chemogeographic evolution of quassinoids in Simaroubaceae. Phytochemistry 30:853-865.
55. PHILLIPSON, J.D., O'NEILL, M.J. 1987. Antimalarial and amoebicidal natural products. In: Biologically Active Natural Products (K. Hostettmann, P.J. Lea, eds.), Clarendon Press, Oxford, pp. 49-64.
56. KOIKE, K., OHMOTO, T., KAWAI, T., SATO, T. 1991. Picrotoxane terpenoids from *Picrodendron baccatum*. Phytochemistry 30:3353-3356.
57. GREGER, H., HOFER, O., KAHLIG, H., WURZ, G. 1992. Sulfur containing cinnamides with antifungal activity from *Glycosmis cyanocarpa*. Tetrahedron. 48:1209-1218.

Chapter Ten

PLANT CELL CULTURES AND SYNTHETIC CHEMISTRY—ROUTES TO CLINICALLY IMPORTANT COMPOUNDS

James P. Kutney

Department of Chemistry, University of British Columbia
2036 Main Mall, Vancouver
British Columbia, Canada, V6T 1Z1

INTRODUCTION

The plant kingdom has provided a wide variety of natural products with diverse chemical structures and a vast array of biological activities, many of which have found applications in the health sciences. For years, synthetic chemists have had the challenge of developing syntheses of such components, but often due to structural complexity the resulting multi-step syntheses rarely find application in the large scale production required in the pharmaceutical drug industry. As a result, starting materials for such drug production, or indeed the final clinical drug, are frequently obtained from tedious and often costly extraction from the living plant. This latter solution is often fraught with well known problems: a) an active agent is present in minute amounts in the plant extract; b) separation of the target compound may be difficult and, in turn, expensive; c) varying concentrations of the target compound depending on seasons during which plant collection is performed; d) desired plant species growing in geographically or politically inaccessible regions, etc.

Appropriate solutions to at least some of the above difficulties are possible by the use of plant cell culture methodology, particularly when such studies are coupled with chemical methods. The advantages of plant cell cultures over living plants in terms of secondary metabolite production are clear: a) growth conditions are laboratory controlled, therefore, reproducible yields of end product are achieved; b) growth parameters such as pH, changes in nutrient media, temperature, etc. can be optimized to achieve metabolite production in yields significantly higher than in the living plant; c) separation of target compounds is much easier due to lower complexity of extract; d) cloning of cell lines provides further optimization for end product yields; e) plant cell cultures are an excellent source of enzymes, much superior to living plants where isolation often leads to enzyme denaturation. With enzyme availability, the opportunity to perform biosynthetic and/or biotransformation experiments related to metabolite production, is clear.

It must be emphasized at the outset that the effectiveness of such a program depends entirely on an interdisciplinary "team" of scientists comprised of the "culture development" group (biologists, biochemists) and the "chemical" group, the latter isolating and characterizing the metabolites via spectroscopic methods coupled with synthesis as required.

The present discussion summarizes studies performed in the author's laboratories in several areas in order to portray the various avenues of research that an interdisciplinary program can achieve.

AVENUES OF RESEARCH

Our research program can be conveniently divided into four areas; the discussion following will briefly present studies within these areas. 1) Studies of the biosynthesis and application of biosynthetic information toward development of efficient syntheses of clinical drugs; 2) Use of plant tissue culture and enzymes derived therefrom, as "reagents" in organic synthesis. Whole cells, crude enzyme extracts or immobilized enzyme systems, can be employed; 3) Use of plant tissue culture to produce higher levels of plant derived-natural products and provide novel compounds for pharmacological screening; 4) Use of well developed cell lines that enable one to separate pharmacological activities exhibited by the complex mixtures that are generally employed in herbal medicine applications.

Studies of the Biosynthesis and Application of Biosynthetic Information Toward Development of Highly Efficient Syntheses of Clinical Drugs

Plant cell cultures can provide excellent sources of relevant enzymes for unravelling biosynthetic pathways and providing information for subsequent development of efficient chemical syntheses of target compounds. The methodology for isolation of the required enzymes follows a procedure outlined in Figure 1. This exemplifies enzyme isolation from cell cultures of *Catharanthus roseus*, the well known plant from which the anti-leukemic drugs, vinblastine (3) and vincristine (NCH3 replaced by NCHO) are presently obtained. We employed this methodology to isolate enzymes from cell culture lines of *Tripterygium wilfordii*, *Podophyllum peltatum*, and *Nicotiana sylvestris*. A brief

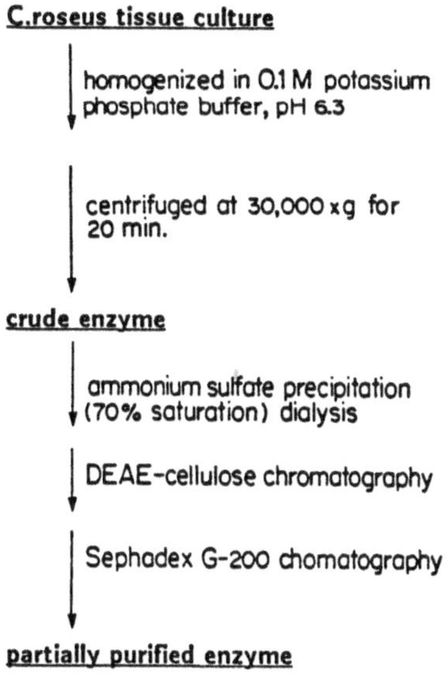

Fig. 1. Isolation of enzymes from cell culture of *C. roseus*.

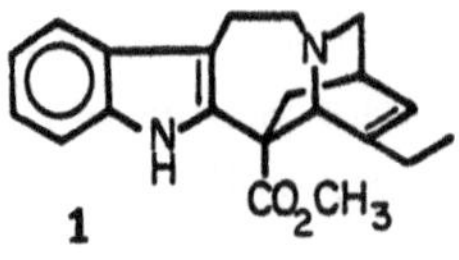

1
catharanthine

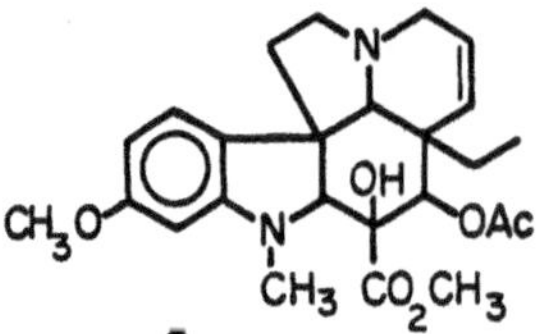

2
vindoline

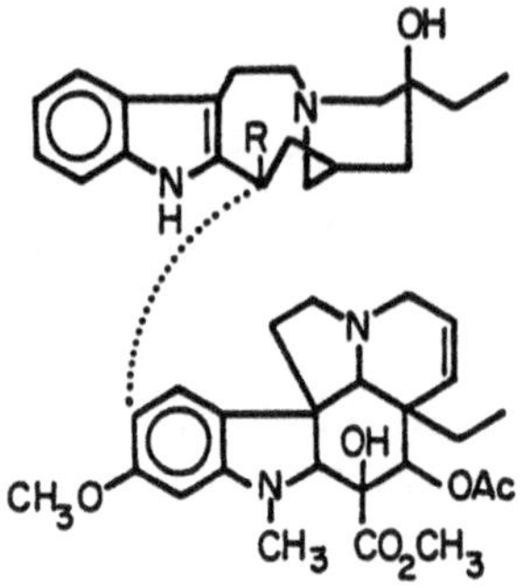

3 R = CO_2CH_3
vinblastine

mention will be made of a highly advanced program in which biosynthetic information derived from experiments with cell cultures of *C. roseus* have formed the basis of an efficient synthesis of the alkaloid vinblastine (**3**).

Detailed studies with crude enzyme preparations (cell free extracts) (Fig. 1) obtained from various ages of *C. roseus* cultures have provided information relating to the manner of enzymatic coupling of catharanthine (**1**) and vindoline (**2**), two major alkaloids in the plant that are presently discarded as waste by-products during extraction of the clinical drugs from the plants. Further studies established how the initially coupled intermediate was subsequently enzymatically elaborated to the vinblastine-vincristine series. Application of this biosynthetic information has allowed us to develop a highly efficient "one-pot" process for the syntheses of **3** and, in turn, vincristine from the waste by-products **1** and **2**. This process of commercial importance has been patented. This program, spanning over more than 10 years, has been the subject of approximately 20 publications and is summarized in several recent reviews.[1-4]

Use of Plant Cell Cultures and Enzymes Derived Therefrom as Reagents in Organic Synthesis

Enzymes derived from plant cell cultures can be utilized as "reagents" in selected biotransformation experiments with synthetic precursors so as to achieve efficient routes to certain clinical drugs. The strategy depends on the postulate that these enzyme systems are sufficiently versatile to accept "foreign" precursors, that is, substrates not normally produced in the plants from which the cultures are derived. Biotransformations of such precursors to desired end

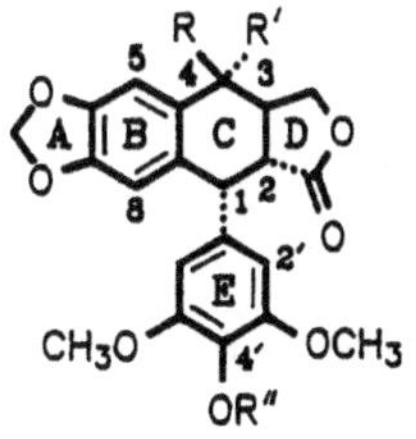

4

Podophyllotoxin : R=H; R' =OH; R'' =CH_3

Epipodophyllotoxin : R=OH; R' =H; R'' =CH_3

Deoxypodophyllotoxin : R=R' =H; R'' =CH_3

4' –Demethylpodophyllotoxin : R=H; R' =OH; R'' =H

4' –Demethylepipodophyllotoxin : R=OH; R' =H; R'' =H

Etoposide : R=CH_3 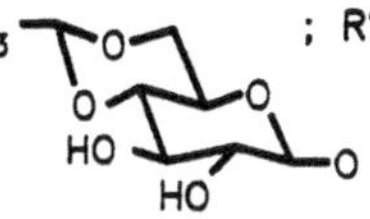 ; R' =H; R'' =H

Fig. 2. The podophyllotoxin family of compounds.

products can be achieved in short term incubation studies and generally in respectable yields. The substrates subjected to such enzymatic transformations can be natural products or synthetic compounds depending on the overall objectives of the research program. We have established that plant cell culture derived enzyme systems possess considerable versatility in biotransforming a variety of substrates, differing widely in chemical structures, to desired target compounds. Experiments employing two cell culture lines *C. roseus* and *Podophyllum peltatum*, and involving substrates relating to the podophyllotoxin family, will be discussed.

The podophyllotoxin family of natural products (**4**, Fig. 2) has received recent considerable attention from both a chemical[5,6] and a pharmacological viewpoint due to the development of the clinical anti-cancer drug etoposide.[7] This drug is effective in the treatment of myelocytic leukemia, neuroblastoma, bladder, testicular, and small-cell lung cancers. The process, under consideration for commercial production, starts with podophyllotoxin, which is chemically converted to 4'-demethylepipodophyllotoxin (**5**), and the latter is subsequently transformed into the clinical drug by the sequence shown in Figure 3.

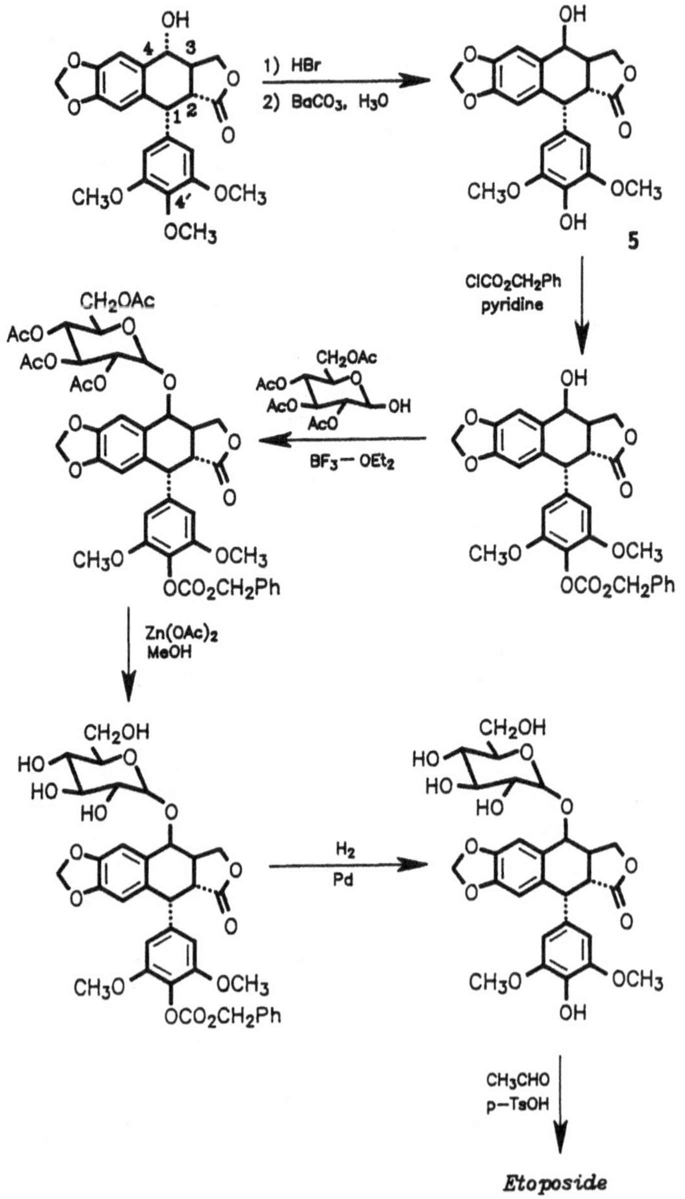

Fig. 3. Commercial production of etoposide from podophyllotoxin.

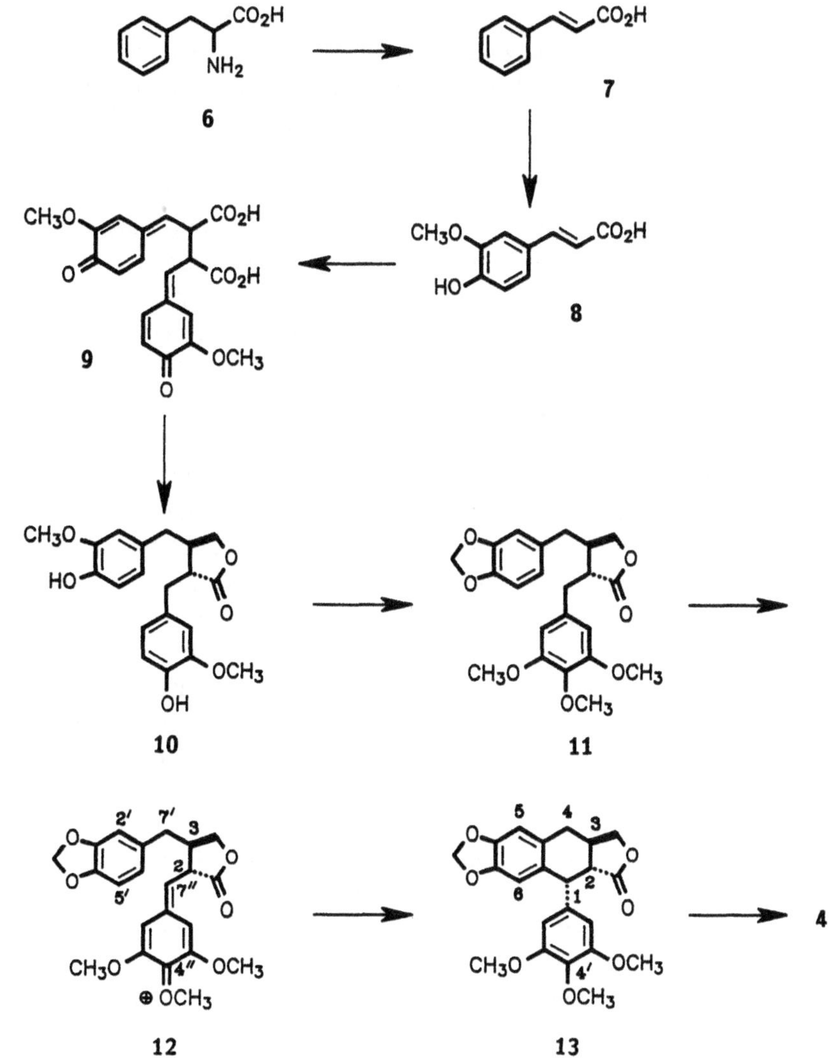

Fig. 4. Proposed biosynthetic pathway of the podophyllotoxins.

Whereas the basic starting material is presently derived from a plant source, *Podophyllum peltatum*, we used an alternative approach that combines synthetic chemistry with plant cell culture methodology. This affords a route to etoposide which need not rely on the living plant as a source of the starting material. The strategy selected was based on the biosynthetic pathway which had been postulated for the podophyllotoxin family. Broomhead *et al.*,[8] on the basis of extensive experiments involving various *Podophyllum* species of plants, has postulated the pathway shown in Figure 4. Although the exact mechanism for the late stages of this pathway remains unclear, a ring closure of an appropriately substituted dibenzylbutanolide, for example, **11**→**12**→**13**, is involved. One possible alternative for such a sequence involves the well known phenol oxidative coupling (originally proposed by D.H.R. Barton) for phenolic natural products. Since such processes are generally considered to involve "peroxidases" within the plant system, it seemed appropriate to consider such enzyme systems in our study.

In earlier experiments with *C. roseus* derived enzymes (Fig. 1) to unravel the possible biosynthetic intermediates for vinblastine-vincristine (**1-3**), we showed that our *C. roseus* cell cultures afford an excellent source of "peroxidases," and it was of interest whether these enzymes could be employed as "reagents" in biotransformation of dibenzylbutanolides of the type shown in Figure 4 to the desired podophyllotoxin system. If successful, such experiments would also reveal enzymatic versatility with plant derived systems in being capable of utilizing "foreign" precursors as appropriate substrates.

A requirement in such a study involved the development of a versatile synthetic route to the requisite dibenzylbutanolides from commercially available starting materials. Figure 5 summarizes the completion of the essential objectives. The route shown employs readily available aldehydes (**14**) (R'=R"=H; R'=H, R"=alkyl; R', R"=methylenedioxy). Excellent overall yields of the corresponding precursors of general structure (**19**) (R'=R"=H; R"=CH_3 or isopropyl, R'=H; R', R"=methylenedioxy) were obtained.

When the synthetic intermediate **16** was treated with Raney nickel, desulfurization was achieved and the precursors, for example, **20**, corresponding to the deoxypodophyllotoxin series, became available. Since Japanese

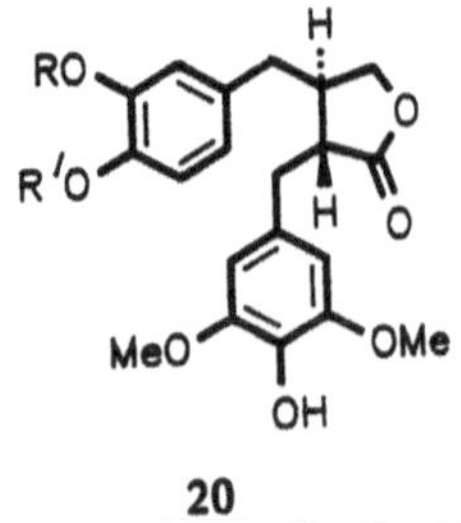

20

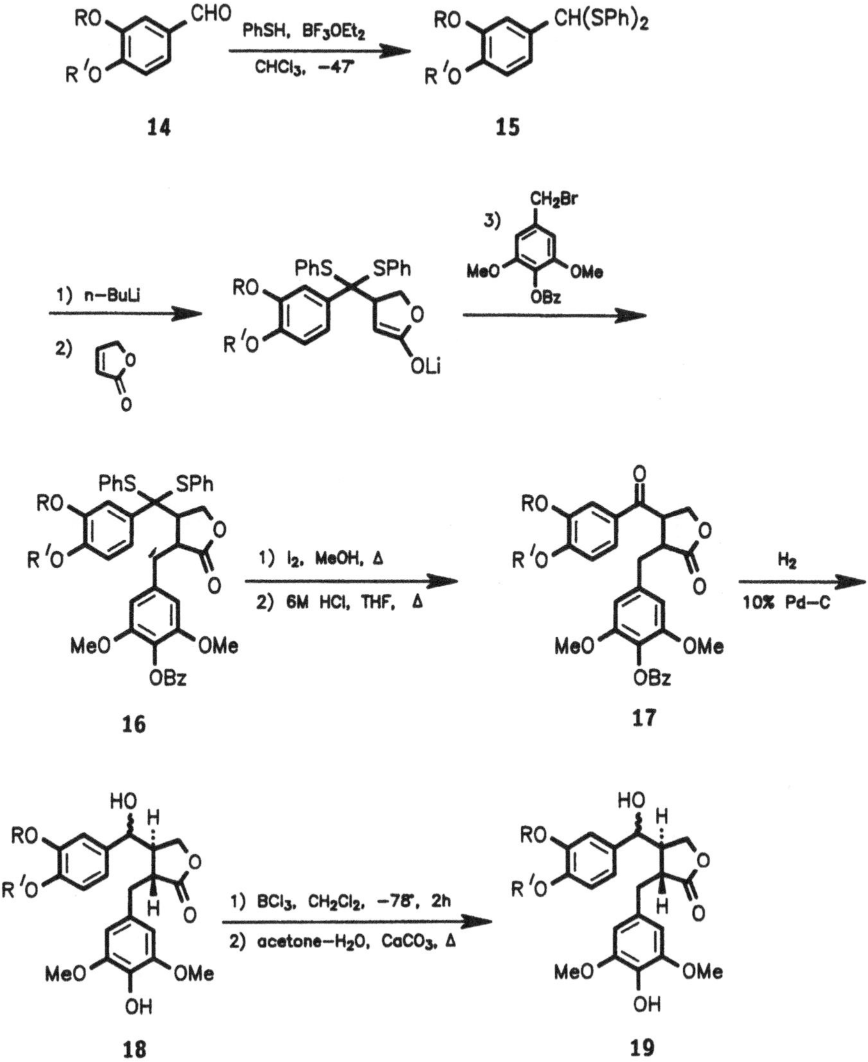

Fig. 5. Synthesis of a 4'-dimethylepipodophyllotoxin precursor.

workers[9] had demonstrated the conversion of deoxypodophyllotoxin to podophyllotoxin by the use of *Penicillium* strains, any successful cyclization of **20** by our plant cell culture derived enzymes would prove of interest in relation to etoposide production.

The initial studies with the deoxy series (general structure **20**) and the crude enzyme preparation obtained from *C. roseus* cell culture (Fig. 1) are now briefly described. A large number of experiments with substrate **21** (Fig. 6) and the crude enzyme preparation (Fig. 1) obtained from our *C. roseus* cell line (AC3) were conducted in order to achieve optimum conditions for the cyclization of **21** to podophyllotoxin analogues, for example, **22**. The reaction parameters

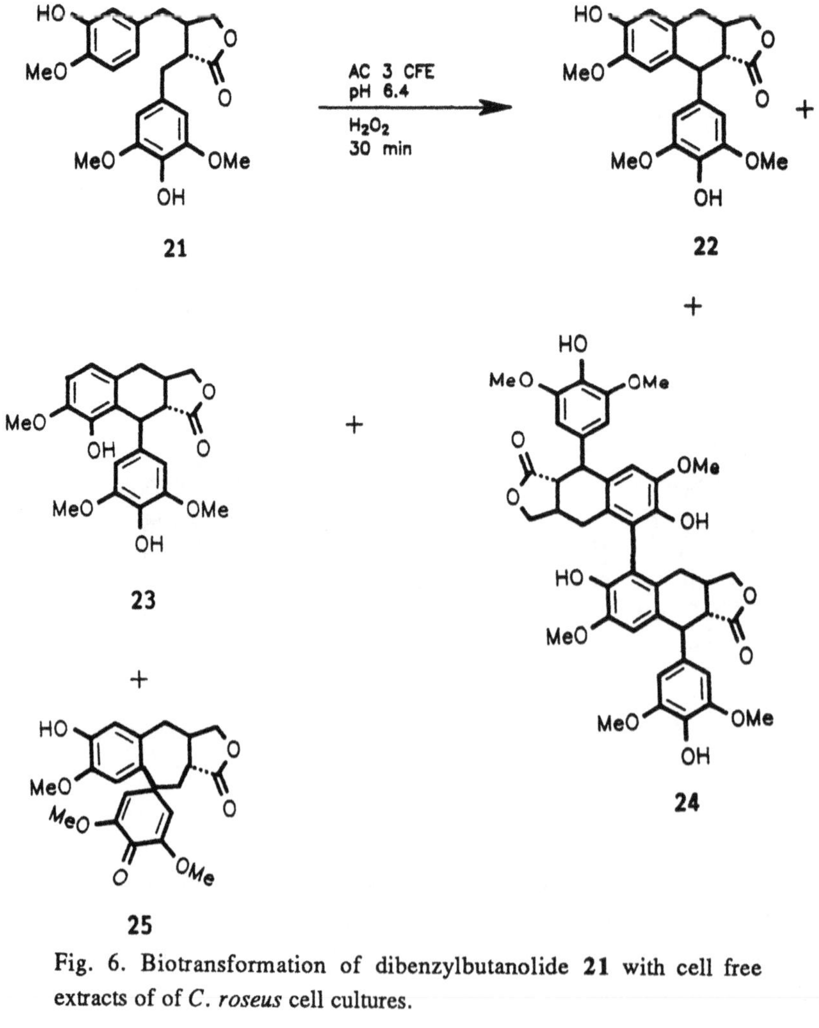

Fig. 6. Biotransformation of dibenzylbutanolide **21** with cell free extracts of of *C. roseus* cell cultures.

(pH, H_2O_2, units of peroxidase enzyme per mmol substrate, reaction time) were evaluated and a summary of the data is provided in Figures 7-11.[12]

One of the important parameters was the pH of the buffer (pH_h) used to prepare the crude enzyme and its effect on the enzymatic activity. Figure 7 summarizes the changes in peroxidase activity and specific activity of the crude enzyme with the pH of the buffer employed. Peroxidase activity and protein concentration measurements were determined according to established procedures.[10,11] A value of 6.3 appeared to be a good compromise between high peroxidase activity and specific activity.

The ratio of peroxidase to substrate was determined using an amount of 2.0 molar equivalents of hydrogen peroxide and conducting the reaction at pH 6.3. As shown in Figure 8, substrate **21** was very nearly consumed within 15 min when crude enzyme corresponding to 400 units peroxidase per millimole of substrate was used. Increasing the peroxidase:substrate ratio or the reaction time (180 min, Figure 9) failed to completely convert **21** to **22**. Instead, the sole effect was to promote the reaction of aryltetralin **22** to **24** (Fig. 6). It was concluded that a peroxidase:substrate **21** ratio of approximately 250 units per millimole of butanolide **21** would give optimum yields of aryltetralin **22**. Higher values promoted dimerization, whereas lower values resulted in incomplete conversion of substrate **21**.

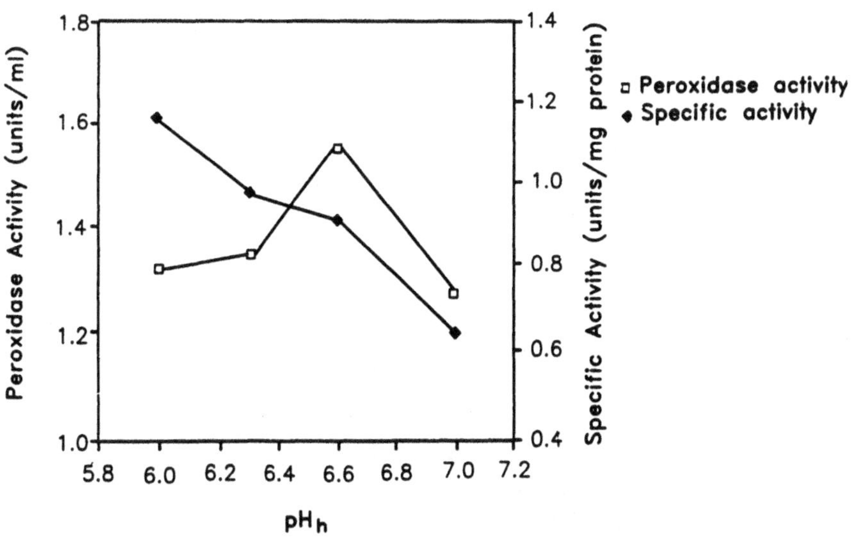

Fig. 7. Effect of pH_h on peroxidase activity and specific activity.

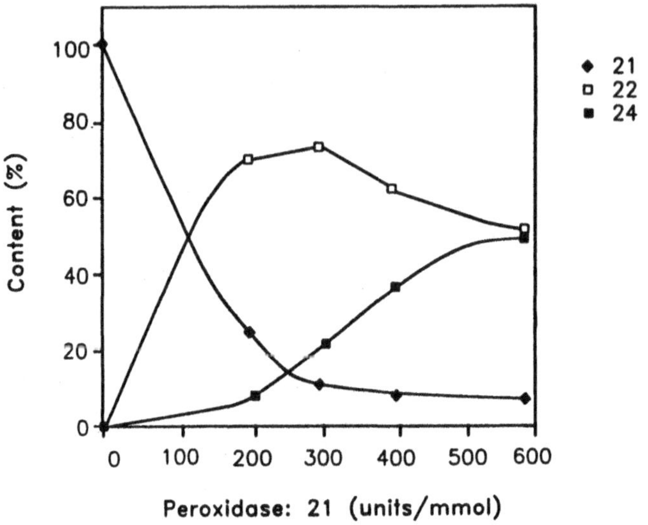

Fig. 8. Effect of peroxidase:substrate ratio on biotransformation on **21**. Reaction time = 15 min.

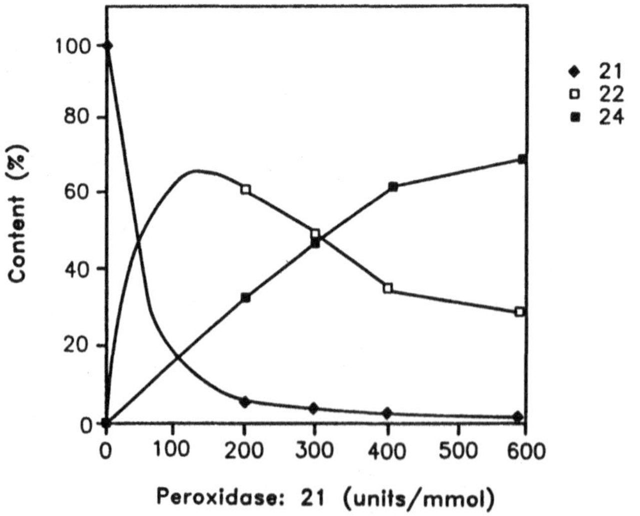

Fig. 9. Effect of peroxidase:substrate ratio on biotransformation on **21**. Reaction time = 180 min.

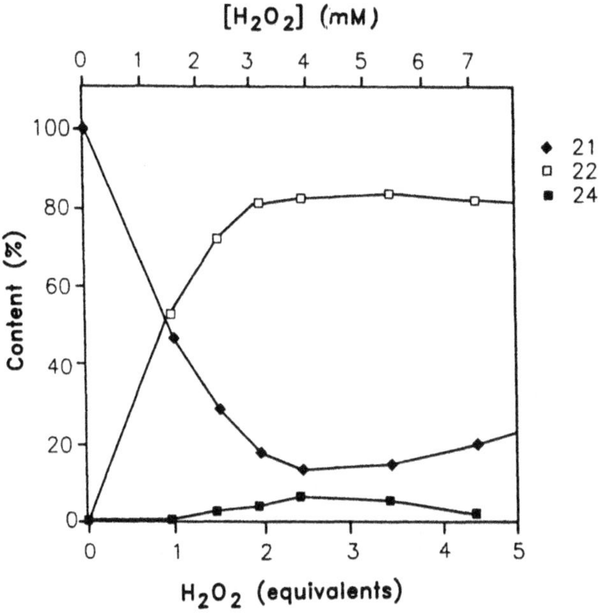

Fig. 10. Effect of hydrogen peroxide concentration on biotransformation of **21**. Reaction time = 15 min.

The optimum hydrogen peroxide:substrate ratio was determined using 250 units peroxidase per millimole of **21** in buffer at pH 6.3. In the absence of hydrogen peroxide, no reaction occurred after 15 min (Fig. 10). A similar result was noted when the reaction was allowed to proceed for 180 min. The yield of **22** and **24** then increased with hydrogen peroxide concentration but remained approximately constant at about 80% when more than 2.0 molar equivalents (3.1 nM) were present. However, these results clearly did not represent the exclusive requirements of the oxidative coupling reactions of **21** and **22** in view of the ability of the crude enzyme alone to consume hydrogen peroxide. This property was demonstrated when the enzyme was stirred for 15 min with hydrogen peroxide *prior* to precursor addition. The yield of **22** was only 28% after one hr (vs 64% for the control, when precursor **21**, hydrogen peroxide and crude enzyme were added simultaneously), but was increased sharply to 50% 1 h after a further 2.0 equivalents of hydrogen peroxide were added. These competing processes may have involved the catalytic decomposition of hydrogen peroxide (catalyzed by peroxidase, catalase or metal ions) or the oxidation of cell material.

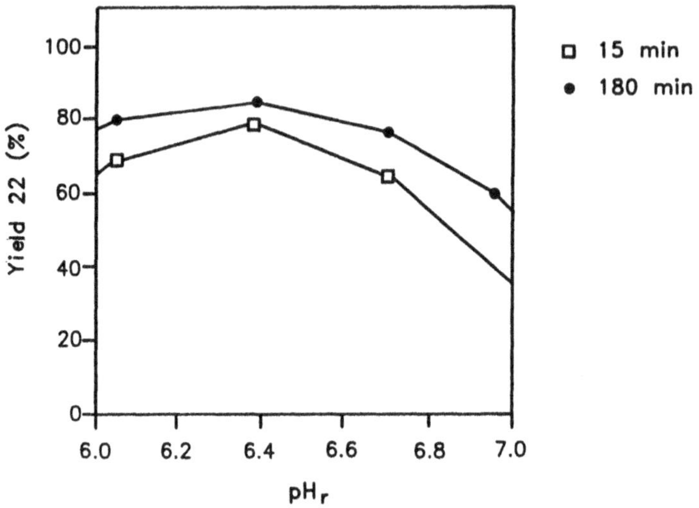

Fig. 11. Effect of pH_r on biotransformation of **21**.

It was concluded that 2.0 molar equivalents of hydrogen peroxide would give optimum yields of **22**. Lower values resulted in incomplete biotransformation of **21** while higher ratios promoted the dimerization of **22** to **24**. There appeared to be a distinct decrease in enzyme activity when more than 3 equivalents of hydrogen peroxide were used.

The optimum pH for biotransformation (pH_r) of **21** to **22** could now be reliably determined with the establishment of the best peroxidase:substrate and peroxide:substrate ratios. Up to this point, a pH of 6.3 had been found to give good results, but it was clearly desirable to identify the optimum value of pH_r more closely.[12] This value (Fig. 11) was found to be about 6.3 based on the yield of **22**, with the optimum conditions established previously. The yield of **22** showed only minor variations over the range pH 6.0-6.6, but decreased markedly at more basic values. When the crude enzyme solution was stored at 4°C, it was quite stable with respect to the biotransformation of **21**. The age of the crude enzyme preparation was thus not a significant factor in the reproducibility of individual experiments.

In summary, the biotransformation of **21** to **22** could be accomplished most efficiently using the conditions described in Table 1. The yields of **22** were consistently 70% or higher in the subsequent larger scale experiments (generally

Table 1. Optimum conditions of *C. roseus* enzyme catalyzed biotransformation of butanolide **21** to aryl tetraline **22** [a]

pH_h	6.3
pH_r	6.3
Hydrogen peroxide (equivalents)	2.0
CFE (units peroxidase per mmol 21)	250
Reaction time (min)	180

[a] pH_h = optimum pH for preparation of extract and for enzymatic activity. (Fig. 7)
pH_r = optimum pH for biotransformation of **21** to **22** (Fig. 11)

1 g of **21**) and the dimer **24** could be completely suppressed. In these latter studies, two minor products **23** (0.53% yield) and the spirodienone **25** (1% yield) could also be detected. It was now clear that the enzymes of *C. roseus* cell cultures, originally developed for the production of indole alkaloids, could tolerate "foreign" substrates of the dibenzylbutanolide family and afford high yields of the desired podophyllotoxin analogues.

The above biotransformation of **21** to **22** in our cell culture extracts was compared with that of commercial horseradish peroxidase. The reaction mixture was complex and the best yields of **22** were 15-19%, thereby revealing that the *C. roseus* derived enzymes were far superior for the present study.

Extension of the above studies to suitable dibenzylbutanolides, which upon cyclization would afford podophyllotoxin analogues possessing the required C(4) hydroxyl function, was now undertaken. For this purpose, compound **26** (Fig. 7), available via the chemical route outlined in Figure 5 and starting with aldehyde **14** (R'=H; R''=isopropyl) was considered. The expected cyclized product **27** (Fig. 12) could be converted through established chemical methods to the C(6)-C(7) dihydroxy system, and, in turn, to the methylenedioxy functionality required for the crucial starting material **5** (Fig. 3) for etoposide synthesis. Indeed, when **26** was subjected to the biotransformation conditions with the crude enzyme preparation obtained from *C. roseus*, a 84% yield of **27** was obtained. Compound **27a** (Fig. 7) was not isolated in these studies.

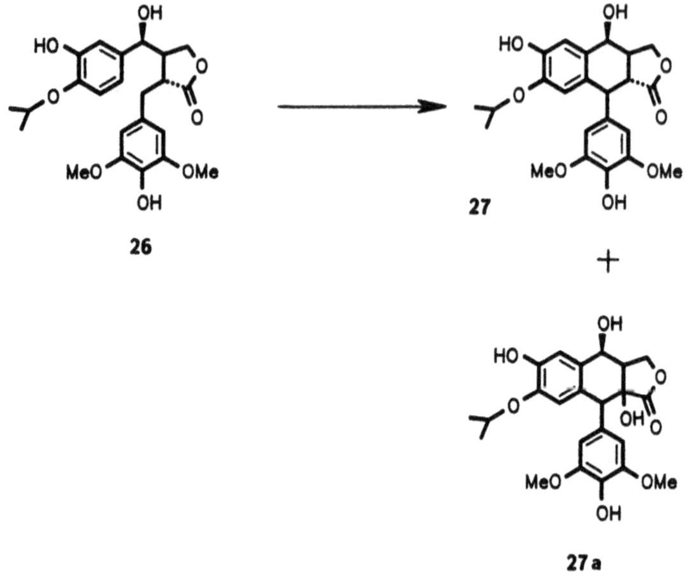

Fig. 12. Biotransformation of dibenzylbutanolide **26** with cell free extracts of *C. roseus* or with whole cells of *P. peltatum*.

While the above studies were underway, we established the first successful cell suspension culture of *P. peltatum*, which afforded podophyllotoxin, deoxypodophyllotoxin, and 4'-demethylpodophyllotoxin (Fig. 2).[13] It seemed that enzyme systems inherent in such cell cultures might be ideally suited for the biotransformation of the above-noted dibenzylbutanolides, since as noted earlier (Fig. 4), closely related intermediates have been implicated in the biosynthesis of the podophyllotoxins.

From a practical standpoint, it would be preferable to obtain successful cyclization reactions, for example, **26**→**27**, with whole cell fermentations rather than by isolating crude enzymes as noted in the above section. We used our new cell line of *P. peltatum* for this purpose. A large number of experiments have been conducted, but the following discussion will focus on a few selected areas to illustrate the successful approach.

One experimental approach involved the addition of dibenzylbutanolide **26** to "growing" cells of the *P. peltatum* cell cultures. Dissolved in a small amount of ethanol, **26** was added to cell cultures of varying ages (5-21 day old) growing in nutrient medium in shake flasks or bioreactors. After an appropriate time of incubation (generally 24-120 h), the fermentation broth and cells were extracted (ethyl acetate) and extracts analyzed for products formed. In general, the major cyclized product **27** was obtained in yields of 50-55%, although conditions for this conversion were not optimized in view of success in the semi-continuous process discussed below. We also found that **27a**, representing a further hydroxylation of **27** (Fig. 7), becomes a significant product especially with longer incubation times.

It is important to recognize that dehydration of **27a** to the corresponding styrene analogue would permit the generation of the desired podophyllotoxin stereochemistry with respect to the orientation of the aromatic unit at C(1) (see **4**, Fig. 2). For example, reduction of the double bond, thus formed, would provide the desired end product. This stereochemical arrangement is important in terms of the clinical efficacy demonstrated by etoposide. Further studies to solve this problem are presently underway.

The above studies indicated that the desired cyclization products **27** and **27a** were extracted almost entirely from the fermentation broth with only insignificant quantities remaining within the cells. It therefore became of interest to determine whether a semi-continuous fermentation process could be established. In this approach, the fermentation broth, after a certain incubation period with **26**, would be withdrawn under aseptic conditions, then new medium and a new batch of **26** added; the process was repeated over a number of cycles. Obviously, if successful, one batch of cell culture could be utilized for successive productions of **27** and/or **27a** and this "biological factory" could afford a highly attractive approach toward multigram quantities of end products. Indeed, a successful approach has been developed and only a brief description of a large number of experiments is provided here.

Inoculum of the *P. peltatum* cell line, grown in Murashige-Skoog (MS) medium[13] in shake flasks for 17 days, is transferred (41 g wet cells/l) to a Microferm bioreactor and cells are grown in this medium (aeration rate, 0.1 liter/liter/min, impeller speed 440 rpm, 26°C) for 2-17 days prior to addition of **26**. After sterilization of an alcoholic solution of **26** via filtration through a 0.22μ membrane, the substrate is added (varying concentration, 0.05-1.0 g/l) to the bioreactor and incubation with the growing cells is varied (24-50 h) during the numerous experiments. In a typical experiment, involving 24 h incubation periods between cycles, the fermentation broth containing the metabolites is withdrawn, new medium (generally at one-tenth to one-half original MS medium) and a new batch of **26** are added, and the process continued for a further 24 h whereupon the procedure is repeated. Although studies are still underway, we have demonstrated that one batch of cells maintains the necessary enzymatic activity to biotransform **26** to **27** and/or **27a** for several months. We have not yet evaluated how long such activity persists since the supply of **26** was depleted, after numerous additions, and the study had to be terminated before enzymatic activity within the bioreactor had been exhausted. In effect, the *P. peltatum* cell line is a "biological factory" in terms of its ability to achieve oxidative coupling of dibenzylbutanolides to podophyllotoxin analogues. The present combined yields of **27** and **27a** are in the range of 50% when **26** is added in 24-48 h cycles and in this particular study mentioned, eleven successive additions were made before the supply of **26** was depleted. Enzymatic activity in this batch of cells was maintained well beyond three months, although the ultimate loss of activity was not determined since the bioreactor contents were harvested and analyzed in order to initiate further studies. The optimum conditions for this process have not yet been established and we anticipate yields in the range of 70%. Finally, it should be noted that an efficient process in removing the isopropyl group in **27** and regeneration of the methylenedioxy function as required in **5** (Fig. 3) has already been completed in our laboratory. In summary, the semi-continuous process, once optimized, will afford a highly attractive route to various analogues of the podophyllotoxin series.

Use of Intact Plant Cells in Culture to Produce Higher Levels of Plant-Derived Natural Products

Since enzyme stimulation within cell cultures can be generally achieved through "artificial" nutrient media changes, it is not surprising that plant cell cultures often produce, in addition to the normal natural products present in the plant, novel metabolites which are not found in plant extracts. The following discussions involving plant cell cultures of the Chinese herbal plant *Tripterygium wilfordii*, illustrate our recent results pertaining to this area.

Tripterygium wilfordii Hook, commonly called Lei Gong Teng (Thunder God vine) or Mang Cao (rank grass) in China, is a perennial twining vine which has been used in herbal medicine for several centuries. A refined extract from the root xylem of this plant, available in Chinese markets as tablets, has been used for the treatment of rheumatoid arthritis, various skin disorders, and more recently as a potential male contraceptive agent.[14] Furthermore, cytotoxic effects of two of the minor plant constituents, tripdiolide (**28**) and triptolide (**29**), isolated by Kupchan[15] have been noted. These data stimulated us to develop cell cultures of *T. wilfordii* in an effort to obtain higher yields of these natural products and, in particular, to attempt to ascertain which chemical structural types present in the tablets used in China, were responsible for the various pharmacological properties noted above.

Earlier studies with our developed cell line of *T. wilfordii* have been published[16,17] so only a summary of the most salient features of these studies will be presented here. Major emphasis is provided on the most recent and as yet unpublished results (Kutney *et al.*, Can. J. Chem. 70, May, 1992). The cell line of *T. wilfordii*, coded as TRP4a, has been stable since its original development in the late 1970's and continues to produce the same spectrum of metabolites. Scale-up of TRP4a in bioreactors, for example, 60 liter Chemapac and other airlift bioreactors, provides several hundred liter quantities of broth and significant amounts of cells from which twenty-one compounds within the diterpene and triterpene families have been isolated and fully characterized. Figure 13 provides a summary of the metabolites isolated. As expected, in addition to providing some metabolites identical to those produced by *T. wilfordii* plants, four completely novel triterpenes (**30**, **31**, **32**, **39**) and two novel diterpenes (**42**, **36**, R=H) have been produced by the TRP4a cell line. As noted below, various approaches to increase metabolite production can be utilized.

A detailed study with the TRP4a cell line of *T. wilfordii* was undertaken in terms of tripdiolide (Td) (**28**) production versus variations in growth media composition. In view of the detailed publication,[17] only a brief summary is provided. Comparison of three basal media (PRL-4, MS, SH), in terms of Td production, is shown in Figure 14. The major differences in media composition are: a) concentration of available nitrogen in the form of NH_4^+ ion or NO_3^- ion. Both are present in much higher concentrations in the MS medium relative to the others; b) concentration of $CaCl_2$ is also higher in MS; c) concentration of thiamine is much lower in MS than in the other two media; d) Glycine is only present in MS. In addition, there are also differences in the concentrations of the micronutrients (e.g. Mn^{2+}, Zn^{2+}, Cu^{2+} and Co^{2+}).

With the realization that the MS medium provides the highest relative yield of Td, further studies with this medium were conducted and the results are summarized in Figure 15. Td production was highest in the medium with 1650 mg/l of ammonium nitrate, whereas biomass was greatest when ammonium nitrate at 850 mg/l was used (Panel A, Fig. 15). Sucrose concentration (Panel B, Fig. 15) is important with higher Td yields at 40 g/l while $CaCl_2$ at a level of 880 mg/l appeared best for Td production (Panel C, Fig. 15). However, Td production was not optimized in this study, even at the 4.0 mg/l yield obtained in the present experiments, since the yield from the living plant is about 36 times greater than that reported by Kupchan.[15]

In a recent, unpublished study with the TRP4a cell line, we have evaluated the approach of "elicitor stimulation" to determine increases, if any, in metabolite production. This approach is based on earlier studies by numerous laboratories involving living plants and phytoalexin production. It is well known that plants produce phytoalexins, secondary metabolites with antimicrobial activity in response to an imposed stress, generally after infection by pathogenic organisms.[18] It seemed reasonable to assume that plant cells

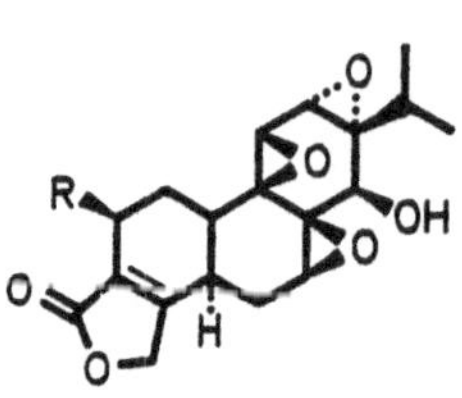

28 R = OH
29 R = H

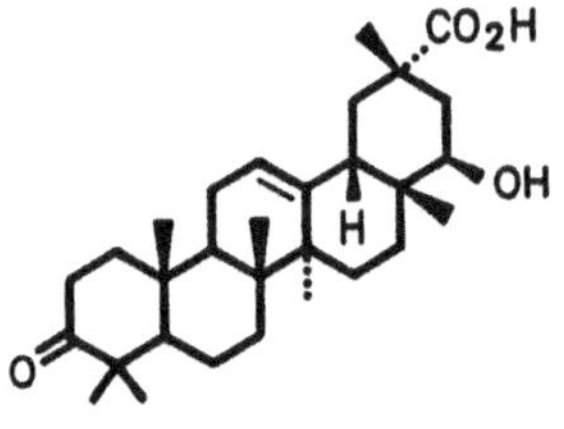

30

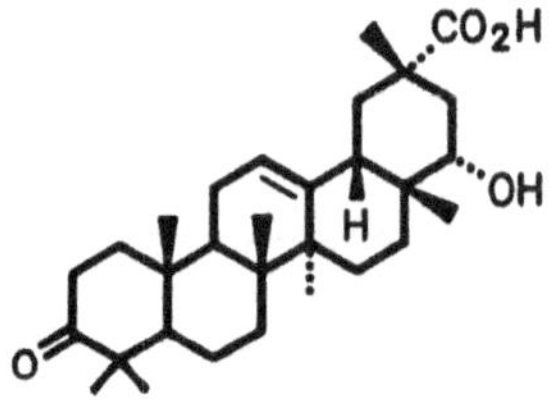

31

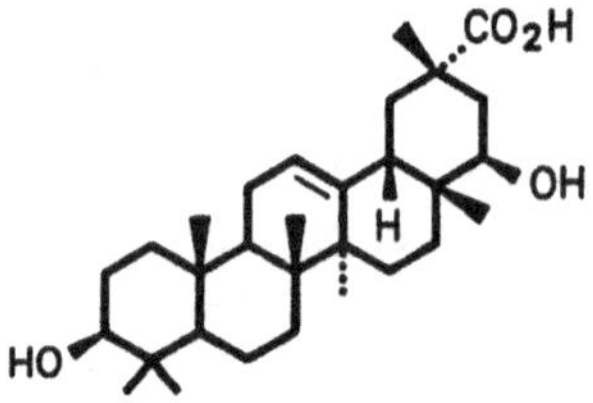

32

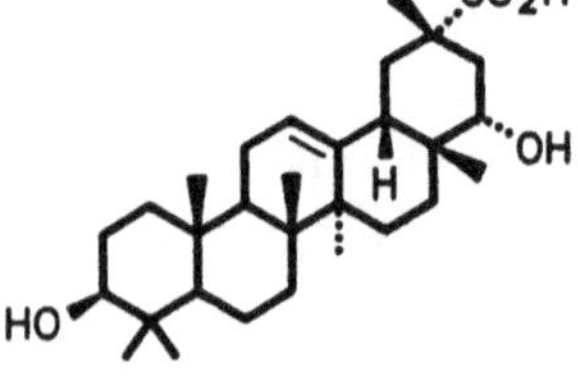

33

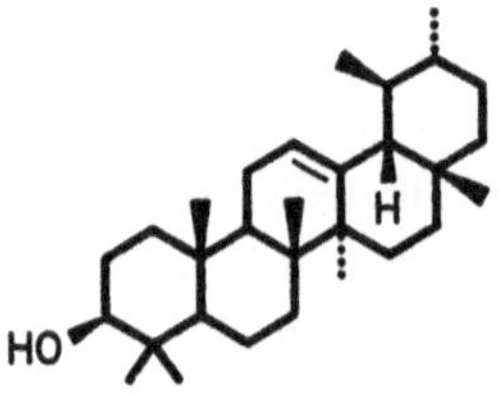

34

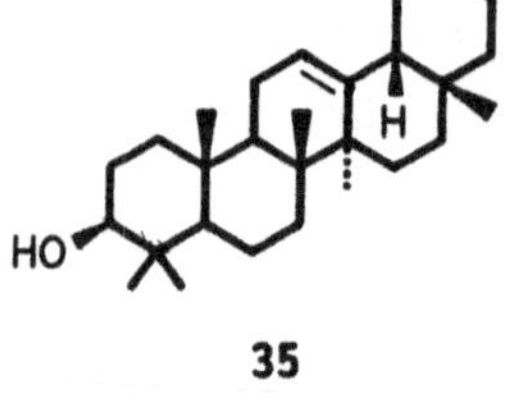

35

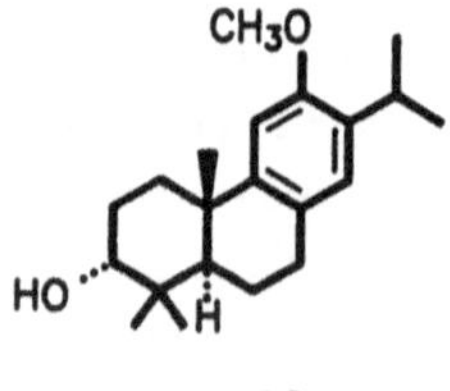

36

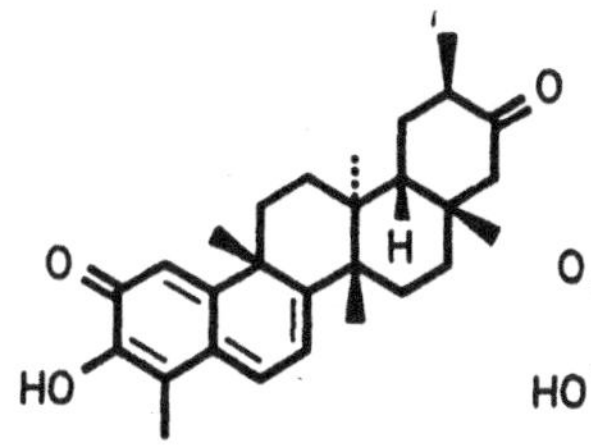

37

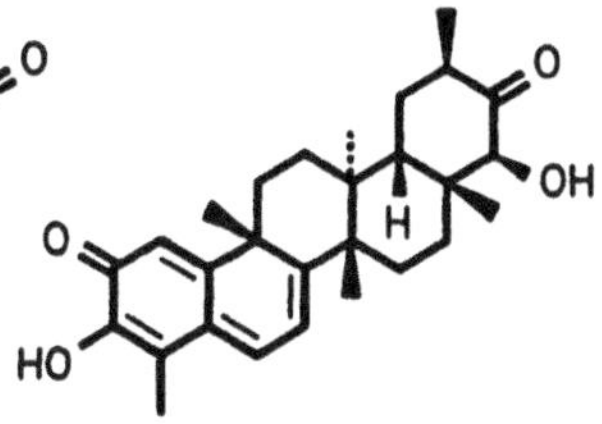

38

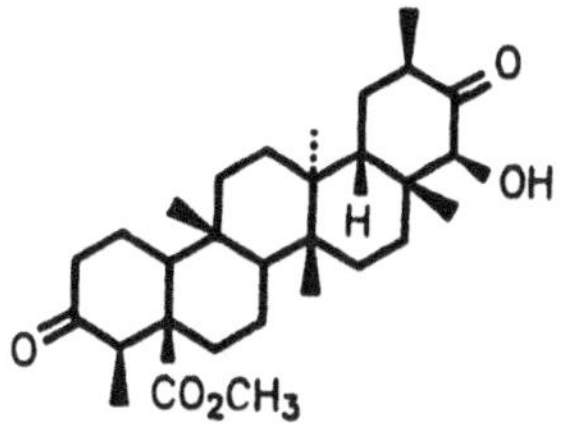

39

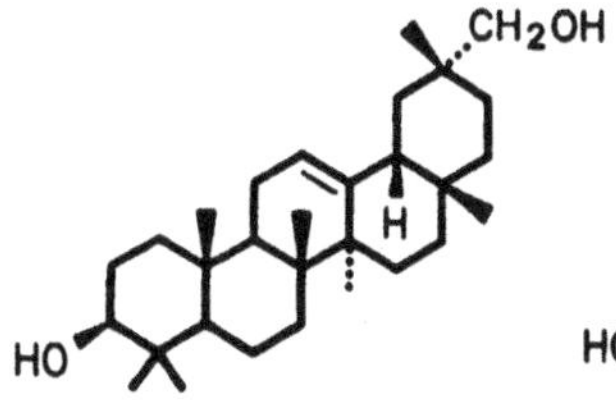

40

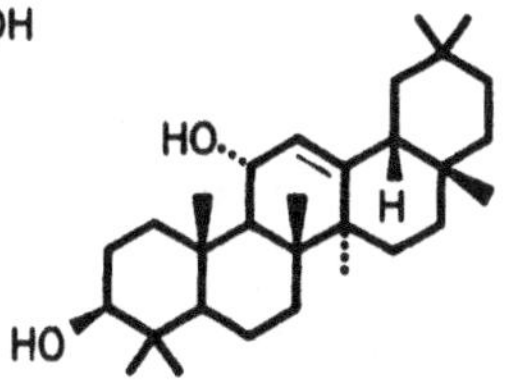

41

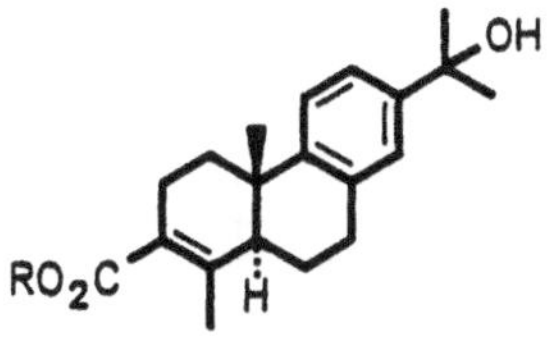

42 R = CH_3
R = H

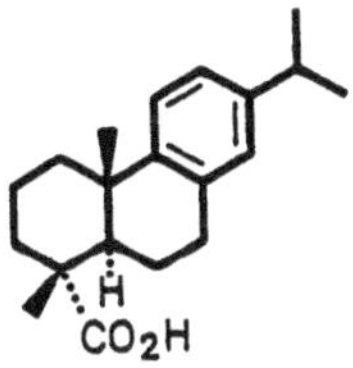

43

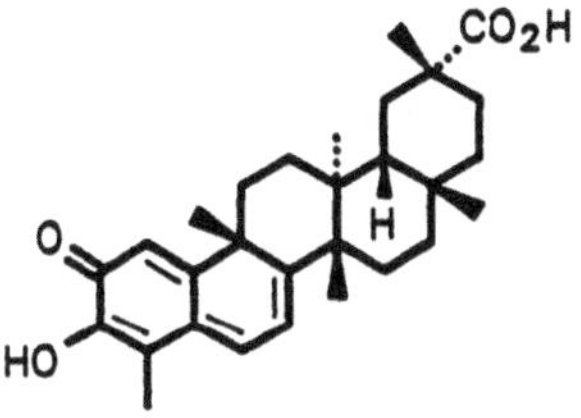

44

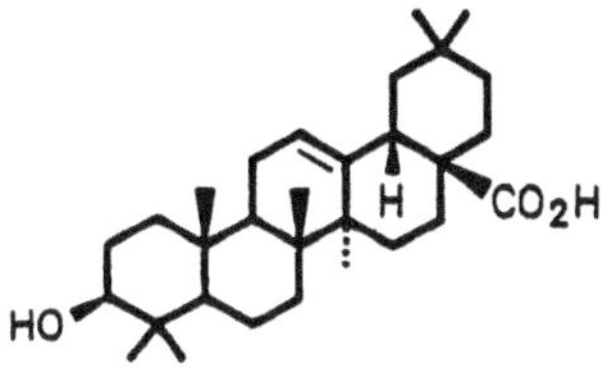

45

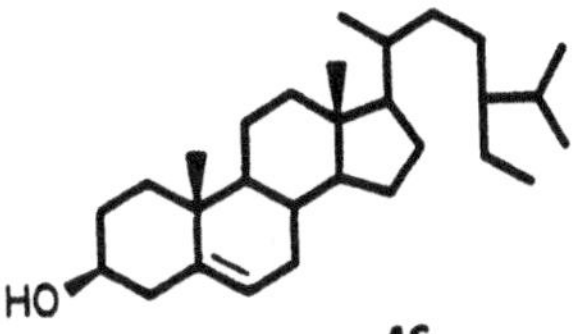

46

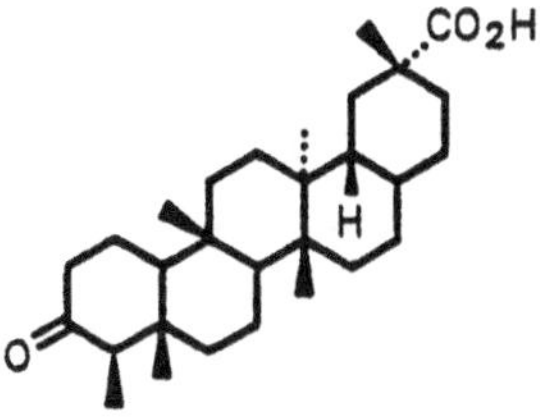

47

Fig. 13. Metabolites isolated from TRP4a cell line of *T. wilfordii*.

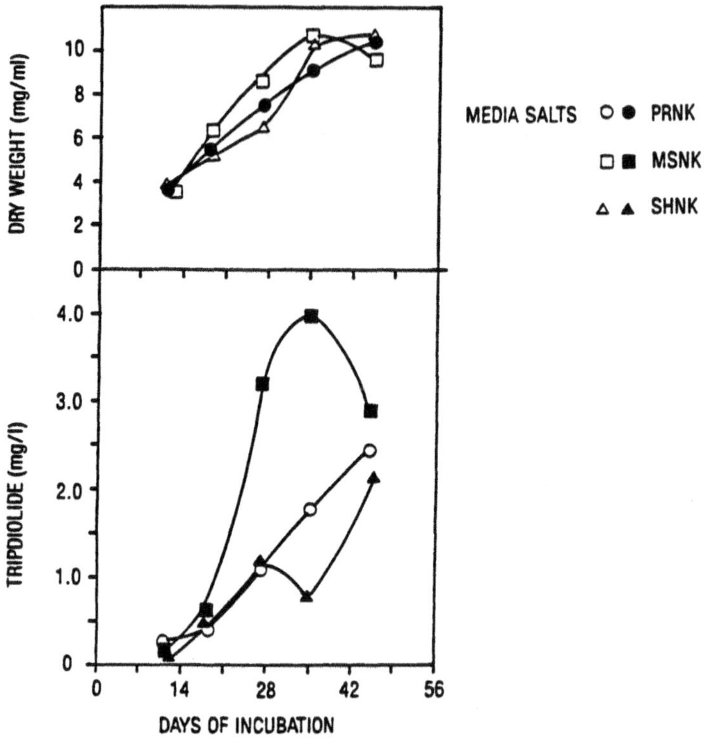

Fig. 14. Comparison of three different basal media for growth and tripdiolide production by TRP4a cell line.

growing in appropriate media may similarly stimulate "phytoalexin" production when exposed to such organisms. If such metabolites were identical and/or structurally related to the target compounds, an effective approach to elevating metabolite yields in such fermentation processes would become available. In order to determine whether elicitation of the desired metabolites (Fig. 13) was indeed occurring upon addition of fungal preparations to growing cells of theTRP4a cell line, a large number of small scale experiments were conducted. Fungal cultures of strains of *Botrytis* sp., *Sclerotinia sclerotiorum*, *Trichoderma viride* were grown to maturity, treated in a homogenizer until the mycelia were fragmented, autoclaved and then added to growing cultures of TRP4a, at various stages of the growth phase. The cultures were then allowed to continue growing for varying time periods (18-72 h). Extraction with ethyl acetate followed by monitoring of tripdiolide and triptolide production by quantitative fluorimetric

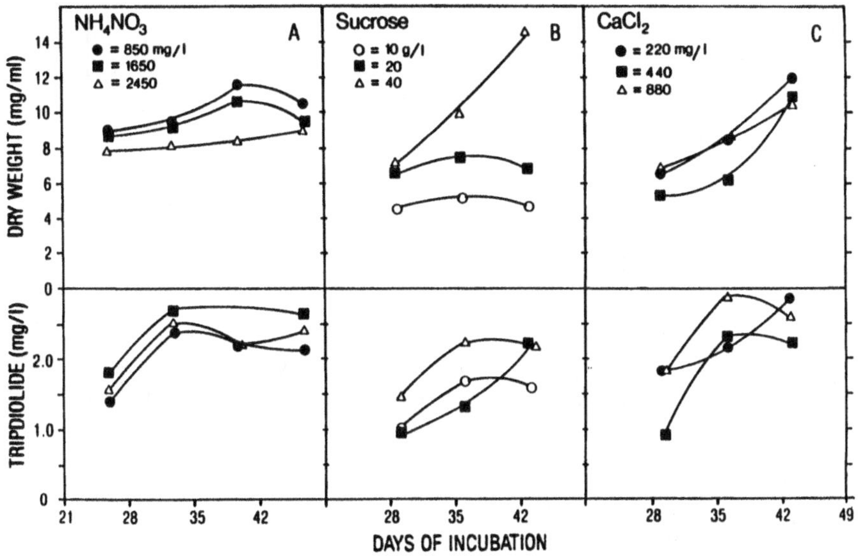

Fig. 15. Effects of tripdiolide and dry biomass production of different medium concentrations of NH_4NO_3 (Panel A), sucrose (Panel B), and $CaCl_2$ (Panel C). TRP4 cell line grown in $MSNA_{0.5}K_{0.5}$ basic medium.

detection and triterpene production by gas liquid chromatography, revealed that *Botrytis* and *Sclerotinia* were indeed effective in elevating triterpene levels in the TRP4a cell line when compared to control cultures to which no fungal preparations had been added. Neither of the diterpene triepoxides **28** or **29** were elicited by addition of any fungal preparations.

Subsequent studies with TRP4a in airlift bioreactors containing 1% elicitor (*Botrytis*) followed by a 72 h incubation period after fungus addition, revealed significant increase in yields of selected triterpenes, specifically the oleanane triterpenes **30-33** (Fig. 13), labelled as A, B, C, D respectively in Figure 16 (based on five separate experiments).

Addition of the fungal preparation to the growing cells of TRP4a was made at different ages of the cell culture and it is clear that "elicitation" of triterpene production varied depending on time of addition. In experiments coded as 256-3 and 259-2 in Figure 16, *Botrytis* (1% concentration) was added after a 10 day initial growth of TRP4a cells in the nutrient medium (refractive index (R.I.)=1.3360), followed by a 72 h incubation period prior to harvest and

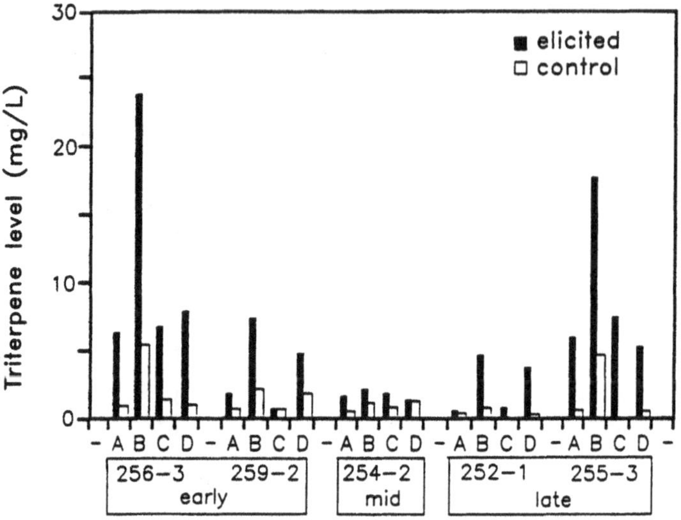

Fig. 16. Triterpene levels from elicitation at various ages of TRP4a cultures. Oleanane triterpenes A-D (**30-33** respectively, Fig. 13. Elicitation with 1% *Botrytis* for 72 h. Controls were grown without *Botrytis*.

extraction of metabolites. In experiment 254-2 addition of fungal preparation was made in the middle of the growth period (14 days of growth, R.I.=1.3355), whereas 252-1 and 255-3 addition of *Botrytis* was in late stages of growth (16 days, R.I.=1.3332; 19 days, R.I.=1.3334 respectively). Refractive index, generally a measurement that reflects consumption of nutrients by the growing cells, was used in our studies to provide a measure of cell development and growth.

In summary, the time course studies shown in Figure 16 demonstrate that maximum triterpene yield was obtained from cultures elicited at the beginning of the rapid growth period (experiment 256-3) and that an approximate five fold increase in production of triterpene **31** (Fig. 13), for example, was observed relative to the control culture. These, and other data not provided here, reveal that elicitation with 1% *Botrytis* produced a steady increase in production levels of the triterpenes **30-33** inclusive for at least six days of growth after addition of the elicitor and usually until the end of culture growth (18-22 days).

The specific mechanism of action of these elicitors is unknown, but based on the well established biosynthetic pathway of isoprenoids, addition of

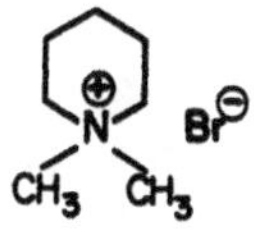

48

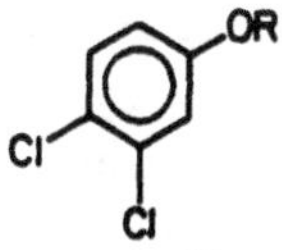

49

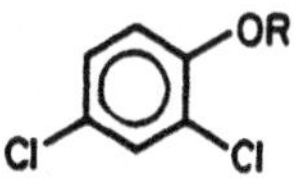

50

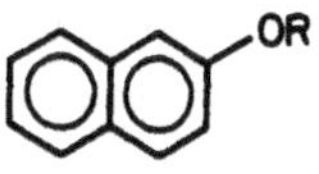

51

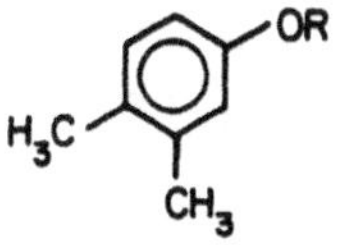

52

$R = CH_2CH_2N(CH_2CH_3)_2$

Botrytis to the TRP4a cell line must stimulate enzymes associated with the late stages of the pathway, such as the cyclization of squalene and subsequent reactions. Since the production of the diterpenes tripdiolide and triptolide is not affected by elicitor addition, the earlier stages of the isoprenoid pathway are not markedly influenced by such elicitors. Although studies to achieve optimum levels of triterpenes have not been completed, this approach may be useful for maximizing yields of metabolite production via plant cell culture methods.

A third approach directed toward an increase in metabolite production involves the addition of "bioregulators" to cell cultures. This method has been studied in some detail with cell cultures of *C. roseus* known to produce various indole alkaloids. They were incubated with a series of aminoethyl phenylethers (**48-52**) which Yokohama[19] had shown stimulated rubber biosynthesis. These results have been published,[20] so only a brief summary is provided.

The five bioregulators **48-52** were added as the free base (except **48** as the bromide), dissolved in methanol-water and filter-sterilized through a 0.45 mm membrane. The cell line of *C. roseus* was initially grown in shake flasks for 7 days, when the cultures were near or at stationary phase of growth and the bioregulators added (concentration 2 mg/l). Incubation periods ranging from 4-31 days were evaluated in terms of biomass production, as measured by cell dry weight, total alkaloid yield, and finally the effect of bioregulators on yields of the alkaloids ajmalicine and catharanthine (**1**), the latter determined by quantitative HPLC measurements. Of the five bioregulators, **48** and **52** were most promising and the data for these are summarized in Figures 17 and 18.

Figure 17 indicates that the total alkaloid yield is increased approximately three-fold after an 18 day incubation with bioregulator **52** and the relative increase depends on periods of incubation. Figure 18 summarizes the data for catharanthine (1), one of the target alkaloids. The increase in **1** is more than ten-fold if an incubation period with **52** of 12-18 days is involved. Although we did not proceed to optimize conditions, it sems clear that bioregulator stimulation of metabolite production can be an attractive approach.

Use of Well Developed Plant Cell Lines to Separate Pharmacological Properties Exhibited by Complex Mixtures

With well developed plant cell lines which can be utilized to provide a constant series of metabolites often in yields substantially higher than the living plant, it becomes advantageous to use this methodology to understand which components of a plant extract are responsible for specific pharmacological effects as claimed in herbal medicine. This approach has been undertaken in one aspect of our program involving the Chinese herbal plant, *Tripterygium wilfordii*.

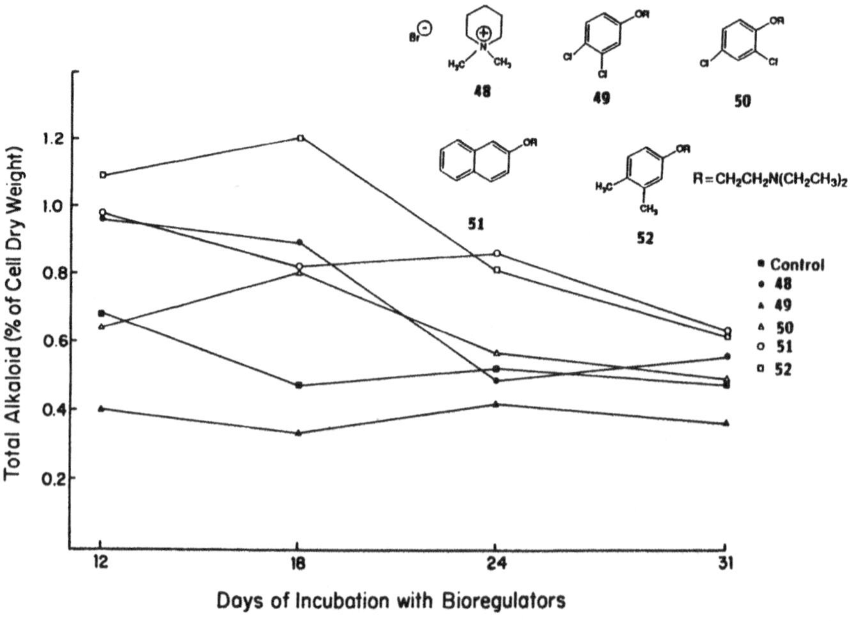

Fig. 17. Effect of bioregulators (**48**) to (**52**) on total alkaloid yield over an incubation period of 12 to 31 days.

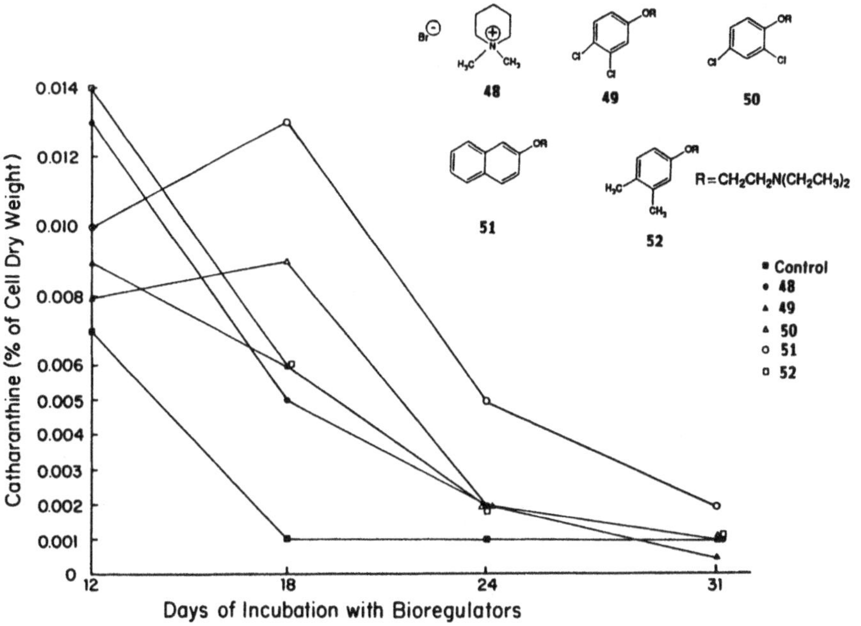

Fig. 18. Effect of bioregulators (**48**) to (**52**) on yield of catharanthine (**1**) over an incubation period of 12 to 31 days.

As noted earlier, extracts of *T. wilfordii* have been administered, in the form of tablets, for clinical treatment of rheumatoid arthritis and skin disorders, and also evaluated as a novel non-steroidal male contraceptive in man. It was of interest to determine which of the natural products present in the tablets are responsible for the different pharmacological effects. In particular, it would be highly desirable to separate the structural requirements responsible for pharmacological activities attributed to arthritis from those required for anti-spermatogenic activity in man.

In order to derive information relative to the above, the TRP4a cell line discussed in the previous section was used to obtain sufficient quantities of the metabolites shown in Figure 13. Some of these metabolites were submitted for pharmacological screening to several laboratories in the People's Republic of China and to pharmaceutical firms in North America. Studies in this area are incomplete, but some understanding of the structural requirements for the above-noted pharmacological effects is emerging.

The diterpenes tripdiolide (**28**) and triptolide (**29**) are highly active, even at nanogram levels, as immunosuppressive agents and the hydroxy acid diterpene **42** (R=H) also show activity at higher dose levels. On the other hand, the various triterpenes shown in Figure 13 lack significant immunosuppressive activity. In a similar manner, **28** and **29** also exhibit anti-spermatogenic activity in tests conducted with rats.

Further studies will be required to put forward a more accurate assessment of these data, but the TRP4a cell line will provide an excellent avenue for further experiments within this area. We already have some significant results involving biotransformation of selected diterpene substrates with TRP4a to provide novel diterpene analogues for a more detailed structure-activity study.

In summary, it appears that the "active" constituents in the tablets being administered in Chinese herbal medicine practice are members of the diterpene family, tripdiolide (**28**) and triptolide (**29**) or closely related analogues. We intend to pursue this avenue of structure-activity relationships in future research.

CONCLUSION

Utilizing specific examples from the author's laboratory, a review of the possible avenues of research in which plant cell culture methods combined with chemistry can afford interesting routes to biologically active compounds, is provided. The areas selected provide examples of the use of plant cell cultures as a source of enzymes for biosynthetic and biotransformation experiments, as "reagents" in organic synthesis, their utilization in production of plant derived natural products used in clinical medicine and as sources of novel metabolites which offer the opportunity to develop systems that exhibit interesting pharmacological activities.

The results presented illustrate the potential of such an interdisciplinary program and, in particular, reveal that various "shortcomings" in the utilization of plant cell cultures sometimes expressed (for example, cell line instability, long term and low level production of metabolites, etc.) may not be justified. With proper selection of methodology, gram scale production of end products, often in short term experiments similar to those with microbial systems, can be achieved. The use of semi-continuous fermentation methods, for example, provides an approach which is of distinct interest in large scale efficient production of commercially important products.

ACKNOWLEDGMENTS

Financial support from the Natural Sciences and Engineering Research Council of Canada, the National Research Council of Canada and Nippon Oil Company, Japan, is gratefully acknowledged. The efforts of the large number of enthusiastic and dedicated research workers whose names appear in the publications from my laboratory in the references section are acknowledged with appreciation by the author.

REFERENCES

1. KUTNEY, J.P. 1987. Studies in plant tissue culture. The synthesis and biosynthesis of indole alkaloids. Heterocycles 617-640.
2. KUTNEY, J.P. 1988. Studies in Plant Tissue Culture. The Biosynthesis of Complex Natural Products. In: Studies in Natural Products Chemistry, Vol. 2, Structure Elucidation (Part A), (A.U. Rahman, ed.), Interscience, Amsterdam, pp. 365-420.
3. KUTNEY, J.P. 1990. Biosynthesis and synthesis of indole and bisindole alkaloids in plant cell cultures - a personal overview. Nat. Prod. Rep. 7:85-103.
4. KUTNEY, J.P. 1991. Plant cell cultures and synthetic chemistry—a potentially powerful route to complex natural products. Synlett. 11-19.
5. JARDINE, I. 1980. In: Anticancer Agents Based on Natural Product Models, (J.M. Cassady, J.D. Douros, eds.), Academic Press, New York, pp. 319-351, and references cited therein.
6. WARD, R.S. 1982. The synthesis of lignans and neolignans. Chem. Soc. Rev. 75-125 and references cited therein.
7. ISSELL, B.F., MUGGIA, F.M., CARTER, S.K. 1984. Etoposide (VP-16) Current Status and New Developments. Academic Press, Orlando.
8. BROOMHEAD, A.J., RAHMAN, M.A.M., DEWICK, P.M., JACKSON, D.E., LUCAS, J.A. 1991. Matairesinol as precursor of *Podophyllum* lignans. Phytochem. 30:1489-1492, and references cited therein.

9. KONDO, K., OGURA, M., MIDORIKAWA, Y., KOZAWA, M., TSUJIBO, H., BABA, K., INAMORI, Y. 1989. Conversion of deoxypodophyllotoxin to podophyllotoxin-related compounds by microbes. Agric. Biol. Chem. 53(3):777-782.

10. SAUNDERS, B.C., HOLMES-SIEDLE, A.G., STARK, B.P. 1964. In: Peroxidase. Butterworth, London.

11. CHANCE, B., MAEHLY, A.C. 1955. In: Methods in Enzymology, Vol. 2, (S.P. Colowick, N.O. Kaplan, eds.), Academic Press, New York, pp. 773.

12. KUTNEY, J.P., HEWITT, G.M., JARVIS, T.C., PALATY, J., RETTIG, S. 1992. Studies with plant cell cultures of *Catharanthus roseus*. Oxidative coupling of dibenzylbutanolides catalyzed by plant cell culture extracts. Can. J. Chem. 70:2115-2133.

13. KUTNEY, J.P., ARIMOTO, M., HEWITT, G.M., JARVIS, T.C., SAKATA, K. 1991. Studies with plant cell cultures of *Podophyllum peltatum* L.I. Production of podophyllotoxin, deoxypodophyllotoxin, podophyllotoxone and 4'-demethylpodophyllotoxin. Heterocycles 2305-2309.

14. QIAN, S.Z. 1987. *Tripterygium wilfordii*, a Chinese herb effective in male fertility regulation. Contraception 36:335-344, and references cited therein.

15. KUPCHAN, S.M., COURT, W.A., DAILY, R.G., GILMORE, C.J., BRYAN, R.F. 1972. Triptolide and tripdiolide, novel antileukemic diterpenoid triepoxides from *Tripterygium wilfordii*. J. Am. Chem. Soc. 94:7194-7195.

16. KUTNEY, J.P., HEWITT, G.M., KURIHARA, T., SALISBURY, P.J., SINDELAR, R.D., STUART, K.L., TOWNSLEY, P.M., CHALMERS, W.T., JACOLI, G.G. 1981. Cytotoxic diterpenes, triptolide, tripdiolide, and cytotoxic triterpenes from tissue cultures of *Tripterygium wilfordii*. Can. J. Chem. 59:2677-2683.

17. KUTNEY, J.P., CHOI, L.S.L., DUFFIN, R., HEWITT, G., KAWAMURA, N., KURIHARA, T., SALISBURY, P., SINDELAR, R., STUART, K.L., TOWNSLEY, P.M., CHALMERS, W.T., WEBSTER, F., JACOLI, G.G. 1983. Cultivation of *Tripterygium wilfordii* tissue cultures for the production of the cytotoxic diterpene tripdiolide. Planta Medica 48:158-163.

18. BRODELIUS, P. 1988. In: Plant Cell Biotechnology, (M.M. Pais, ed.), Springer-Verlag, Berlin, pp.195-209, and references cited therein.

19. YOKOHAMA, H., HAYMAN, E.P., HSU, W.J., POLING, S.M., BAUMAN, A.J. 1977. Chemical bioinduction of rubber in Guayule plants. Science 197:1076-1077.

20. KUTNEY, J.P., AWERYN, B., CHATSON, K.B., CHOI, L.S.L., KURZ, W.G. 1985. Alkaloid production in *Catharanthus roseus* (L.) Don cell cultures. XIII. Effects of bioregulators on indole alkaloid biosynthesis. Plant Cell Rep. 4:259-262.

Chapter Eleven

UTILITY OF THE PHYTOCHEMECO DATABASE

Stephen M. Beckstrom-Sternberg

National Germplasm Resources Laboratory
Agricultural Research Service—U.S. Department of Agriculture
Beltsville, Maryland USA 20705-2350

INTRODUCTION

The phytochemical/ecogeographic database (PHYTOCHEMECO) of the National Germplasm Resources Laboratory (NGRL), U.S. Department of Agriculture, is a unique blend of phytochemical, taxonomic, ecological, geographic and climatic aspects, and yields information and germplasm databases in a UNIX minicomputer environment. The phytochemical database contains information about 16,332 chemical compounds found in 1,164 plants of economic importance, including quantitative data where available, as well as localization and specific activities (1,517 activities of 4,329 chemicals) of the compounds.[1,2] The taxonomic, ecological and yield databases contain plant names of 8,771 taxa, growing locations of 6,242 taxa, and crop yields of 239 taxa, respectively. The geographic and climatic[3] databases contain specific information about more than 18,000 world-wide locations, including latitude, longitude, elevation, soil type, rainfall, temperature and lifezones. The Germplasm Resources Information Network database (GRIN) lists accessions of plant clonal and seed material available for study through the National Plant Germplasm System.

The data are accessed with standard structured query language (SQL), which is maintained in a relational database management system. The data are structured into tables which can be interconnected through one or more common fields, thereby preventing duplication of entries and facilitating rapid and precise querying of the data. These disparate databases have been interconnected to rapidly address questions previously considered excessively time-intensive.

During the week at the Agricultural Research Service (ARS) of the USDA, the scientific community or the public may pose questions directly to the NGRL staff, which are readily addressed through the PHYTOCHEMECO database. Examples of six questions posed during the course of a typical week are explored below using this database; they are but a sampling of those potentially addressed:

1. Are antileukemic compounds correlated with plant lifezones?
2. What specific antihypertensive compounds are found in *Citrus* species and what other plants contain these compounds?
3. What plants are highest in pectins?
4. Do any anticholeric plants grow in Peru?
5. What plants contain greater than 100 ppm selenium?
6. What is the value of the chemical constituents of a hectare of sunflowers?

ANTILEUKEMIC COMPOUNDS AND CORRELATED LIFE ZONES

To make meaningful and predictive ecological distinctions among climatic data, lifezones have been defined on the basis of temperature and rainfall ranges[4] (Table 1). The question of correlating lifezone and chemical activity is most easily viewed in the following way: Are there chemical activities which predominate in plants found in certain lifezones? As an example, antileukemic compounds were chosen.

A search of the PHYTOCHEMECO database found 372 species containing 523 unique chemicals with antileukemic activities. Of these, 139 plants containing 20 antileukemic chemicals had lifezone data. Table 2 lists the

Table 1. Climatic life zones and their corresponding ranges of annual average temperature and annual average rainfall

Symbol	Lifezone	Annual Average Temperature (^{o}C)	Annual Average Rainfall (mm)
Tx	Tropical Desert Scrub	> 24	125—250
Tt	Tropical Thorn Woodland	> 24	250—500
Tv	Tropical Very Dry Forest	> 24	500—1,000
Td	Tropical Dry Forest	> 24	1,000—2,000
Tm	Tropical Moist Forest	> 24	2,000—4,000
Tw	Tropical Wet Forest	> 24	4,000—8,000
Tr	Tropical Rain Forest	> 24	> 8,000
Sx	Subtropical Desert Scrub	18—24	125—250
St	Subtropical Thorn Woodland	18—24	250—500
Sd	Subtropical Dry Forest	18—24	500—1,000
Sm	Subtropical Moist Forest	18—24	1,000—2,000
Sw	Subtropical Wet Forest	18—24	2,000—4,000
Sr	Subtropical Rain Forest	18—24	> 4,000
Wx	Warm Temperate Desert Bush	12—18	125—250
Wt	Warm Temperate Thorn Steppe	12—18	250—500
Wd	Warm Temperate Dry Forest	12—18	500—1,000
Wm	Warm Temperate Moist Forest	12—18	1,000—2,000
Ww	Warm Temperate Wet Forest	12—18	2,000 -4,000
Wr	Warm Temperate Rain Forest	12—18	> 4,000
Cx	Cool Temperate Desert Scrub	6—12	125—250
Cs	Cool Temperate Steppe	6—12	250—500
Cm	Cool Temperate Moist Forest	6—12	500—1,000
Cw	Cool Temperate Wet Forest	6—12	1,000—2,000
Cr	Cool Temperate Rain Forest	6—12	> 2,000
Bx	Boreal Desert	3—6	< 125
Bd	Boreal Dry Scrub	3—6	125—250
Bm	Boreal Moist Forest	3—6	250—500
Bw	Boreal Wet Forest	3—6	500—1,000
Br	Boreal Rain Forest	3—6	> 1,000
Pd	Subpolar Dry Tundra	< 3	< 125
Pm	Subpolar Moist Tundra	< 3	125—250
Pw	Subpolar Wet Tundra	< 3	250—500
Pr	Subpolar Rain Tundra	< 3	> 500

Table 2. Antileukemic chemicals found in the PHYTOCHEMECO database

aloe-emodin	genistein
astragalin	heliotrine
β-sitosterol	isofraxidin
β-sitosterol-β-D-glucoside	methyl-gallate
colchicine	piceatannol
cucurbitacin-I	pinocembrin
demecolcine	reserpine
deoxypodophyllotoxin	ricin
digallic acid	tricin
emetine	ursolic acid

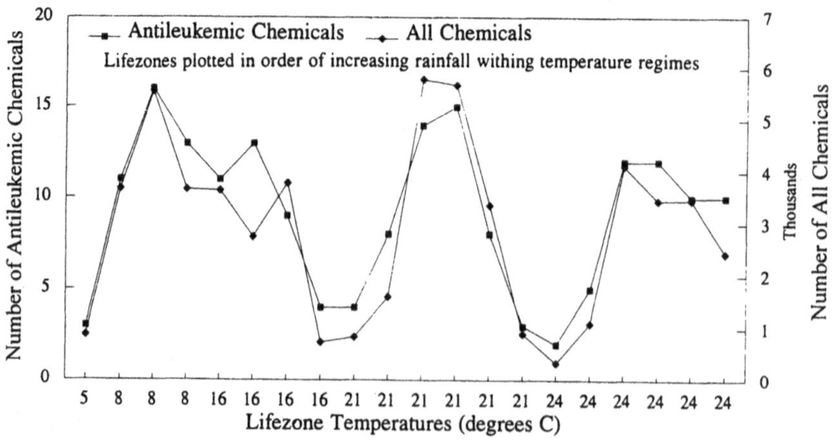

Fig. 1. Antileukemic and all chemicals in different lifezones as a function of temperature. For list of lifezone abbreviations see Table 1.

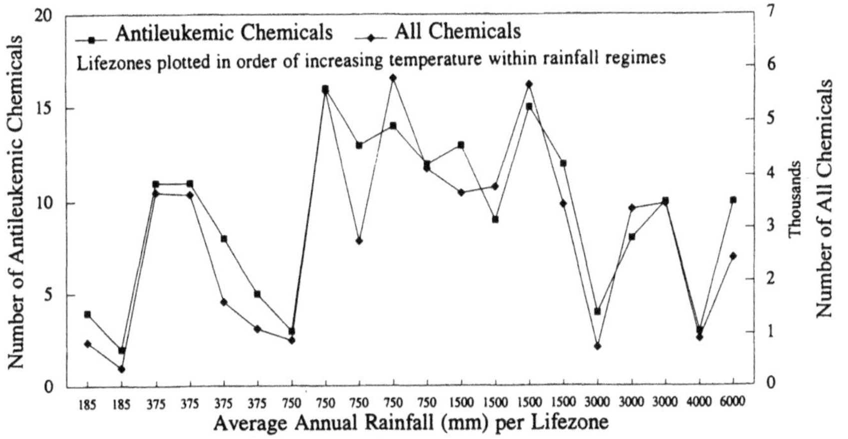

Fig. 2. Antileukemic and all chemicals in different lifezones as a function of rainfall. For list of lifezone abbreviations see Table 1.

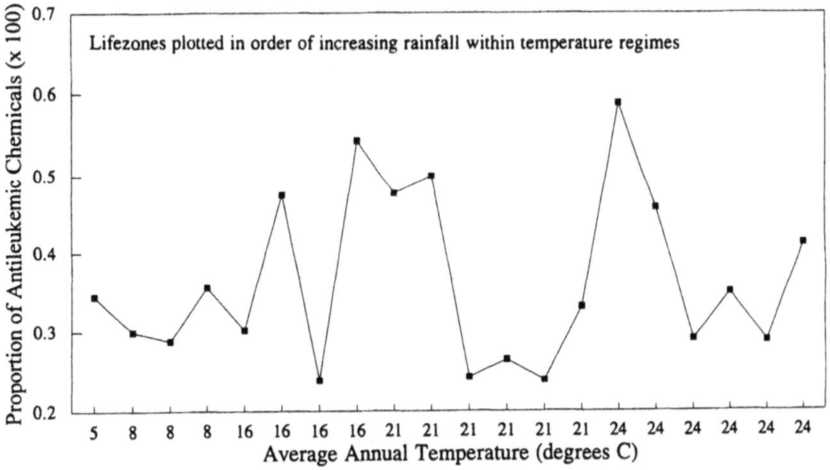

Fig. 3. Proportion of antileukemic chemicals per lifezone as a function of temperature. Proportions were calculated by dividing number of antileukemic chemicals by total number of chemicals, per lifezone, and multiplying by 100. For list of lifezone abbreviations see Table 1.

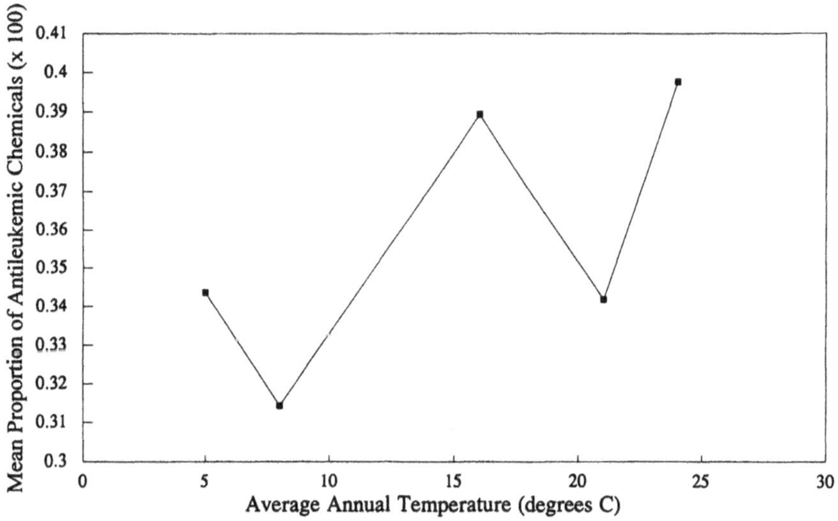

Fig. 4. Mean proportion of antileukemic chemicals as a function of temperature. Lifezones were grouped on the basis of temperatures, and means and proportions were calculated.

20 unique antileukemic phytochemicals. The distribution of antileukemic chemicals among lifezones was not correlated with either temperature (Fig. 1) or rainfall (Fig. 2), but rather with the pattern of distribution of all chemicals among lifezones. To nullify the effect of sampling, proportions were calculated by dividing the number of antileukemic chemicals per lifezone by the total number of chemicals per lifezone. These values were plotted as functions of temperature (Figs. 3 and 4) and rainfall (Figs. 5 and 6). As shown above, the correlations with temperature and rainfall were non-existent. Proportionally, the driest (Fig. 6) and hottest (Fig. 4) areas had the highest number of antileukemic chemicals, indicating,in the absence of other data, that hot, arid lifezones might conceivably be preferred starting points for antileukemic phytochemical prospecting. However, the lack of correlation with either temperature or rainfall, coupled with the small sample of 20 out of a total of 523 antileukemic chemicals, lends mimimal support to this conclusion, and demonstrates the need for more balanced chemical coverage of the lifezones, as well as a much expanded ecological coverage of the plants contained in the phytochemical database.

ANTIHYPERTENSIVE CHEMICALS IN *CITRUS* AND OTHER PLANTS

All *Citrus* species contain the two antihypertensive chemicals, potassium and tryptophan (Table 3). A query of the PHYTOCHEMECO database revealed that 179 other plants contained greater than 16,360 ppm of potassium, which is the highest level found in the grapefruit (*Citrus paradisi*). Table 4 lists the first 48 species in descending order of potassium content. Likewise, 99 other plants were found to contain significantly greater amounts of tryptophan than the 680 ppm found in the orange (*C. sinensis*). Table 5 lists the first 48 species in descending order of tryptophan content.

In addition to potassium and tryptophan, the PHYTOCHEMECO database reveals fourteen antihypertensive compounds found in plants other than *Citrus* (Table 6), of which the genus *Raufolfia* contains five. The result obtained for *Rauvolfia* may indicate synergy for this activity and illustrates the potential for the PHYTOCHEMECO database to search for instances where synergies may be occurring, warranting further investigation.

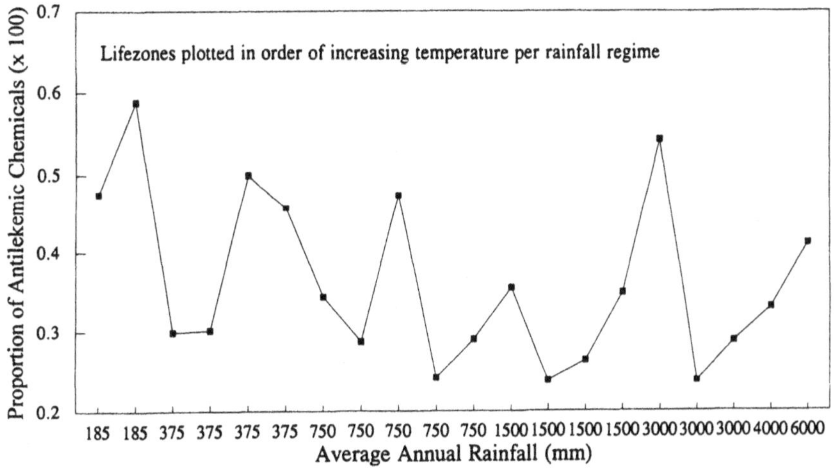

Fig. 5. Proportion of antileukemic chemicals per lifezone as a function of rainfall. Proportions were calculated by dividing number of antileukemic chemicals per lifezone by total number of chemicals and multiplying by 100. For list of lifezone abbreviations see Table 1.

PLANTS HIGHEST IN PECTINS

Pectin is a source of dietary fiber and as such has been reported to have cancer-preventive, antiulcer and cholesterol-lowering activities.[2] However, it has also been found to reduce the bioavailability of β-carotene in humans.[5]

A search of the PHYTOCHEMECO database found that marshmallow (*Althaea officinalis*) and carrot (*Daucus carota*) roots, hops (*Humulus lupulus*), passionfruit (*Passiflora edulis*) and eggplant (*Solanum melongena*) fruits, sunflower (*Helianthus annuus*) plants, and flax (*Linum usitatissimum*) seeds have the highest concentrations of pectin (Table 7).

Table 3. Potassium and tryptophan as hypertensive chemicals in fruits of various *Citrus* species

	Amount (ppm)	
	Low	High
POTASSIUM		
Citrus paradisi	1,300	16,360
Citrus medica	–	15,500
Citrus limon	–	14,700
Citrus aurantium	7,020	13,800
Citrus sinensis	1,400	13,772
Citrus reticulata	1,200	13,127
Citrus aurantifolia	820	9,533
Citrus aurantium	–	1,020
TRYPTOPHAN		
Citrus sinensis	90	680
Citrus reticulata	60	484
Citrus aurantifolia	30	255
Citrus paradisi	20	220

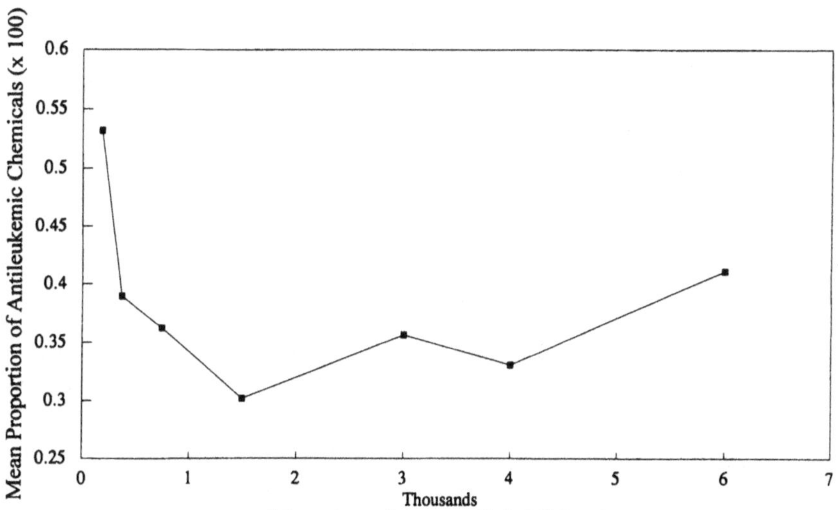

Fig. 6. Mean proportion of antileukemic chemicals as a function of rainfall. Lifezones were grouped on the basis of rainfall, and means and proportions were calculated.

Table 4. Species other than *Citrus* that contain high levels of potassium

		Amount (ppm)	
		Low	High
Lactuca sativa	Leaf	2,900	121,800
Cichorium endivia	Leaf	2,915	96,000
Vigna mungo	Seed	1,710	89,790
Chenopodium album	Leaf	–	87,100
Raphanus sativus	Root	2,215	85,700
Brassica pekinensis	Leaf	74,100	81,900
Portulaca oleracea	Herb	4,880	81,200
Avena sativa	Plant	2,000	78,900
Chrysanthemum coronarium	Bud	5,710	76,745
Anethum graveolens	Plant	8,700	76,450
Taraxacum officinale	Root	12,000	75,000
Amaranthus sp.	Leaf	2,630	73,503

Table 4 cont'd

Plant		Amount (ppm) Low	High
Cucumis sativus	Fruit	1,465	72,500
Brassica chinensis	Leaf	1,804	69,143
Spinacia oleracea	Plant	2,060	69,077
Borago officinalis	Leaf	4,700	67,210
Rheum rhabarbarum	Petiole	2,510	66,400
Nasturtium officinale	Herb	3,300	66,000
Aralia cordata	Leaf	3,100	65,950
Beta vulgaris	Leaf	4,380	61,798
Lycopersicon esculentum	Fruit	780	58,800
Phaseolus vulgaris	Fruit	1,960	58,500
Apium graveolens	Petiole	2,689	57,800
Lepidium sativum	Leaf	6,060	57,170
Apium graveolens	Root	3,900	56,360
Asparagus officinalis	Shoot	2,210	55,200
Houttuynia cordata	Plant	—	54,300
Petroselinum crispum	Plant	4,425	53,833
Colocasia esculenta	Leaf	4,370	51,774
Anthriscus cerefolium	Leaf	47,400	51,200
Beta vulgaris	Root	3,033	50,000
Lycium chinense	Leaf	5,180	49,808
Ipomoea aquatica	Leaf	1,500	49,200
Brassica oleracea	Flower	3,300	49,080
Coriandrum sativum	Leaf	5,600	48,177
Lycopersicon esculentum	Leaf	–	47,000
Perilla frutescens	Leaf	6,500	46,429
Daucus carota	Root	3,000	46,360
Glechoma hederacea	Plant	32,000	46,000
Corchorus olitorius	Leaf	4,440	45,500
Momordica charantia	Fruit	2,700	45,000
Vigna unguiculata	Shoot	4,550	44,520
Prunus domestica	Fruit	1,677	44,200
Cucumis melo	Fruit	3,018	44,000
Hordeum vulgare	Plant	19,880	44,000
Brassica napus	Root	2,200	43,850
Triticum aestivum	Plant	1,200	43,500
Raphanus sativus	Leaf	5,000	43,478

Table 5. Species other than *Citrus* that contain high levels of tryptophan

Plant	Part	Amount (ppm) Low	High
Oenothera biennis	Seed	2,400	16,000
Helianthus annuus	Seed	7,037	15,900
Psophocarpus tetragonolobus	Seed	7,620	8,313
Lablab purpureus	Seed	880	7,255
Nasturtium officinale	Herb	300	6,000
Psophocarpus tetragonolobus	Tuber	2,520	5,915
Sinapis alba	Seed	5,260	5,628
Cucurbita foetidissima	Seed	1,840	5,472
Psophocarpus tetragonolobus	Leaf	1,160	5,011
Cicer arietinum	Seed	1,850	4,970
Sesamum indicum	Seed	2,010	4,969
Phaseolus vulgaris	Sprout Seed	440	4,731
Spinacia oleracea	Plant	390	4,632
Cucurbita pepo	Seed	4,310	4,630
Trigonella foenum-graecum	Seed	2,800	4,300
Basella alba	Leaf	280	4,060
Corchorus olitorius	Leaf	300	4,000
Brassica nigra	Leaf	270	3,976
Vigna radiata	Sprout Seed	370	3,886
Allium schoenoprasum	Leaf	310	3,875
Asparagus officinalis	Shoot	300	3,871
Juglans cinerea	Seed	3,660	3,786
Prunus dulcis	Seed	3,580	3,745
Amaranthus sp.	Leaf	310	3,729
Cichorium intybus	Leaf	180	3,672
Valerianella locusta	Plant	260	3,610
Portulaca oleracea	Herb	160	3,400
Brassica oleracea	Flower	260	3,360
Colocasia esculenta	Leaf	480	3,345
Arachis hypogaea	Seed	2,540	3,321
Vigna unguiculata	Seed	340	3,293
Brassica oleracea	Leaf	200	3,278
Vigna unguiculata	Seed	3,000	3,276
Lupinus albus	Seed	2,890	3,225
Brassica chinensis	Leaf	150	3,206

Table 5 cont'd

Plant	Part	Amount (ppm) Low	High
Brassica oleracea	Leaf	290	3,115
Phaseolus lunatus	Seed	900	3,024
Avena sativa	Seed	–	3,000
Triticum aestivum	Seed	1,000	3,000
Hordeum vulgare	Seed	1,000	3,000
Vicia faba	Seed	560	2,950
Pistacia vera	Seed	2,830	2,944
Phaseolus vulgaris	Seed	2,560	2,877
Vigna radiata	Seed	2,600	2,859
Lens culinaris	Seed	2,510	2,826
Foeniculum vulgare	Fruit	2,530	2,775
Papaver somniferum	Seed	2,550	2,735
Bertholletia excelsa	Seed	2,600	2,690

ANTICHOLERIC PLANTS IN PERU

With the outbreak of cholera in Peru, the question of whether anticholeric plants grow in Peru is of significance. A search for anticholeric chemicals and plants reveals both shortcomings and strong points of the PHYTOCHEMECO database. Unfortunately, there are too few plants which have locality data. Additionally, the chemical data are presently limited to 1,164 plants. Furthermore, chemicals applicable as vibriocides (compounds which kill the bacterial genus, *Vibrio*, responsible for cholera) are not listed as anticholeric.

However, since lifezone information is available for 422 plants found concurrently in the phytochemical database, it is possible to predict which plants could be grown in locations where they may not be currently growing.[6] Furthermore, it is a simple matter to broaden the search to include vibriocides.

Using these additional criteria, two compounds were found: berberine, an anticholeric chemical found in two plants which can be grown in Peru; and menthol, a vibriocide found in at least 13 plants which grow in Peru (Table 8). Since menthol is widespread in occurrence, it could also be found in virtually any plant in the mint family.

Table 6. Antihypertensive compounds in various plants, other than *Citrus*

AJMALINE		RHOMITOXIN	
Pausinystalia johimbe	Bark	*Rhododendron molle*	Plant
Rauvolfia tetraphylla	Plant		
Rauvolfia serpentina	Root	SALSOLINE	
Rauvolfia vomitoria	Root	*Anabasis aphylla*	Plant
BERBAMINE		TETRAMETHYLPYRAZINE	
Berberis vulgaris	Plant	*Capsicum annuum*	Fruit
Mahonia aquifolia	Root	*Glycine max*	Seed
Berberi vulgaris	Plant	*Camellia sinensis*	Leaf
COREXIMINE		TRACHELOGENIN	
Annona muricata	Plant	*Cnicus benedictus*	Fruit
Coptis spp.	Rhizome	*Arctium lappa*	Plant
Asimina triloba	Plant		
Annona montana	Plant	TRACHELOSIDE	
		Carthamus tinctorius	Seed
RAUBASINE			
Rauvolfia serpentina	Plant	TYROSINASE	
		Allium sativum	Bulb
RESCINNAMINE		*Lablab purpureus*	Seed
Rauvolfia serpentina	Root	*Solanum tuberosum*	Plant
Rauvolfia vomitoria	Root	*Taraxacum officinale*	Root
RESERPILINE		VERATRAMINE	
Rauvolfia tetraphylla	Leaf	*Veratrum album*	Root
Rauvolfia vomitoria	Root	*Veratrum viride*	Root
Rauvolfia serpentina	Root		
RESERPINE			
Rauvolfia serpentina	Root		
Vinca minor	Leaf		
Catharanthus roseus	Plant		
Rauvolfia tetraphylla	Root		
Rauvolfia vomitoria	Root		

Table 7. High pectin contents of various plants

Plant	Plant Part	Amount (ppm) Low	High
Althaea officinalis	Root	110,000	350,000
Lagenaria siceraria	Fruit	8,400	210,000
Daucus carota	Root	100,000	188,000
Limonia acidissima	Fruit	30,000	160,000
Humulus lupulus	Fruit	120,000	140,000
Passiflora edulis	Pericarp	24,000	140,000
Rosa canina	Fruit	–	110,000
Solanum melongena	Fruit	–	110,000
Helianthus annuus	Plant	–	110,000
Linum usitatissimum	Seed	53,900	100,000
Spondias dulcis	Fruit	–	97,600
Cucurbita spp.	Fruit	6,000	94,935
Solanum tuberosum	Leaf	21,000	73,000
Malus domestica	Fruit	1,400	66,585
Crotalaria juncea	Stem	–	64,100
Glycyrrhiza glabra	Shoot	–	58,000
Plumeria acutifolia	Leaf	–	56,000
Ficus carica	Fruit	18,000	50,000
Viburnum opulus	Fruit	–	50,000
Ficus carica	Leaf	–	48,400
Rosa rubiginosa	Seed	–	46,000
Musa x paradisiaca	Fruit	7,000	40,000
Punica granatum	Pericarp	20,000	40,000
Prunus domestica	Fruit	8,000	40,000

Table 8. Anticholeric (containing berberine) and vibriocidic (containing menthol) compounds in plants grown in Peru

Plant	Part	Amount (ppm)
ANTICHOLERIC		
Andira inermis	Bark	–
VIBRIOCIDES		
Brassica oleracea	Plant	-
Cymbopogon nardus	Plant	-
Helianthus annuus	Flower[a]	235,000
Mentha aquatica	Leaf	3,570
arvensis	Leaf	1,000
longifolia	Shoot	20
pulegium	Plant	3,780
spicata	Leaf	375
x piperita	Leaf	156
x rotundifolia	Leaf	80
Ocimum basilicum	Plant	4

[a] in essential oil of flower

Table 9. Plants containing high levels of selenium

Plant	Part	Amount (ppm)
Bertholletia excelsa	Seed	3-497
Silybum marianum	Plant	171
Hibiscus sabdariffa	Flower	143
Nepeta cataria	Plant	123
Elytrigia repens	Plant	102
Polygonum multiflorum	Root	74
Agathosma betulina	Leaf	70

Table 9 continued

Plant	Part	Amount (ppm)
Barosma betulina	Leaf	70
Cymbopogon citratus	Plant	62
Cypripedium pubescens	Root	49
Valeriana officinalis	Root	44
Caulophyllum thalictroides	Root	35
Berberis vulgaris	Root	34
Myrica cerifera	Bark	34
Cnicus benedictus	Plant	34
Althaea officinalis	Root	33
Rhodymenia palmata	Plant	33
Cimicifuga racemosa	Root	32
Cucurbita pepo	Seed	32
Smilax spp.	Root	31
Juglans nigra	Fruit	30
Mentha pulegium	Plant	25
Rumex crispus	Root	25
Stevia rebaudiana	Leaf	25
Rubus idaeus	Leaf	25
Panax ginseng	Root	25
Hordeum vulgare	Stem	24
Ruscus aculeatus	Root	24
Aloe vera	Leaf	23
Viburnum opulus	Bark	23
Urtica dioica	Leaf	22
Echinacea spp.	Root	21
Carthamus tinctorius	Flower	20
Crataegus oxycantha	Fruit	20
Larrea tridentata	Plant	19
Chondrus crispus	Plant	18
Gentiana lutea	Root	18
Vitis vinifera	Stem	18
Chrysanthemum parthenium	Plant	17
Fucus vesiculosus	Plant	17
Humulus lupulus	Fruit	17

PLANTS CONTAINING GREATER THAN 100 PPM SELENIUM

Selenium is an essential micronutrient with antioxidant, cardio-protective, and cancer-preventive functions .[2,7,8] A search of the phytochemical database retrieved only five plants having greater than 100 ppm selenium, and only three of these in edible portions of the plant (Table 9). The seed of the Brazil nut (*Bertholletia excelsa*) contained the highest amount of this essential element (497 ppm). It should be noted however that this high selenium content is unusual and that the mean value for 529 nuts sampled was actually 29.6 ppm (Palmer *et al* 1982).[9] Other edibles with high selenium content (Table 9) were hibiscus flowers (*Hibiscus sabdariffa*) and catnip (*Nepeta cataria*).

The variable selenium content of Brazil nuts illustrates a weakness of the database. All data entered into the PHYTOCHEMECO database are from published sources. However, it is not possible to validate these data since this would require performing duplicate analyses of each study. In addition, the data obtained are often from only one source or possibly only one plant. Genetic variation among individuals, coupled with the variation of inter- and intra-lab technical competence and consistency further confuses the overall variability of the data. With this in mind, it would be unwise to apply these data in an uninformed manner. These data are best utilized as a stimulus for further study, not as final, conclusive evidence.

THE VALUE OF THE CHEMICAL CONSTITUENTS OF A HECTARE OF SUNFLOWERS

Are there alternative uses for traditional crops? A unique perspective on the value of sunflowers is an assessment of their chemical constituent value. The resulting non-traditional use of sunflowers would be to supply the chemical market with a new source of phytochemicals.

This question is of course limited by a number of factors. Since yield data are only available for sunflower seeds, the value is limited to only the seed portion of the plant. Unfortunately, this omits some of the costlier chemicals, such as chlorophyll. Furthermore, the value is limited to chemicals for which quantitative information as well as pricing information is available. The pricing of the chemicals is determined by its current (1992) market value purchased from the Sigma Chemical Company under the current supply constraints. If chemicals were available from plants in the quantities proposed, the prices would

Table 10. Value of the chemical contents of sunflower seeds produced on one hectare of land

Chemical	Low Amount (g/ha)	High Amount (g/ha)	Price ($/g)	Low Value ($/ha)	High Value ($/ha)
Arginine	22,860	129,150	20.65	472,059.00	2,666,947.50
β–Carotene	<1	<1	10.50	0.01	4.96
Caffeic acid	11,440	22,835	2.01	2,894.40	5,698.35
Chlorogenic acid	11,710	44,100	26.60	45,486.00	1,173,060.00
Choline	11,080	11,890	0.02	21.60	37.80
Citric acid	11,710	22,993	0.02	34.20	59.85
Glycine	113,860	24,255	0.02	277.20	485.10
Lecithin	347	33,623	173.10	59,979.15	627,054.75
Linoleic acid	80,100	551,250	6.60	528,660.00	3,638,250.00
Myristic acid	450	11,418	0.07	31.50	99.23
Niacin	43	76	0.02	0.86	1.51
Oleic acid	32,400	280,350	6.30	204,120.00	1,766,205.00
Palmitic acid	112,825	50,400	0.23	2,949.75	11,592.00
Riboflavin	2	5	0.13	0.23	0.61
Stearic acid	88,100	44,100	5.00	40,500.00	220,500.00
Total				1,357,013.90	10,109,996.66

undoubtedly fall. Perhaps the most important caveat is that it may be far more costly to extract the chemicals than they are worth at market prices.

Nevertheless, the value of chemicals from the seeds of one hectare of sunflowers ranged from $1.4 million to $10.1 million, which is based on yields of 900 to 1,575 kg/ha, respectively (Table 10).

CONCLUSION

These practical questions demonstrate the varied application of the database as well as its shortcomings. The PHYTOCHEMECO database is unique in its coupling of several previously unconnected sources of data, and therefore allows researchers to address questions connecting environment with phytochemistry and pharmacology. The database is particularly strong in its rapid access time, simple structure, thorough coverage of the chemicals contained in 1,164 plants and their chemical activities, and depth of ecological coverage.

Areas for improvement include:

1. A much broader taxonomic coverage of plant growing regions and chemical constituents;
2. The need for quantitative data on many compounds;
3. The need for information on how individual chemicals are metabolized in the human body;
4. The need to insure robustness of the data, including some index of validity.

ACKNOWLEDGMENTS

I gratefully acknowledge the critical comments and suggestions of James A. Duke, Ed Bird, Calvin Sperling (USDA), Jean-Paul Hofmann, and Leigh Anderson (Large Scale Biology Corporation).

REFERENCES

1. DUKE, J.A. 1992. Handbook of Phytochemical Constituents of GRAS Herbs and Other Economic Plants. CRC Press. Boca Raton, FL.

2. DUKE, J.A. 1992. Handbook of Biologically Active Phytochemicals and Their Activities. CRC Press. Boca Raton, FL.
3. WERNSTEDT, F.L. 1983. World Climatic Data. Climatic Data Press, Lemont, PA.
4. DUKE, J.A. 1978. The quest for tolerant germplasm. Chap. 1. In: Crop Tolerance to Suboptimal Land Conditions, (G.A. Jung, ed.). ASA Special Symposium 32. American Society of Agronomy. Madison, Wisconsin, pp. 1-61.
5. ROCK, C., SWENDSEID, M. 1992. Plasma β-carotenc response in humans after meals supplemented with dietary pectin. Am. J. Clin. Nutr. 55: 96-9.
6. BECKSTROM-STERNBERG, S.M., DUKE, J.A. 1992. Sister-site selection for the neem tree using the ORACLE® climate-crop data base. In: Proceedings of BARC Poster Day 1992, (S.M. Hubbard, ed.). ARS-102, U.S. Department of Agriculture, Agricultural Research Service, p. 59.
7. LEVANDER, O.A. 1987. A global view of human selenium nutrition. Ann. Rev. Nutr. 7:227-50.
8. MINO, Y., OTA, N. 1990. Inorganic chemical approaches to pharmacognosy. VIII. Determination of selenium in crude drugs by energy-dispersive X-ray fluorescence spectrometry. Chem. Pharm. Bull. 38:2208-2211.
9. PALMER, I., HERR, A., NELSON, T. 1982. Toxicity of selenium in brazil nuts to rats. J. Food Sci. 47:1595-1597.

INDEX

GPSR Compliance

The European Union's (EU) General Product Safety Regulation (GPSR) is a set of rules that requires consumer products to be safe and our obligations to ensure this.

If you have any concerns about our products, you can contact us on ProductSafety@springernature.com

In case Publisher is established outside the EU, the EU authorized representative is:

Springer Nature Customer Service Center GmbH
Europaplatz 3
69115 Heidelberg, Germany

Batch number: 10761599

Printed by Printforce, the Netherlands